Human Physiology in Extreme Environments

Human Physiology in Extreme Environments

Hanns-Christian Gunga

Professor
Center for Space Medicine and
Extreme Environments
Institute of Physiology
CharitéCrossOver (CCO)
Charité University Medicine Berlin
Berlin
Germany

With contributions from

Oliver Opatz

Postdoctoral Research Associate
Center for Space Medicine and
Extreme Environments
Institute of Physiology
CharitéCrossOver (CCO)
Charité University Medicine Berlin
Berlin
Germany

Alexander Christoph Stahn

Postdoctoral Research Associate
Center for Space Medicine and
Extreme Environments
Institute of Physiology
CharitéCrossOver (CCO)
Charité University Medicine Berlin
Berlin
Germany

Mathias Steinach

Postdoctoral Research Associate
Center for Space Medicine and
Extreme Environments
Institute of Physiology
CharitéCrossOver (CCO)
Charité University Medicine Berlin
Berlin
Germany

AMSTERDAM • BOSTON • HEIDELBERG • LONDON
NEW YORK • OXFORD • PARIS • SAN DIEGO
SAN FRANCISCO • SINGAPORE • SYDNEY • TOKYO
Academic Press is an imprint of Elsevier

Academic Press is an imprint of Elsevier
32 Jamestown Road, London NW1 7BY, UK
525 B Street, Suite 1800, San Diego, CA 92101-4495, USA
225 Wyman Street, Waltham, MA 02451, USA
The Boulevard, Langford Lane, Kidlington, Oxford OX5 1GB, UK

British Library Cataloguing-in-Publication Data
A catalogue record for this book is available from the British Library

Library of Congress Cataloging-in-Publication Data
A catalog record for this book is available from the Library of Congress

ISBN: 978-0-12-386947-0

For information on all Academic Press publications
visit our website at http://store.elsevier.com/

Typeset by SPi Global, India

Printed and bound in United States of America
15   16   17   10 9 8 7 6 5 4 3 2 1

Working together
to grow libraries in
developing countries

www.elsevier.com • www.bookaid.org

Dedication

This book is dedicated to Luise, Leonard, Maxim, and Arthur.

Berlin, May 2014

Contents

6. Cold Environments

Mathias Steinach and Hanns-Christian Gunga

Preface

This book deals with humans in extreme environments. It originated from talks and field studies with my academic teacher Prof. K. Kirsch in the gold mines of Tarkwa in Ghana's tropical rain forest more than 20 years ago. The contents of the book are based on lectures and seminars to medical students at the Institute of Physiology at the Free University of Berlin[1] at the beginning of the 1980s. Meanwhile, additional courses were given with support of the Deutschen Akademischen Austauschdienst (DAAD, Germany) and the State Administration of Foreign Experts (China) at the Faculdad de Medicina de Chile (Santiago, Chile) and at the Key Laboratory for Space Biosciences and Biotechnology at the Northwestern Polytechnical University (Xi'an, China), respectively. Our experience is based on numerous studies in the laboratory and the field, including studies conducted in Africa, North and South America, Antarctica, and in space.

Several different kinds of books on this topic have been published during the last decades [1–3]. However, some cover only one topic, such as Edholm's *Man: Hot and Cold* [1]. Other excellent books are extremely comprehensive, such as volumes I/II of the *Handbook of Physiology: Environmental Physiology* [4], Pandolf's "Human performance physiology and environmental medicine at terrestrial extremes" [5], or Auerbach's outstanding handbook *Wilderness Medicine* [6]. Therefore, we felt that the present book should be "as broad as possible and as focused as necessary" [7],[2] aiming primarily at medical students and interested laymen. Thus, our work follows most closely of Edgar Folk's classical *Textbook of Environmental Physiology* in its second edition [8] or more recently the publications by Piantadosi [9] "The biology of human survival life and death in extreme environments" and Cheung's "Advanced environmental exercise physiology" [10]. However, we thought that for an initial introduction to the topic some additional aspects should be addressed such as evolutionary, anthropological, and methodological issues.

---

1. In 2003 the Medical Faculty of the Free University of Berlin and the former Medical Faculty of the Humboldt University, the Charité, were unified under the common name Charité University Medicine Berlin.
2. It would seem that Arnold Schönberg faced similar problems when he tried to summarize the historical and aesthetic developments of music in the last centuries, guided by the captivating principle that "a universal education should be another virtue of the specialist" [7, p. 7].

FIGURE 1   Nathan Zuntz (1847-1920) at his office in Berlin at the Königlich Landwirtschaftliche Hochschule around 1915. *Courtesy Institute of Physiology, Charité University Medicine Berlin.*

In an era when knowledge is expanding at an ever-increasing pace and in-struction increasingly takes place in the form of modules that actually demand a considerable basis of theoretical knowledge, an integrative approach to observ-ing fundamental physiological processes is even more critical to avoid getting lost in details and losing sight of the larger picture. This integrative approach, a working philosophy of not relying on "armchair postulation" [11], but instead testing and verifying laboratory results in the field, as well as the research on humans in extreme environments, is not new for us in Berlin. One hundred years ago, Nathan Zuntz (1847-1920) was already a key figure in the history of research on the physiology of humans in extreme environments (Figure 1).

His books *Höhenklima und Bergwanderungen in ihrer Wirkung auf den Menschen* (*The Effect of High-Altitude Climate and Mountain Hiking on Human Beings*) [12] as well as *Studien zu einer Physiologie des Marsches* (*Studies on the Physiology of Marching*) [13] must be considered classic publications in the his-tory of applied physiological sciences and on a par with Paul Bert's *La pression barométrique* (Barometric pressure) [14] and Adolph's *Man in the Desert* [11].[3] Adolph (1895-1986) led the Rochester Desert Unit in the United States and con-ducted several studies of water requirements and thermoregulation in humans un-der contract with the Office of Scientific Research and Development in the United States in close cooperation with the famous Harvard Fatigue Laboratory [19]. The latter was conceived by Lawrence J. Henderson in 1926, started operating in the fall of 1927 under the directorate of David Bruce Dill, and became a leading center

---

3. For further details, please compare Refs. [15–18].

FIGURE 2 Otto Gauer (1909-1979) at the age of 65 in Berlin. *Courtesy Institute of Physiology, Charité University Medicine Berlin.*

of human adaptation to extreme environments dealing with exercise, exhaustion, heat, cold, and high altitude adaptation [20]. Otto Gauer (1907-1979) (Figure 2), the famous gravitational physiologist and former director of the Institute of Physiology of Berlin, was fascinated by this kind of research.

He obviously developed a deep familiarity with it, especially during his time at Brooks Air Force Base in the United States during the 1950s and early 1960s, where he personally met and worked with members of these two important research schools and laboratories. This collaboration definitely paved his way to studies of humans in extreme environments (Figure 3).

Gauer taught the scholars K. Kirsch (1938-) and L. Röcker (1940-) in Berlin, who continued and broadened this research, especially in the fields of cardiovascular, blood, exercise, and space physiology. This kind of research has been and remains primarily a domain of English-language scholarship, as recently documented by Tipton's (*History of*) *Exercise Physiology* [17].

In Germany, this field of study does not enjoy the same level of recognition among physiologists, in spite of the fact that the research field "extreme environments" is distinguished by an entertainment value that should not be discounted. This is no small advantage, even, or perhaps especially, with regard to university-level education. The lack of recognition may result from the fact that research on human adaptation to extreme environments is often inherently limited to descriptive work, without the possibility of providing concrete explanations. What some fail to recognize, however, is that over the long term, research that initially provided purely descriptive observations of humans under extreme conditions has since resulted in significant new approaches and ways of thinking that are of essential importance to research. One outstanding current example of this process is the research on non-osmotic storage of Na^+, primarily in the skin, conducted as a part of the investigations into the effects of long-term

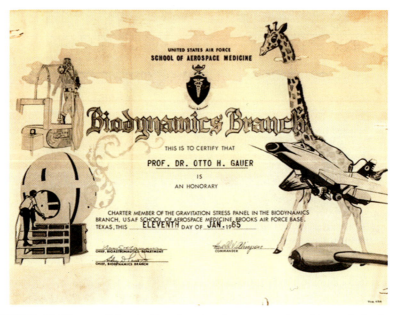

FIGURE 3 Certificate verifying that Otto Gauer is an honorary member of the gravitational stress panel dated January 11, 1965. *Courtesy Institute of Physiology, Charité University Medicine Berlin.*

isolation on humans to prepare for space missions [21]. Furthermore, working in extreme environments, especially when conducting physiological studies in the field, usually requires high-performance, miniaturized, noninvasive, and easy-to-use devices. The potential applications for such devices are not limited to extreme environments; rather, they offer great utility in the many various everyday applications on earth. For several decades, our working group has devoted considerable attention to the development of miniaturized, noninvasive tools and procedures for human physiological measurements in close cooperation with companies active in this field. These technological developments are possible only thanks to substantial and, most importantly, continuous financial support provided to our research projects by the *Deutsches Zentrum für Luft- und Raumfahrt* (DLR, German Aerospace Center) in Bonn-Oberkassel, namely Dr. P. Preu, Prof. G. Ruyters, Dr. P. Gräf, and Dr. H.-U. Hoffmann. We must also not forget that it was small to mid-sized industrial firms as well as major corporations that created the conditions for this research institution to continue its work over the long-term in Berlin by establishing and funding the Nathan Zuntz Professorship for Space Medicine and Extreme Environments at the Freie Universität Berlin in 2003 and currently the Nathan-Zuntz-Förderkreis e.V. In addition, we received financial support from the Deutsche Gesellschaft für Technologische Zusammenarbeit (GTZ, Eschborn) and the Bundesministerium für Bildung und Forschung (BMBF, Bonn).

In summary, this book was enabled by work done in the following three major areas: (i) classical education and teaching at a major German university, including the history of science; (ii) planning, implementation, and evaluation of our own research in human physiology under sometimes extreme conditions in the field and the laboratory, including research conducted with German and international partners; and (iii) close cooperation with small and mid-sized corporations in planning, developing, and producing new tools and processes. Therefore, some of the material in following chapters is based on or complied from previous manuscripts or publications that have been published in seminar handouts (scripts) of regular courses given at the Charité, published separately in journals, derived from works on guidelines by expert councils, or have been the essence of long-lasting research groups, such as the DFG Research Group 533 as well as material from classical text [22–24] and handbook chapters [25,26] written by the authors in previous years.

Finally, this book would not have been possible without the continuous support of L. Roecker and K. Kirsch and my scientific team members at the Department, namely Dr. O. Opatz, Dr. A. Stahn, and Dr. M. Steinach, co-authors of some chapters, and Dr. A. Werner, S. Thomas, as well as the large technical support by B. Bünsch, B. Himmelsbach, E. Hofmann, F. Kern, and finally A. Sommer, who prepared with great long-lasting enthusiasm the illustrations for this book. Thanks to all of them.

REFERENCES

[1] Edholm OG. Man, hot and cold. London: E. Arnold; 1978.

[2] Sloan A. Man in extreme environments. Springfield, IL: Charles C. Thomas; 1979.

[3] Kamler K. Surviving the extremes: a doctor's journey to the limits of human endurance. New York: St. Martins Press; 2004.

[4] Fregly MJ, Blatteis CM, editors. Environmental physiology. Oxford, New York: Oxford University Press; 1996.

[5] Pandolf KB, Sawka MN, Gonzalez RR. Human performance physiology and environmental medicine at terrestrial extremes. Indianapolis, IL: Brown & Benchmark; 1988.

[6] Auerbach PS, editor. Wilderness medicine. Philadelphia. Mosby Elsevier; 2007.

[7] Schneider F. Arnold Schönberg: Stil und Gedanke. Leipzig: Reclam Verlag; 1989.

[8] Folk GE. Textbook of environmental physiology. 2nd ed. Philadelphia: Lea & Febiger; 1974.

[9] Piantadosi CA. The biology of human survival life and death in extreme environments. Oxford, New York: Oxford University Press; 2003.

[10] Cheung SS. Advanced environmental exercise physiology. Champaign, IL: Human Kinetics; 2010.

[11] Adolph EF. Physiology of man in the desert (Reprint 1969). New York: Hafner Publishing Company; 1947.

[12] Zuntz N, Loewy A, Müller F, Caspari W. Höhenklima und Bergwanderungen. Berlin, Leipzig, Wien, Stuttgart: Deutsches Verlagshaus Bong; 1906.

[13] Zuntz N, Schumburg W. Studien zur Physiologie des Marsches. Berlin: Verlag von August Hirschwald; 1901.

[14] Bert P. Pression barometrique. Paris: Masson; 1878.

[15] Fishman A, Richards DW, editors. Circulation of the blood—men and ideas. New York: Oxford University Press; 1964.

[16] West JB. High life: a history of high-altitude physiology and medicine. Oxford, New York: Oxford University Press; 1998.

[17] Tipton CM. Exercise physiology. People and ideas. Oxford, New York: Oxford University Press; 2001.

[18] Gunga H-C. Nathan Zuntz. His life and work in the fields of high altitude physiology and aviation medicine. Amsterdam: Elsevier, Academic Press; 2009.

[19] Horvath SM, Horvath EC. The Harvard fatigue laboratory. Its history and contributions. Englewood Cliffs, NJ: Prentice-Hall, Inc.; 1973.

[20] Folk GE. The Harvard fatigue laboratory: contributions to World War II. Adv Physiol Educ 2010;34:119–27.

[21] Titze J, Machnik A. Sodium sensing in the interstitium and relationship to hypertension. Curr Opin Nephrol Hypertens 2010;19:385–92.

[22] Hierholzer K, Schmidt RF. Pathophysiologie des Menschen. Weinheim: VCH; 1991.

[23] Klinke R, Pape HC, Silbernagl S, editors. Lehrbuch der Physiologie. 5th ed. Stuttgart, New York: Thieme; 2005.

[24] Speckmann E-J, Hescheler J, Köhling R, Rintelen H, editors. Physiologie. München, Jena: Elsevier, Urban & Fischer Verlag; 2013.

[25] Ley W, Wittmann K, Hallmann W, editors. Handbook of space technology. Chichester, West Sussex: Wiley and Sons; 2009.

[26] Preedy VR, editor. Handbook of anthropometry. Physical measures of human form in health and disease. Berlin: Springer; 2012.

Chapter 1

Introduction

Hanns-Christian Gunga

Professor, Center for Space Medicine and Extreme Environments, Institute of Physiology, CharitéCrossOver (CCO), Charité University Medicine Berlin, Berlin, Germany

1.1 UNIVERSE

According to latest calculations, the age of the universe is around 13.8 billion years [1]. It probably came into existence by the "Big Bang." The presence of possible antecedent universes is beyond our knowledge. Immediately after the Big Bang, in the so-called *Planck Epoch*, the four presently known elemental forces—gravitation, strong and weak nuclear energy, and electromagnetism— began to separate, and an explosive inflation of the universe ensued. Thereafter, electrons, quarks, and radiation developed, and about 10^{-5} s after the Big Bang the first protons and neutrons came into existence. In the course of this first second after the Big Bang and the subsequent cooling processes due to the explosive inflation of the universe, known matter originated. Presumably, in addition, matter and energy forms were created that currently still elude our access and exact description (Figure 1.1). The important fact is that this fraction of matter and these energy forms constitute the predominant part in the universe. According to recent calculations, merely 4% of the mass of the universe can be allotted to the matter that we can see and analyze, the so-called "baryonic matter," which represents the construction material of atoms, 73% being constituted by "dark energy" on the one hand and 23% by "dark matter" on the other hand, leaving unclear what this matter actually consists of [2]. One should be at least aware of this fact when in later chapters the chemical composition of the human body will be discussed in detail. Among these particles of hypothetic matter are the so-called *weakly interacting massive particles* (WIMPs) [3]. These are heavy, inert particles that hardly interchange with the visible world as we see it. According to recent concepts, even more complex super-WIMP structures are supposed to exist in this exotic world of dark matter possessing its own forms of power and light not perceptible by us. According to the concepts of theoretical physicists and astronomers, these exotic particle

Human Physiology in Extreme Environments. http://dx.doi.org/10.1016/B978-0-12-386947-0.00001-0

1

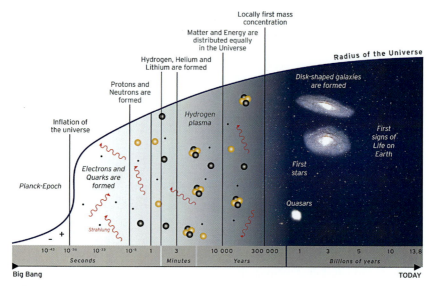

FIGURE 1.1   Emergence of the universe from the *Planck-Epoch* 13 billion years ago; its sudden explosive inflation, and the forming of electrons, quarks, protons, neutrons, hydrogen, helium, and mass concentrations that became the first stars and galaxies until today.

forms originated immediately after the Big Bang, only to be partly destroyed 1 ns afterwards by clashes of particles of dark matter. Only after further expansion, and thus cooling of the universe (age of the universe >10 ns), the amount of WIMPs now calculated theoretically remained in the universe because, due to too low temperature and density, no new WIMPs could be formed in the universe and because the probability of clashes of dark matter among each other gradually decreased drastically [2]. Approximately three minutes after the Big Bang, the first elements, hydrogen and helium, came into existence, and in the following 10,000 years an almost equal distribution of matter and energy in the universe occurred. Until today, the universe has consisted predominantly of these two elements—92.9% hydrogen and 6.9% helium—and the remaining known elements in the universe, the origination of which shall be described in detail later, amount to merely 0.2% of the atoms in the universe (Table 1.1). Due to slight differences in the distribution of the atoms, about 300,000 years later accumulations of matter, and then the first stars appeared. As far as we know right now, the oldest object in the universe to be verified is a quasar aged 13 billion years [5] (Figure 1.1). According to our current state of knowledge, the first galaxies were formed approximately 3-5 billion years after the Big Bang. The unequal distribution of galaxies in the universe, generated by extremely small differences in density in the distribution of matter directly after the Big Bang in the inflationary state of the universe, is still perceptible by the differences in temperature. The temperature differences were recently impressively demonstrated [6] (Figure 1.2). The differences found in the distribution of temperatures are only

TABLE 1.1 Atoms in the Universe—Percent atom abundance of most abundant atoms in the Universe.

Atoms	% Abundance	SEM
H	**92.969**	0.0
He	**6.901**	0.311
O	0.068	0.003
C	0.040	0.002
N	0.009	0.0006
All others	0.011	0.0005
Total	100.000	

(Adapted from [4])

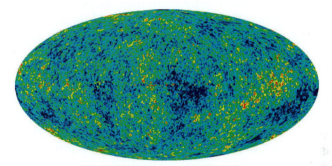

FIGURE 1.2 The Cosmic Microwave Background temperature fluctuations. The image is a Mollweide projection of the temperature variations over the celestial sphere. The average temperature is 2.275 K, and the colors represent the tiny temperature fluctuations. Red regions are warmer and blue regions are colder by about 0.0002°K. *(Adapted from http://wmap.gsfc.nasa.gov/media/101080) [7].*

around <0.0002 K, but they allow conclusions regarding the mass differences in the early universe and enable a hypothesized "frozen image" of the state of the universe directly after its formation following the Big Bang.

1.2 GALAXIES

First, there were no galaxies as we know them now as more or less disc-shaped assemblies of stars, but as loose clouds of galaxies. About 11 billion years ago, gravity began to form the first galaxies as we know them today (Figure 1.1). With more and more mass accumulation, especially toward the center of a galaxy, a black hole is created where even light cannot escape. Our own galaxy, the Milky Way, has such a super massive black hole today.

This structure is not fixed and cannot be seen alone. Instead, galaxies in themselves form clusters, and these clusters even form bigger super clusters, which again are concentrated into filaments that connect the different super clusters, forming a kind of gigantic web cluster. For a long time an open question had been what keeps galaxies from falling apart, forming instead those clusters and super clusters that keep them together. Today, astronomers are convinced it is the "dark matter," which, as mentioned above, is matter that exists approximately six times more often in the universe than that baryonic matter that we know. Indirect hints to the existence of matter have come from so-called "gravitational lenses" that deviate light on its way through the universe. However, the dark matter has a probably even stronger counterpart, the "dark energy" in the universe, an unknown force that pulls the galaxies, clusters, and other structures apart from each other. For the existence of the Earth and ourselves, it is crucial to recognize that our solar system is actually placed in a very special part of our own galaxy. We are far enough away from the center of the Milky Way with its black hole and highly radioactive zones—an environment that makes life forms as we know them impossible—and not too far away from the center so that we have a fairly stable place in the outer part of the galaxy, the so-called "habitable zone" of the galaxy, circulating around the center in about 225-250 million years, the cosmic or galactic year [8–10].

Our galaxy, the Milky Way, is a medium-sized spiral galaxy in the galaxy cluster Local Group. It has a diameter of about 100,000 light-years,[1] and consists of approximately 200 billion fixed stars like our own sun [11–13]. A fixed star of the size of our sun produces its energy by nuclear fusion processes. At about 5 million Kelvin, four hydrogen atoms (^{4}H), also called protons (^{1}H), fuse to one helium atom (^{4}He), a very stable nucleus (alpha particle) with two protons and two neutrons. Each second a star like our sun, which has 1.3 million times the volume of the Earth, converts about 600 million tons of hydrogen nuclei into helium nuclei, thereby dissipating heat, which enables life on Earth, but losing as well about 4 million tons of mass per second. As intermediate products of this nuclear fusion process, the elements deuterium (^{2}H) and helium 3 (^{3}He) also originate.[2] The hydrogen supplies in our sun should suffice for approximately another 5 billion years (http://www.enchantedlearning.com/subjects/astronomy/sun/ [15]; http://solar-center.stanford.edu/about/ [16]). In the distant future, when almost all of the hydrogen in the nucleus of the sun has been transformed into helium, the temperature in the nucleus of the sun will increase to more than 100 million Kelvin. New nuclear fusion processes will arise, leading to the generation of new elements that should turn out to be particularly significant for the creation of life, because from the fusion of three ^{4}He atoms

---

1. One light-year is the distance that light travels in 1 year, about 9.46 trillion kilometers.
2. This isotope of helium gives rise to far-reaching speculations as to the final solution of urgent energy problems on Earth [14].

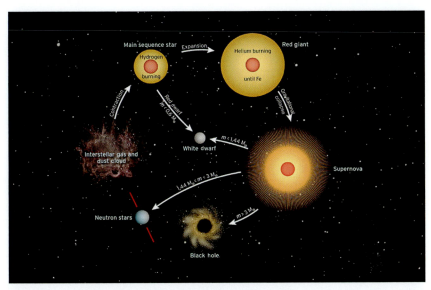

FIGURE 1.3   The life cycle of stars with a similar or larger mass as our sun. Depending on the mass of the object ($M \sim$ sun mass) several pathways can be predicted. *(Adapted from http://marvin. sn.schule.de/~erzkoll/projekte/astro1/sterne.html) [18].*

the element carbonate (^{12}C) originates, and from the fusion of ^{12}C and ^{4}He in the nucleus of the sun, the element oxygen (^{16}O), which is vitally important for us, originates [17]. During these nuclear processes, the sun swells, forming a so-called red giant. In the last act, the red giant explosively ejects its hydrogen-rich external shell, which is enriched with elements such as oxygen and carbon, and in the first instance becomes a white dwarf then, after further cooling, a black dwarf. Whether a star develops within its cosmologic life story to become a red dwarf, a white dwarf, a neutron star, a supernova, or even a black hole depends essentially on its mass. In a schematic representation, Figure 1.3 shows these possible actions in the history of development of a star, depending on its mass.

1.3 PLANETARY SYSTEM

Our own sun and planetary system were formed about 4.56 billion years ago [19]. According to the 2006 findings, Pluto is no longer counted among the planets of our solar system, which contains eight planets and at least two larger belts, the Asteroid belt between the planets Mars and Jupiter and the Kuiper Belt beyond the planet Neptune (Figure 1.4). The heliopause, in general agreement the influence of the gravity of the sun, has been postulated as the outermost limit of our solar system, but this still needs to be verified. The heliopause lies about 23 billion kilometers away from the sun [20] and according to recent data on August 25, 2012, Voyager 1 had crossed this heliopause and entered interstellar space, making it the first spacecraft to do so. It is necessary to record

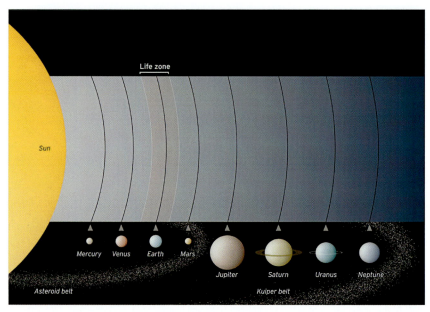

FIGURE 1.4   Our solar system with its eight planets, the Asteroid belt between Mars and Jupiter, and the Kuiper belt in the outer solar system. According to recent findings, icy comets that hit the Earth frequently during its early phase brought the water on Earth and originate from this Kuiper belt (not in scale).

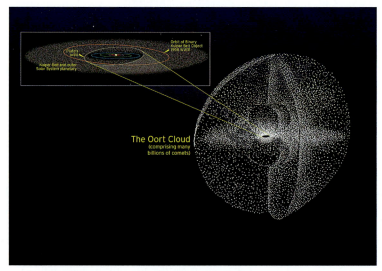

FIGURE 1.5   The dimensions of the Oort cloud, which is the major source of icy comets in our solar system [21].

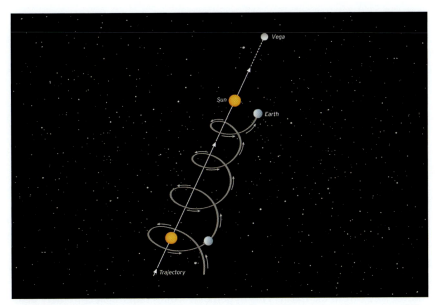

FIGURE 1.6   The solar system is moving with a velocity of 20 km/s on a trajectory toward Vega in the Lyra.

at this point that objects at the edge of this heliopause, such as, for example, Oort's cloud (Figure 1.5), probably exerted a decisive influence on the development of life on Earth (see Section 1.4). At the same time, our solar system is not spatially constant, but the sun rather moves, together with its planets, moons, and comets, as a cosmic unit toward the star Vega in the Lyra at a velocity of approx. 20 km/s (Figure 1.6) *(adapted from http://www.rasnz.org.nz/Stars/Lyra. htm) [22]*.

1.4   EARTH-MOON SYSTEM

The Earth, in comparison to other planets of our solar system, is accompanied by a relatively large Moon. For the development of life on Earth, the importance of the Earth-Moon system is not always fully recognized [23]. Therefore its presumable origin and impact on life on Earth will be briefly discussed.

1.4.1   Origin of the Earth-Moon System

For about 100 years this question has been intensively discussed, and the main hypotheses presented included: fission from the Earth, capture, co-accretion, and a giant impact [23,24]. Today, the giant impact hypothesis is the most accepted among scientists because it can explain concomitantly several curiosities of the Earth-Moon system (see below) [25], although recently some doubts have been raised about this theory again [26].

1.4.1.1 Fission Hypothesis

George Darwin (1845-1912) proposed this hypothesis in 1879 [27]. He thought that the ancient Proto-Earth ejected a piece of its mass by an accelerated rotation approximately 4.5 billion years ago. It was assumed that the scar of this event on our planet would be the Pacific Ocean. However, this theory has been disproven because (i) with an age of 200 million years, the oceanic crust of the Earth is much too young in comparison to the Moon's crust, which is several billion years old, based on the analysis of material from the APOLLO mission; and (ii) this hypothesis also cannot explain the angular momentum of the Earth-Moon system.

1.4.1.2 Lunar Capture Hypothesis

This hypothesis assumes that the Moon was captured by the gravitational field of the Earth. The strongest argument against this theory is the fact that the Earth and Moon share almost identical oxygen isotope ratios [23,24,28]. Furthermore, such a close encounter in the early phase of our solar system would more likely lead to a collision or an alteration of the trajectory, which would move the Moon forever away from the Earth. This hypothesis would require an extremely dense atmosphere around the Proto-Earth, which could have slowed down the velocity of the Moon, altering its trajectory into one orbiting around the Earth.

1.4.1.3 Co-accretion Hypothesis

This hypothesis assumes that Earth and Moon originated together from the primordial accretion disk of the solar system [23]. However, this theory cannot explain the relatively small iron core of the Moon (25% of its radius) as compared to the Earth (50%) [24,29].

1.4.1.4 Giant Impact Hypothesis

This hypothesis assumes that the Moon derives from a giant collision of a Mars-like protoplanet with the proto-Earth during the early accretional phase of the solar system. This theory can explain (i) the relatively small iron core of the Moon and the large one of the Earth because during the collision process, iron material from the core of the Moon was shifted/injected to the Earth's core; (ii) the similar oxygen isotopes on Earth and Moon; (iii) possibly the angular momentum; and (iv) the lack of volatiles in the lunar samples due to the enormous energy of the collision, which completely evaporated the material; this would require an initially molten Moon, although some researchers argue that this depletion of volatile elements on the Moon is not as complete as theoretically would have been expected due to high temperatures occurring in the process of collision [28,30]. Thirty years ago Binder's theory of an initially totally molten Moon [30,31] was heavily controversial [32]. Recently it has again attracted intensive research [33–35].

1.4.2   Earth-Moon System and Development of Life on Earth

Regardless of the exact origin of the Earth's Moon—excluding the first theory by George Darwin—presumable scenarios 2, 3, and 4 indicate that the Earth has had a relatively large companion with a small iron core for approximately 4.5 billion years. Let us summarize how this might have influenced the development of life on Earth. First of all, (i) this giant impact nearly destroyed the Earth shortly after it was born about 4.5 billion years ago. (ii) This giant collision with another Mars-like protoplanet led to a "transfusion" of iron to the core of the Earth and a large mass ejection of pre-Earth's crust material into the orbit, which afterwards circulated closely around our protoplanet, followed by an aggregation of orbiting material that finally formed our Moon (initially totally molten Moon theory) [31,36]; (iii) The new existence of the Moon stabilized the 23° axis of the Earth, which (iv) gave rise to tides and the development of certain seasons in higher and lower latitudes known as spring, summer, fall, and winter. During those prehistoric times, day-and-night cycles were much shorter and, due to the close distance (i.e., gravitational forces), the Moon "departed" much more slowly (2 cm/year) from the Earth than today (3.5 cm/year). We have to assume that 2 billion years ago the Moon was still only at a distance of about 40,000 km—today approximately 384,500 km—orbiting the Earth 3.7 times per day and thereby inducing high and low tides that were a thousand times stronger and continuously slowed down the rotation of the Earth and the Moon [37]. In a second step, the thermodynamic up- and downstream in the liquid iron core of the Earth induced the formation of an early magnetic field. These streams acted and still act today as important UV shielding, which otherwise would have meant that any organic life forms would be destroyed. The Earth thereby kept water in its atmosphere.

1.4.3   Earth

Soon after the solar system was built 4.56 million years ago the Earth was also created. Today, the surface of the Earth is about $510 \, km^2$; one-third, $149 \, km^2$, is covered by land, the other two thirds, $361 \, km^2$, by water. Our own body consists of about 60% water, and our blood can contain up to 90% water. Some other organisms can be even up to 99% water [38–40]. Therefore, it is an important question where all the water on Earth comes from. Definitely, in the early phase as a protoplanet 4.5 billion years ago, the Earth's surface was too hot to retain any water [41]. It took approximately 200-600 million years until the Earth's crust was cool enough so that water would not evaporate, because the chemical formation of water depends on hydrogen bonds keeping the water molecules together. The earliest fossil documents of water are reported from Greenland and Australia (http://www.fossilmuseum.net/Tree_of_Life/Stromatolites.htm) [42]. First theories assumed that the water originated from the Earth's crust [39], but astrophysicists argued for a long time that there was still too much water because reasonable amounts of water in our solar system are found beyond Mars, in the Asteroid belt or Kuiper belt (Figure 1.4). Astrobiologist John Oró hypothesized as early as 1961 [43] that the water originated from comets that were composed of nearly 90% water and came from far away in the solar system, the Oort cloud. They hit the Earth's surface frequently and in massive numbers during the so-called "early bombardment" phase. This Oort cloud with its numerous comets marks the outer limits of our solar system (see Figure 1.5). One of the most famous of those comets, Halley's Comet, originates from this cloud. In 1986, when the European Giotto Mission made an isotope analysis of Halley's Comet, it surprisingly revealed that the comet's water had a completely different isotope composition than the water in the Earth's oceans. Soon, other sources of water were under discussion such as the Asteroid belt. Until very recently, researchers analyzed the composition of the comet 103P/Hartley 2, which originates from the Kuiper belt [44] (Figure 1.4), where approximately more than 70,000 objects larger than 100 km can be found consisting mainly of water [45]. Based on data from the Herschel telescope (spectrograph), this comet from the Kuiper belt showed the exact isotope composition of water as we know it from Earth. Therefore, it is currently thought that the so-called trans-Neptunian objects could have been a major source of the water on Earth—and ourselves.

1.5 LIFE

The elements of carbon, hydrogen, oxygen, nitrogen, sulfur, and phosphorus derived from those catastrophic events in the universe in the past, as described in the previous sections, form the material on which life is based on Earth[3] [17,23,46]. As very recently pointed out by Mora et al. [47], this diversity of life is one of the most striking aspects of our planets, but, somehow surprisingly, this diversity is far from completely described in spite of 250 years of taxonomic classification. Today about 1.2 million species are catalogued in a central database. Over 1 million of these belong to the arthropods (especially insects, spiders, and crustaceans); 46,500 species of vertebrates have been found—the smallest one being only 16 mm long and a few g in mass [48] to a 30 m long blue whale with a body mass of >100 million g—the largest in numbers among them being fish with approximately 20,000, and mammals, with 3700 species including humans. Theoretically, it is predicted by Mora et al. [47] that in total 8.7 million (±1.3 million SE) eukaryotic species globally exist, of which ~2.2 million (±0.18 million SE) are marine. According to Mora et al. [47], this means that only 14% of the species existing on Earth and 9% of the species in the ocean are described. Probably even these estimations are underestimating

---

3. Although already nearly 15 years old, a very informative, dense overview on the topic can be found in the ESA special papers published by the Exobiology Working Group in 1997 [46].

the real number of life forms. Most recently, Bhanoo [49] reported on a newly discovered deep-sea ecosystem in the southern ocean near Antarctica where 23 new animal species live in the hot, dark environment around hydrothermal vents, which are as hot as 720 °F (382 °C). At these extremely high temperatures no animal can survive. Most of the animals found were living in a so-called interface zone that shows water temperatures from 50 to 68 °F (10-20 °C), but some of them, the procaryotes, can even survive temperatures >110 °C [50,51]. However, one has to keep in mind that the temperature range where Earth's life forms (procaryotes, eucaryotes) can exist is extremely narrow if we compare it with the temperature range that could be observed in the universe, in general (Figure 1.7). Today about 700 species living near hydrothermal vents have been described, most of them living close to the vents. In contrast to those previously described, hydrothermal faunas in the Pacific, Atlantic, and Indian Oceans, the researchers found no worms, mussels, and shrimp near the Antarctic hydrother-mal vents in this deep-sea ecosystem. But similar to those, because of the com-plete darkness at these deep-sea vents at 8000 ft, creatures use chemicals such as hydrogen sulfide continuously emitted to the environment by these vents instead of light energy from the sun. They are thereby able to maintain their energy requirements for their biochemical processes [52]. Specifically, these hydrothermal vents deserved more and more consideration in recent years as possible ecosystems where life could have started on our planet because the three major requisites for first organic molecules—carbon, water, and energy—can be found there ubiquitously. The crucial switch from organic chemistry to real life forms is characterized according to de Duve [53] by the following seven points: Every living system must be able to (i) produce its components by means of the material available in its environment, that is, synthesize them;

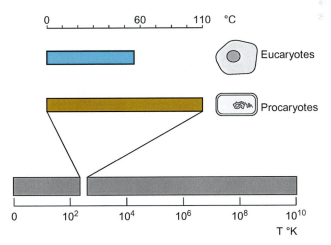

FIGURE 1.7   The wide range of temperatures in K that can be found the universe in contrast to the very narrow range where active life of Procaryotes and Eucaryotes can exist in K [46]. (*Courtesy of ESA*).

(ii) absorb energy from its environment and convert it into different forms of vital work; (iii) catalyze the manifold chemical reactions required for its activities; (iv) control its biosynthetic and other processes by means of information in order to safeguard its true reproduction; (v) isolate itself in such a way that the exchange with the external environment is controlled precisely; (vi) control its activities in such a way that its dynamic organization is maintained in spite of environmental changes; and finally, (vii) reproduce itself.

1.5.1 Palaeontological Aspects

1.5.1.1 First Life Forms

The major steps from the formation of the Earth approximately 4.5 billion years ago include a stable hydrosphere, prebiotic chemistry, pre-RNA, RNA world, the occurrence of the first DNA/protein life, and last, a universal common ancestor 3.6 billion years ago (Figure 1.8). This universal common ancestor includes all the three domains of life so far known on Earth: archaea, bacteria, and eucaryota (Figure 1.9). It seems that in the tree of life, archaea, which belong to the procaryota (showing no real nucleus and no cell organelles), are closest to this universal common ancestor so far. Currently, the oldest known rock formations are found in Greenland and Labrador (3.82-3.65 billion years old); however, highly metamorphosed rocks, better preserved rocks with an age of 3.5-3.0 billion years, are described from the Pilbara craton in western Australia and the Barberton region in South Africa and Swaziland [41]. For years it has been widely accepted that oxygen-releasing cyanobacteria may have been already present at 3.45 billion years ago [55–58], although recently these findings have raised increasing concerns mainly due to the high rate of metamorphism of the material and the related difficulties in making unambiguous descriptions and sound assumptions about the earliest signs of life forms on Earth [41].

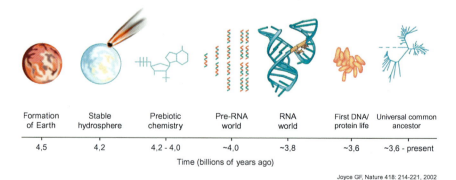

Formation of Earth	Stable hydrosphere	Prebiotic chemistry	Pre-RNA world	RNA world	First DNA/ protein life	Universal common ancestor
4,5	4,2	4,2 - 4,0	~4,0	~3,8	~3,6	~3,6 - present

Time (billions of years ago)

Joyce GF, Nature 418: 214-221, 2002

FIGURE 1.8   A more detailed schematic diagram of the presumable steps leading from prebiotic-chemistry, over pre-RNA and pre-DNA to the first self-replicating life forms on Earth and their time of occurrence. *(Adapted from [54]).*

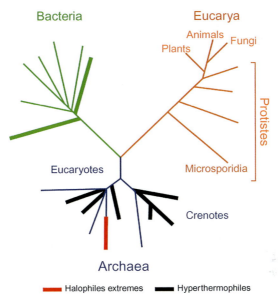

FIGURE 1.9 Phylogenetic tree of the three domains Archaea, Bacteria, and Eucaryota (plants, animals, fungi), indicating that halophiles extremes and hyperthermophiles belong mainly to Archaea. *(Adapted from [46]).*

Nevertheless, according to these sources, the current state of research hypothesizes that (i) hyperthermophiles are most likely among the early life forms on Earth [51,59,60] because based on geological-mineralogical data, volcanic and hydro-thermal rocks are the predominant rocks to be found in the time span between 3.5 and 3.0 billion years ago; and (ii) anaerobes dominated the early life forms using as an energy source anaerobic photoautotrophy [61,62] or chemosynthetic metabolism; and finally (iii) a hyperthermophilic and anaerobic environment as an endolithic origin of life 3.5 billion years ago currently cannot be excluded.

If it is still assumed, as indicated by Figure 1.10, that life originated 3.5 billion years ago, then it took another billion years, from 2.6 billion years, until cyanobacteria and other oxygenic autophototrophs could be frequently found at different places around the Earth. Their global, massive photosynthesis subsequently caused a continuous increase in oxygen partial pressure. From an evolutionary point of view, there came a silent, longtime of ~2 billion years with no visible further main steps in the evolution of life on Earth—probably because the organisms had to deal with the complex toxic side effects of an increasing oxygen partial pressure (Figure 1.10)—then, surprisingly rapidly between 570 and 530 million years ago, a rich fauna developed, consisting of worms, molluscs, arthropods, and early vertebrates (our own ancestors), indicating that the design of vertebrates goes back to the early evolutionary phase of life-forms on Earth, as documented by fossils (Figure 1.10). The ever-increasing number of fossil specimens led palaeontologists to name this period the

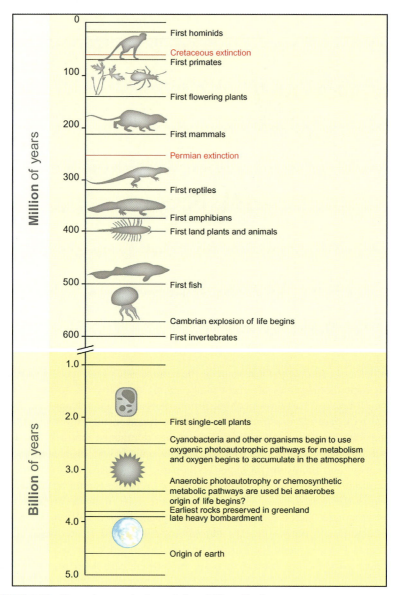

FIGURE 1.10   The major steps in the evolution of life on Earth.

"Cambrian explosion" [63,64]. A famous location where these organisms were found is the Burgess Shale in Canada and more recently several places in South China [65] as well. In both places fossils of the first vertebrate, *Pikaia*, were found. Another true vertebrate was the *Arandaspis*, which existed at approximately the same time as *Pikaia* about 500 million years ago, and had neither

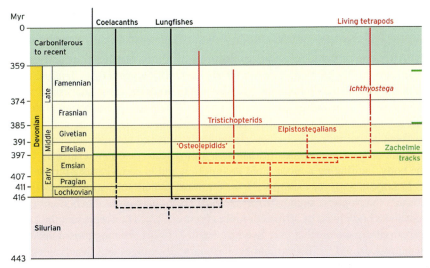

FIGURE 1.11   Simplified evolutionary tree of the living and fossil lobe-finned fishes and tetrapodes. *(Adapted from [69]).*

jaws nor fins. From the first fish a larger group branched off approximately 450 million years ago—the cartilaginous fish—which evolved with a skeleton of cartilage instead of bone. Today's sharks and rays belong to this group. In the cartilaginous fish group, the skeleton and the respiratory apparatus in a number of species developed further, whereby a portion of the stomach evolved into an air bladder or lung [64,66–68]. *Dipterus*, an early example of these lungfishes, showed both gill and lung breathing. The latest research indicates that the living tetrapods share with these lungfish and the Crossopterygians a common ancestor that probably lived during the Silurian Age (Figure 1.11). The transitional walk from water to land presumably took place in the freshwater ponds or rivers that, depending on the season, tended to dry up. About 400 million years ago, organisms with ossified front fins would have been capable of maneuvering shorter distances on riverbanks. Some tracks from Poland (Zachelmie fossils, Figure 1.11) [69], and recent fossils from Scotland [70] document this step from water onto land. From this stage of evolution, it was only a small step to the first amphibians, which were much better adapted to life between land and water. One of the best-known exponents of these early amphibians is *Ichthyostega* (Figure 1.11). Its bony skeleton displayed some of the following characteristics: massive vertebral column, fishtail, and a seven-fingered hand. The next 100 million years, from 350 to 250 million years ago, was dominated by amphibians, which increasingly adapted to survival on the land but continued to return to an aquatic milieu to lay their eggs. The first amphibians initially lived off arthropods, spiders, and scorpions that found their way from water to land slightly earlier, approximately at the same time that the first plants inhabited the land,

400 million years ago (Figure 1.10). Finally, about 270 million years ago, the reptiles, which evolved from the amphibians, were the first true vertebrates to live permanently out of water.

After one of the most severe extinctions of life on Earth, about 250 million years ago during the Perm age, for nearly 200 million years from Triassic times until the end of the Cretaceous age, dinosaurs were the most successful vertebrates on Earth, invading land, water, and air. Their abrupt extinction at the end of the Cretaceous age 65 million years ago might be related to the giant impact in Yucatan that induced obvious global environmental changes. Principally, dinosaurs differ from reptiles in that their limbs were positioned directly below the body, facilitating a more rapid locomotion and in some cases a bipedal one, as in *Lagosuchus*, for example. However, most remarkably, some terrestrial dinosaurs (sauropods) such as *Brachiosaurus* or *Supersaurus* reached enormous size and a body mass of 40 tons or more (Figure 1.12). They represent the largest terrestrial organisms that ever lived on Earth [71–74]. The gigantism occurring in sauropods and the conceptual evolutionary advantages and disadvantages of being big are the focus of very recent research, especially in view of metabolism, growth rates (up to 10 kg/day), and the cardiovascular system (Figure 1.13) [71,76–80]. As mentioned before, during the Mesozoic time, the dinosaurs were the dominating vertebrate life on Earth; however, during the Cenozoic times, the mammals took over, later showing giant forms as well, being driven by diversification to fill the ecological niches left by the dinosaurs, cooler environmental global temperature, and greater land areas—factors that might have triggered gigantism in sauropods as well [81]. The origin of mammals is assumed to start

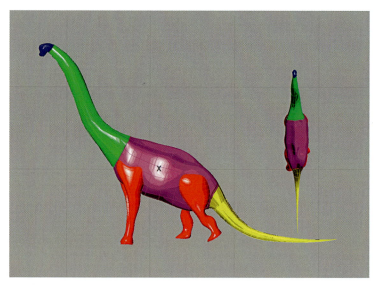

FIGURE 1.12   3-D reconstruction of *Brachiosaurus brancai* by photogrammetry and laser scanning. The center of gravity of this nearly 40,000 kg dinosaur is marked with an "x" [71].

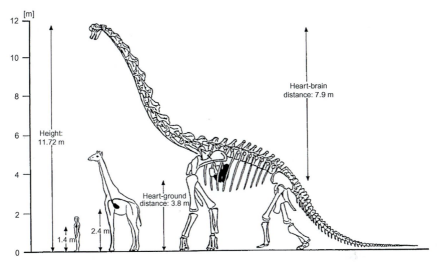

FIGURE 1.13 Hydrostatic distances acting on the heart and the cardiovascular system of man, giraffe, and *Brachiosaurus brancai* [75].

during Triassic/Jurassic times 160 million years ago [82]. From them early primates originated. These early primates pave the way for the human lineage, which will be described in detail in the following chapter.

1.5.2 Anthropological Aspects

Today, only one species, *Homo sapiens sapiens*, is living on Earth. This is an exceptional situation. Only in the last 25,000 years has this been the case in human evolution, because *Homo neanderthalensis* died out at that time. In contrast, during the last 4 million years, when modern humans developed, several different members of the genus *Homo* evolved separately and lived together in close coexistence, probably mixing up with each other now and then, here and there or dying out, as did *Homo neanderthalensis*. The following section will deal (i) with the major steps in human evolution and (ii) the high degree of diversity in the early genus *Homo*; in addition, it is a goal to demonstrate (iii) the environmental constraints that led finally (iv) to different kinds of anatomical, biochemical, physiological, and recently even social-cultural adaptations.

1.5.2.1 The Origin of Primates

The ancestry of primates is still doubtful. Approximately more than 70 million years ago, early primates evolved from shrew-like insectivores, the so-called *Prosimii*. They paved the way later for the anthropoids or higher primates (monkeys, apes, humans) when the climate on Earth got colder and colder in the northern hemisphere. This happened in North America and Europe about 30 million years ago and split the anthropoids into the New World monkeys (*Platyrrhini*) and the Old World monkeys (*Cercopithecoidea*) and hominoids, apes and humans (*Catarrhini*). Very recent findings from Saudi Arabia revealed the divergence of apes and Old World monkeys, the split hominoid-cercopithecoid, which happened between 29 and 24 million years ago (Figure 1.14) [83]. However, those anthropoids showed the following characteristics cited according to Lambert [84]: tail—long, short, or absent; prehensile hands and feet; thumbs at least partly opposable to fingers in most; big toe turns in for use in grasping in most; flat nails on toes and fingers (not claws); only two nipples on chest; large rounded cranium; large, forward-facing eyes; face protrusion variable; and 30-36 vertical teeth, canines mostly large. Furthermore, in contrast to the *Prosimii*, they had a different placenta, which indicates longer pregnancies. Among those early anthropoids we find those that were mainly swinging from tree to tree and still walking on

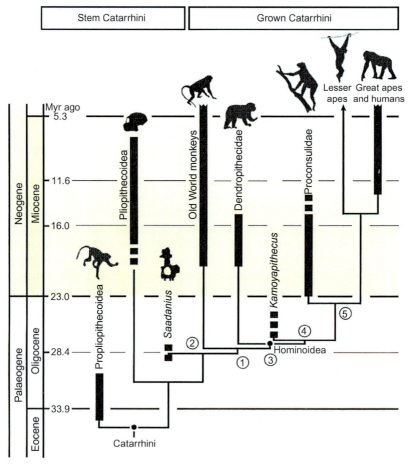

FIGURE 1.14 Presumed relationship of a new oligocene primate (*Saadanius*) and the divergence of apes and Old World monkeys in an overview. *(Adapted from [83]).*

all fours, such as Proconsul, approximately 20 million years ago. *Dryopithecus* (mid-late Miocene) and Proconsul (early-mid Miocene) belong to the so-called *Proconsulidae* (Figure 1.14), which show a mixture of ape and monkey features untypical for hominids, but they might perhaps be the ancestors of apes and man 22-15 million years ago, found in Europe and East Africa [85]. What gave rise to them was probably once again a changing climate [86].

During the Oligocene, the climate in the northern hemisphere had already become cooler and cooler. This development fostered (i) the change of the tropical rainforest in East Africa into wide savannas during the Miocene period, and (ii) the formation of new land bridges between Arabia and Africa (Ethiopia) because of the falling sea level. The primates had a better survival rate and were more able to cope with the increasing shortness and new kinds of food now available in the open savanna (more bark and leaves of evergreens), which gave rise to the anthropoids. During the 19 million years lasting from the Miocene epoch—from 24 to 5 million years ago—the climate in the East African Rift Valley became more and more arid. Tropical or subtropical forests changed into open savannas. It can be assumed that the related heat stress combined with water, salt, and food shortages [87–89] placed a decisive selective genetic pressure on the early primates living for several million years under those challenging environmental conditions. It has been estimated that due to the strict vegetarian type of paleolithic nutrition (fruits, leaves) that the salt (NaCl) intake might have been only about 0.6 g/day [90], and even in the middle and late Pleistocene salt intake may have been only 1.9 g/day [91]. In this context Watanabe et al. [91] presented the interesting hypothesis that probably the various uricase mutations that took place during last the 20 million years of hominoid evolution [92] are linked with this shortage of NaCl and the salt-sensitivity in humans. Watanabe et al. [91] assumed that the development of salt sensitivity in humans may be related to environmentally driven mutations of the urate oxidase (uricase) gene that made it nonfunctional, leading thereby to >2 mg/dL serum uric acid concentrations and significantly higher uric acid levels in hominoids (apes and humans) and certain New World monkeys as compared to other mammals (<2 mg/dL) [93]. The evolutionary advantage for early primates of high uric acid levels might have been that, apart from its function as a potent antioxidant [94], the higher uricase concentrations may act as a decisive factor to maintain blood pressure under low-sodium availability [91]. In addition, the changing climate probably gave rise to species with (i) a larger body mass, (ii) a shift from night to day activity, and (iii) better vision, which outranked smell [84]. In the same time span, when the *proconsulidae* evolved in Europe and Africa, the *pongidae* with *Sivapithecus*, *Ramapithecus*, and *Gigantopithecus* were living in Europe, Asia, and China. Today, the great apes include *Pongo* (orangutan), *Pan* (chimpanzee), and *Gorilla* (gorilla). But genetic studies have shown that because gorilla, especially chimpanzee, and modern man, differ in about

4.4% they are taken as one subfamily called Homininae [95]. As shown most remarkably in Figure 1.15, the evolutionary split between ape and man occurred—based on those recent molecular studies [95]—only 4-5 million years ago (Figure 1.16). Again, dramatic climate changes fostered evolutionary developments. About 3.5 million years ago the worldwide temperature

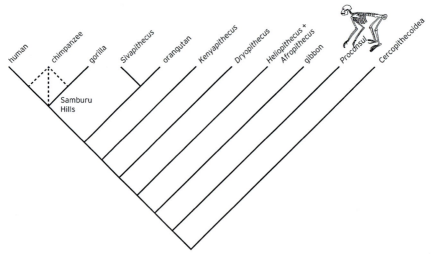

FIGURE 1.15   The major steps in the evolution of life and oxygen concentration on Earth from the origins 4.5 billion years ago to the present. *(Adapted from [71]).*

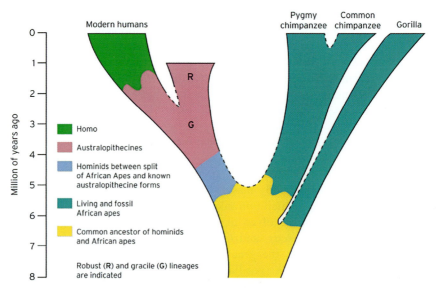

FIGURE 1.16   The relationship of hominids to modern humans and African apes. *(Adapted from [96]).*

dropped, and the Antarctic ice cap as we know it now was formed. These ice caps in the polar regions of the Earth kept so much water that the sea level sank and rainfalls dwindled, leading to an increasingly dry, cold climate with a typical wide spreading of savannas that decreased the availability of water, salt, and food items (fruits, seeds) for our ancestors and changed seeds to tough, hard, fibrous ones. The relatively rapid harsh changes in climate were combined in East Africa with a geological uplift of the East African Rift Valley Region (Figure 1.17), adding to the thermal and nutritional stresses, a hypobaric-hypoxic challenge [98].

This change definitely deserves a closer look. The observation that, based on paleoclimatological and geological data, rapid environmental changes occurred

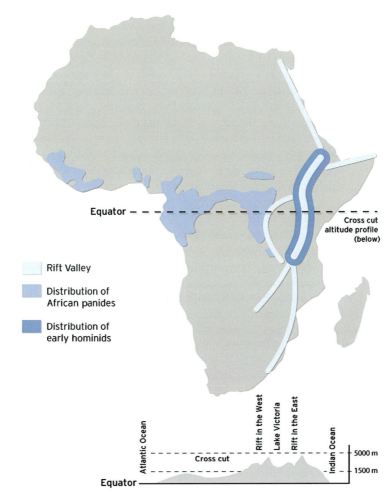

FIGURE 1.17   The Rift Valley, human evolution, and the role that altitude might have played. *(Adapted from [97]).*

from the Miocene on is interesting because they led to a dryer and high altitude environment in the East African Rift Valley system. Biomedical researchers have long known that there are quite a few mechanistic similarities in human physiology between adaptations for endurance performance and for hypoxia tolerance [99–102]. But how specific evolutionary pathways of complex physiological systems such as hypoxia tolerance or endurance performance have evolved within our phylogeny has remained largely unexplored due to methodological problems. Some years ago we approached this question from different perspectives to try to understand whether our ancestral phenotype is an adaptation for hypoxia tolerance and for endurance performance [98]. We summarized that comparative studies on animals and pinnipeds revealed that (i) some physiological-biochemical characteristics considered necessary in diving animals are conserved in all pinnipeds. These traits were presumably formed largely by negative selection (any mutations affecting them are not survivable). This kind of mutation includes diving apnea, bradycardia, tissue hypoperfusion, and hypometabolism of hypoperfused tissues [103]. (ii) A few other anatomical and physiological-biochemical characters are obviously more malleable and are correlated with long-duration diving and prolonged foraging at sea, such as spleen weight, blood volume, and red cell mass. The larger these anatomical and physiological-biochemical adaptations are, the greater the diving capacity, that is, diving duration, will be. As we pointed out earlier [98], because the relationships between diving capacity and any of these traits (enlarged spleens, blood volume, red cell mass) are evident, it is reasonable to conclude that these three traits extending diving duration thereby increase survival chances. These characters evolved presumably by positive selection to ensure prolonged diving times. Finally, (iii) the evolutionary physiology of the diving response can thus be described in terms of the degree of development to adaptable versus conservative characters of diving characters (such as bradycardia). Therefore, from an evolutionary point of view, it is important to study how these patterns change through time and how the patterns are lineage-specific. According to Hochachka and Somero [104], the timeline for response can be divided into three major categories: (i) acute, (ii) acclimatory, and (iii) genetic or phylogenetic. Figure 1.18 gives a diagrammatic summary of the formally defined relationships between time and physiological responses. As can be seen from this diagram, several crucial steps can be identified in this process, and it must be emphasized here that all the different steps may be altered gradually through evolutionary time. In Figure 1.19 the main "downstream consequences" of continuous hypobaric-hypoxic exposure are given as an example: (i) Sensing mechanisms are needed that tell the organism when the problem occurred and how dangerous it might be; (ii) this information has to be transmitted to various other parts or compartments of the body so that adequate countermeasures can be initiated by the body; then (iii), the genetically specified physiological and biochemical sensing systems will initiate by specialized signal transduction pathways either adequate acute or (iv) in larger scale more complex acclimatory responses. These acute/acclimatory responses or acclimations

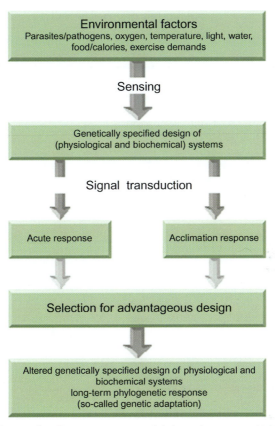

FIGURE 1.18 Acute and acclimatory processes and their sensing systems. *(Adapted from [98]).*

require some fraction of the organism's lifetime from milliseconds (cellular signal transduction) to days/months/years (metabolism). Both responses will lead the organism or finally—from a phylogenetic viewpoint—the species to the selection of an advantageous design (genetic adaptation) by using the same signal transduction pathways as for acute responses. A key role in this signal transduction pathway is played by the up- and downregulation of the hypoxic inducible factor (HIF-1 alpha). Here, more generally, the acute and acclimation response of lowlanders to hypobaric hypoxia should be taken as a typical example of how signal transduction pathways in different organs and systems initiate an adequate response to an environmental stress. These different approaches of the whole organism to the hypoxic stress are diagrammatically shown in Figure 1.19 and will be briefly described now. The acute and acclimation processes start with (i) the carotid body, which senses that arterial pO_2 falls [105], initiating increased ventilation to compensate for the drop in arterial pO_2. This mechanism, the so-called "hypoxic ventilatory response" (HVR) starts, although there might be a

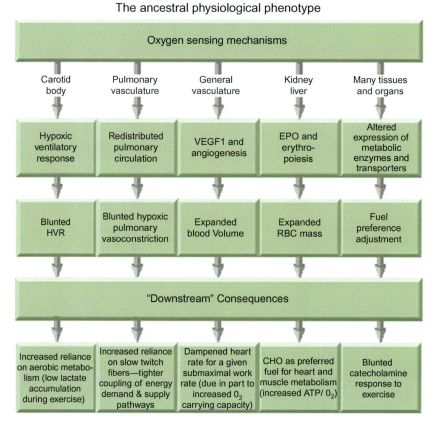

FIGURE 1.19 An integrated view of the impact of continuous hypobaric-hypoxic exposure on organs or organ systems in humans. *(Adapted from [98]).*

risk of developing a respiratory alkalosis. (ii) O_2 sensors in the vasculature of the lung initiate a vasoconstriction in those parts of the lung that show a ventilation/perfusion mismatch, the so-called Euler-Liljestrand-Reflex, ensuring thereby a more efficient oxygen uptake [106]. (iii) The decrease of pO_2 in other tissues will locally be sensed by O_2 sensors as well, and the vasculature in these tissues will release the endothelial growth factor (VEGF) to initiate angiogenesis, thereby ensuring a better microcirculation of these tissues stressed by hypoxia, especially in the heart, the skeletal muscles, and the brain [107–109]. (iv) O_2 sensors in the kidneys [108,110] and the liver [111] sense the drop in pO_2 as well and start the production of erythropoietin, which triggers an increase in red blood cell production in the bone marrow, thereby ensuring a higher oxygen transport capacity in the blood [108,112–115]. Finally, (v) tissue-specific O_2-sensing mechanisms and signal transduction pathways will induce an alteration in metabolism to ensure a better use of the energetic resources under hypobaric-hypoxic conditions [99,116,117]. This kind of response usually takes a few days.

However, these acute responses, in cases of severe hypobaric conditions might actually not be efficient enough to counteract the detrimental effects of a severe fall in pO_2 on the whole body. Therefore, with continued exposure of lowlanders to hypoxia, acclimation processes will start [98] that include the following main adaptations: (i) increased O_2 affinity of the oxygen sensors at a biochemical level so that any change will be sensed earlier, producing an exaggerated hypoxic ventilator response to a given hypoxic stimulus [99,105,116]; (ii) exaggerating the Hering-Breuer reflex to further improve oxygen uptake in the lung [106,118]; (iii) high angiogenic [106] and (iv) erythropoietic activity will still be maintained [108,115], the latter leading eventually to excessive erythrocytosis, the so-called Monge's disease [119]; finally, (v) metabolic pathways will give preference to carbohydrate metabolism during physical exercise under hypoxic conditions [120]. In summary, we can assume that several million years of harsh environmental conditions in the East African Rift Valley favored an endurance and hypoxic adaptation.

As shown in Figure 1.20, it took the human lineage from *Proconsul* to humans several million years to progress from quadrupedal to a bipedal locomotion. In human lineage *Ardipithecus ramidus*, at 4.4 million years, is currently the oldest known hominid from Aramis in Ethiopia, which also lived in the East African Rift Valley system. Its precursor has not yet been identified (Figure 1.21). There is good evidence that *Ardipithecus ramidus* at least partly used bipedal locomotion, a hallmark in human evolution in conjunction with increasing brain size and remodeled teeth [121,122]. Only 200,000 years later *Australopithecus anamensis* from Kenya joined the human lineage, and

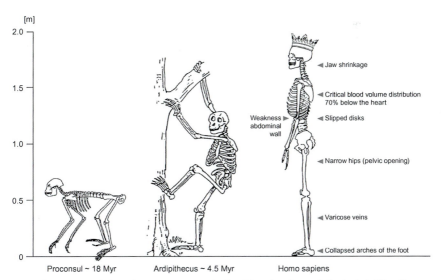

FIGURE 1.20 Evolution of *Homo sapiens* and related functional-morphological pitfalls of human upright posture.

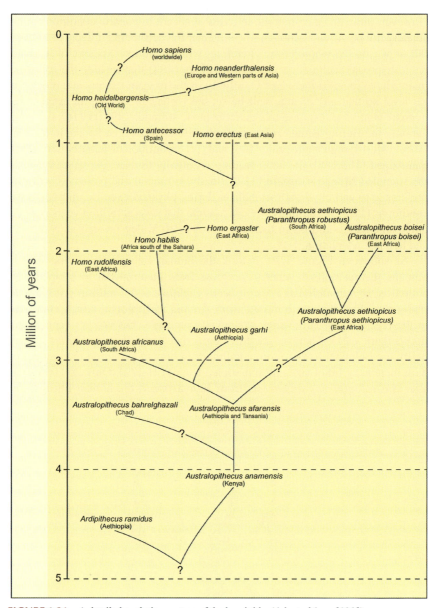

FIGURE 1.21   A detailed evolutionary tree of the hominids. *(Adapted from [121]).*

then followed *Australopithecus bahrelghazali* and *Australopithecus afarensis* (Figure 1.21) from which 3.8-million-year-old fossil footprints are known (Laetoli) [123], clearly indicating that this early australopithecine was upright walking, a kind of locomotion that differs from what we know from apes because it afforded a decisive remodeling of hips and lower limbs [124]. Undisputedly, this is a hallmark in human evolution because this evolutionary step enabled free use of the hands and the rise in the use of tools, which have been identified at 2.5 million years ago. An energetically advanced bipedal style of locomotion, upright posture combined with increasing vision capabilities, and free hands definitely stimulated and challenged brain functions in early australopithecines. The different selective advantages of a bipedial locomotion are summarized in Figure 1.22. The loss of hair and the development of sweat glands as well as emissary veins in the skull indicate that, although we have to deal with a dryer, higher, and colder overall climate, the thermal stresses on the body during the search for food (plants, fruits, nuts), or increased metabolic heat production due to hunting prey were inevitable, and had to be counterbalanced by these physiological adaptations in the late phases of human development.

In the following 2 million years, the changing climate and morphology in the different regions of North, East, and South led obviously to a rapid differentiation of australopithecines (Figure 1.21). *Australopithecus afarensis* from Ethiopia, Laetoli, and Tanzania lived from >4 to 2.5 million years ago, *Australopithecus africanus* in the south of Africa from about 3 to <2.5 million years ago. Both were quite small (up to 1.4 m) with a body mass of 30-70 kg and a brain size of about 400-500 ml [121]. In contrast, *Australopithecus* (*Paranthropus*) *robustus* and *Australopithecus* (*Paranthropus*) *boisei*, which evolved from *Australopithecus* (*Paranthropus*) *aethiopicus* (2.8 million years ago), were much taller and heavier. *Australopithecus boisei*, which lived from 2.6 until 1.2 million years ago, showed a body height up to 1.4 m, a weight between 40 and 80 kg, and brain size of about 530 ml, still much less than that

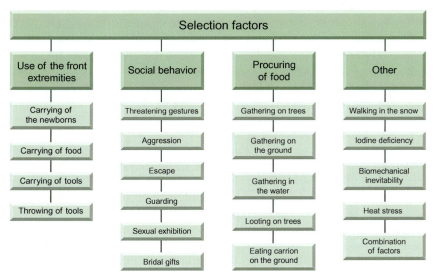

FIGURE 1.22 An overview of the evolutionary advantages of bipedal locomotion in humans. *(Adapted from [125]).*

of most members of the genus *Homo*, which have a brain volume of >700 ml. *Australopithecus robustus* lived from 2.1 million years ago and died out presumably 1.5 million years ago, was up to 1.7 m in height, and had a body mass of 70 kg and a brain size of 410-530 ml. The different body builds might be related to their different lifestyles, but this is hardly disputed among paleoanthropologists [139]. Whereas some researchers assume that the gracile australopithecines, *Australopithecus afarensis* and *Australopithecus africanus*, were eating fruit, seeds, and meat, some argue they did not. It seems undisputed that the robust forms of the australopithecines, *Australopithecus robustus* and *Australopithecus boisei*, based on the morphologies of their skull and teeth, were eating mainly fruits and hard seeds. However, whereas the two robust australopithecines died out about 1 million years ago, the smaller but probably smarter, brainier gracile australopithecus gave rise to modern, bipedal walking *Australopithecus garhi* (2.6 million years old, Ethiopia), *Homo rudolfensis* (1.8 million years old, East Africa), and finally the first stone tool-making humans, *Homo habilis* (Figure 1.21). *Homo habilis* is commonly regarded as an intermediate between australopithecines and *Homo sapiens*, which lived at least from 2.4 until 1.6 million years ago in the East African region Omo (Shungara), East Turkana (Koobi Fora), Olduvai Gorge, Uraha, and in Sterkfontein, South Africa. It showed a body height of 1.5 m, a body weight of 50 kg, and a brain size of about 700 ml. However, especially the great variation in morphology of skulls of *Homo habilis* found in Koobi Fora (East Africa), are complicating the picture of early human evolution for anthropologists. Some researchers argue therefore that *Homo habilis* was not the only human-like inhabitant of the East African region around 2 million years ago [122]. Also disputed is whether a bulge in the left skull in the area of the broca, the area for speech in the central nervous system in humans today, can be taken as first signs of the use of speech by *Homo habilis* because the larynx seems not be designed to create as many sounds as ours. However, there is no dispute that about 1.8 million years ago, a new type of early humans appeared, named *Homo ergaster* in East Africa and *Homo erectus* in East Asia [121]. *Homo ergaster* lived for about 1.6 million years until 200,000 years ago (when this species exactly died out is disputed among scientists) was up to 1.8 m in height, about 70 kg in weight, and showed a brain size of 800-900 ml, considerably above *Homo habilis* (700 ml). This species was the first to spread via land bridges from northern Africa to Arabia and from there quite rapidly to Southeast Asia (Figure 1.21). They made tools from wood, invented (1 million years after the first use of pebble tools in Olduvai) the hand axe, which assumes the ability to comprehend and think through a coherent plan of action to produce it. *Homo ergaster* were nomadic hunters and gatherers, used fire, and were obviously eating a mixed diet. The emergence of our own species, *Homo sapiens*, occurred by about 400,000 years ago. Sometimes these early humans are lumped together into the so-called group of "archaic *Homo sapiens*." To this archaic *Homo sapiens* group probably belong *Homo heidelbergensis* and *Homo rhodesiensis* in Africa. Both were as robust and muscular as *Homo ergaster* but showed a definitely larger brain, with 1100-1300 ml, than *Homo ergaster* and some anatomical signs at the base of the skull that could be interpreted as first signs of a voice box of a modern type [122]. It might well be that during interglacial warm periods *Homo ergaster* invaded Europe and evolved there when isolated during the following ice periods into such species as *Homo antecessor* (800,000 years old, northern Spain) and finally *Homo heidelbergensis*, the remains of which were found all over the old world from 600,000 to 200,000 years ago, which gave rise to *Homo neanderthalensis* (Figure 1.21). In this period dramatic climatic changes happened. The coincidence of marked steps in human evolution and dramatic climatic changes are again more striking. *Homo neanderthalensis* had brains that were even larger than the brains of the present *Homo sapiens* (1200-1750 ml) and were, or became, perfectly cold-adapted when environmental temperatures dropped significantly in Europe from 120,000 to 70,000 years ago and remained cold, interrupted by only short interglacial warm periods. Most researchers believe that these modern humans left Africa again about 150,000-200,000 years ago over a land bridge between Africa and Arabia and met *Homo neanderthalensis* in the Middle East [121].

Palaeo-climatological studies have revealed that temperatures during the summer were not higher than 10 °C. The heavy body build of *Homo neanderthalensis* and its technical skills to build shelters, hunt, gather, and prepare a rich variety of different kinds of food by use of fire enabled *Homo neanderthalensis* to survive under environmental conditions becoming continuously harsher. During the Würm Age, 70,000-30,000 years ago, when Northern Europe was in a strong grip of the Ice Age, Neanderthal humans clearly dominated the field and their remains were found all over Europe, even in Asia. Some of them even started 50,000 years ago to inhabit the Tibetan Plateau (4000 m) [126,127] and Figure 1.23 [98]. However, they disappeared rapidly in a time span of only 10,000 years when about 40,000-30,000 years ago the first real early modern humans, the so-called *Cro Magnons*, appeared from the southeast (*Homo sapiens sapiens*). Very recent studies have revealed that *Homo neanderthalensis* and these first humans had a long period of coexistence of about 60,000 years, and obviously even mixed with each other [140], so that even today about 5% of our genome can be traced back to *Homo neanderthalensis* [128]. About 25,000 BCE *Homo neanderthalensis* died out in Northern Europe, and approximately 13,000 years later, the earliest man reached North America over a land bridge between Siberia and Alaska at 12,000 BCE. It then only took about 1000 years for these humans to reach New Mexico (Clovis culture) [129] and an additional 1000 years to find their way down to Patagonia [130]. This means that these humans moved south with an expansion speed of ~13 km/year. Obviously, they preferred not only to live along the coastal areas in North and South America but also invaded extreme altitudes as well, as in case of the Quechua and Aymaras in Bolivia, Peru, and Chile (Figure 1.23) [98]. A major pitfall for the possible development of agriculture in America compared with Europe/Asia Minor was, as pointed out by Jared Diamond [130], the extinction

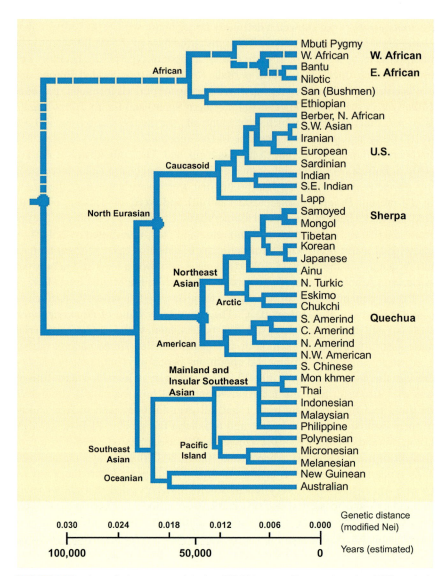

FIGURE 1.23   An evolutionary tree of the last 150,000 years of humans based on DNA analysis. *(Adapted from [98]).*

of the American megafauna on the double continent that occurred after the 22nd Ice Age 11,000 years ago (Clovis culture).[4] As a dramatic consequence, no large animals were later available for domestication on the double continent. In Asia Minor, in contrast, animals available for domestication by 8000 BCE were cattle,

---

4. In contrast to this, on the Australian continent the megafauna was extinguished by humans [131].

sheep, goats, horses, reindeer, as well as water buffalos, yaks, dromedaries, and camels. The advantages of domesticating animals in order to keep livestock are obvious: (i) Meat was available, and for those ethnic groups who kept a high level of enzymatic activity of lactase in adolescence due to a special mutation of chromosome 2 [132], milk, butter, cheese, and yoghurt were available as well.[5] (ii) Multiple calories could be obtained within one life span of a domesticated animal instead of only those available once after immediate slaughtering. (iii) The better nutrition achieved in peasant societies led to higher birth rates compared to the lifestyle of hunters and gatherers. And finally, (iv) a higher density of population led to the differentiations and specialization of new occupations, including those of writers who fostered the documentation of knowledge. These advantages paved the way for the evolution of agriculture in Asia Minor starting with the domestication of plants (emmer) 8500 BCE in the Fertile Crescent, 6500 BCE in Greece, and 5000 BCE in Germany in conjunction with the building and establishing of larger, complex societies concentrated in the first Neolithic cities such as Catal Hüyük in Southern Anatolia (Turkey) [134].

Once again it can be seen that environmental and nutritional aspects played a very distinctive and decisive role in human evolution. Let us therefore finally briefly summarize the major three evolutionary steps that started 25 million years ago with gathering food, followed approximately 23 million years later by hunting and by agriculture only 10,000 years ago, respectively. The first long phase (gathering) was characterized by a high food diversity, very low caloric input, salt shortage, and because of an increasingly arid climate, a long-lasting foraging time combined with long-distance running, that is, relatively high exercise levels [135]. The hunting phase started about 2 million years ago and led to a decrease in biodiversity and an increase in caloric intake, but salt was still difficult to find. Finally, 10,000 years ago, as the agriculture phase started, biodiversity of different food items still decreased, caloric intake dramatically increased because of the constant availability of sugars and fatty food items, combined with an approximately 15 g/day salt intake as compared to the 1.5 g/day in our ancestors in the Pleistocene mentioned before. The year-round, ubiquitous high caloric and high salt intake combined with low physical activity causes—at least in the Western and especially North American countries in the last 30 years—a dramatic increase in the incidence of cardiovascular and metabolic diseases, namely the metabolic syndrome (diabetes mellitus, impaired glucose tolerance, insulin resistance, dyslipidemia, obesity) [136–138]. Over

---

5. Today, lactose intolerance—also called lactase deficiency or/and hypolactasia—is very rare to find in Northern Europe (Sweden 2%, Denmark 5%) and common in the population of China (94%) and Southeast Asia (98%). Recent anthropological studies revealed that the autosomal-dominant mutation of the LCT (Lactase Gen)-allele of chromosome 2 (Gene ID 3938), which is responsible for keeping a high enzymatic activity of lactase, occurred after the domestication of cattle had started http://de.wikipedia.org/wiki/Laktoseintoleranz [133].

25 million years it seems that our human ancestors had difficulties in permanently assuring adequate fluid, food, and salt supply. Because body weight, body composition, and physical fitness are crucial parameters for human performance in extreme environments, a short overview on this topic will be given in the following chapter before the specific adaptation of humans to extreme environments will be discussed.

REFERENCES

[1] De Bernardis F, Melchiorri A, Verde L, Jimenez R. The cosmic neutrino background and the age of the Universe. J Cosmol Astropart Phys 2008:1–9.

[2] Feng J, Trodden M. Der verborgene Plan des Kosmos. Spektrum der Wissenschaft 2011;1:38–46.

[3] Feng J. Dark matter candidates from particle physics and methods of detection. Ann Rev Astron Astrophys 2010;48:495–545.

[4] Gilbert DL. Evolutional aspects of atmospheric oxygens and organisms. Handbook of physiology, vol. II. Oxford University Press; 1996. p. 1059–94.

[5] Mortlock DJ, Warren SJ, Venemans BP, Patel M, Hewett PC, McMahon RG, et al. A luminous quasar at a redshift of $z = 7.085$. Nature 2011;474:616–9.

[6] Fumagalli M, O'Meara JM, Prochaska JX. Detection of pristine gas two billion years after the Big Bang. Science 2011;334:1245–9.

[7] Seven Year Microwave Sky. http://wmap.gsfc.nasa.gov/media/101080; 2012.

[8] Morris M. The Milky Way. The world book encyclopedia, World Book Incorporated; 2002, p. 551.

[9] Goodwin S. Hubble's universe: a new picture of space. Constable; 1996.

[10] Kerrod R. Encyclopedia of science: heavens, stars, galaxies, and the solar system. New York: MacMillan Reference USA; 1997.

[11] Pananides NA, Arny T. Introductory astronomy. Addison Wesley Longman Publishing Co.; 1979.

[12] Coble CR, Murray EG, Rice DR. Earth science. Englewood Cliffs, N.J.: Prentice-Hall; 1987.

[13] Karachentsev ID, Kashibadze OG. Masses of the Local Group and of the M81 Group estimated from distortions in the local velocity field. Astrophysics 2006;49:3–18.

[14] Schätzing F. Der Schwarm. Köln: Kiepenheuer & Witsch; 2004.

[15] The Sun. http://www.enchantedlearning.com/subjects/astronomy/sun/; 2012.

[16] About the Sun. http://solar-center.stanford.edu/about/; 2012.

[17] Gilbert DL. Evolutionary aspects of atmospheric oxygen and organisms. Handbook of physiology, environmental physiology, Wiley; 2011.

[18] Sterne. http://marvin.sn.schule.de/~erzkoll/projekte/astro1/sterne.html 2012.

[19] Bouvier A, Wadhwa M. The age of the Solar System redefined by the oldest Pb-Pb age of a meteoritic inclusion. Nat Geosci 2010;3:637–41.

[20] Gurnett DA, Kurth WS. Radio emissions from the outer heliosphere. Space Sci Rev 1996;78:53–66.

[21] Oort cloud. http://en.wikipedia.org/wiki/Oort_cloud; 2012.

[22] Lyra, a constellation for August. http://www.rasnz.org.nz/Stars/Lyra.htm; 2012.

[23] Canup RM, Righter K. Origin of the Earth and Moon. Tucson, AZ: University of Arizona Press; 2000.

[24] Münker C, Pfänder JA, Weyer S, Büchl A, Kleine T, Mezger K. Evolution of planetary cores and the Earth-Moon system from Nb/Ta systematics. Science 2003;301:84–7.

[25] Lang KR. The Cambridge guide to the solar system. Cambridge: Cambridge University Press; 2011.

[26] Zhang J, Dauphas N, Davis AM, Leya I, Fedkin A. The proto-Earth as a significant source of lunar material. Nat Geosci Lett 2012;http://dx.doi.org/10.1038/ngeo1429.

[27] Darwin GH. On the precession of a viscous spheroid and on the remote history of the Earth. Philos Trans R Soc Lond A 1879;170:447–538.

[28] Mackenzie D. The big splat, or how the Moon came to be. John Wiley & Sons; 2003.

[29] Jordan TH. Structural geology of the Earth's interior. Proc Natl Acad Sci 1979;76:4192–200.

[30] Binder AB. On the origin of the Moon by rotational fission. Moon 1974;11:53–76.

[31] Binder AB, Gunga H-C. Young thrust-fault scarps in the highlands: evidence for an initially totally molten moon. Icarus 1985;63:421–41.

[32] Solomon SC, Chaiken J. Thermal expansion and thermal stress in the Moon and terrestrial planets: clues to early thermal history. Proc Lunar Sci Conf 7th 1976;3229–43.

[33] Watters TR, Robinson MS, Beyer RA, Banks ME, Bell JFI, Pritchard ME, et al. Evidence of recent thrust faulting on the moon revealed by the lunar reconnaissance orbiter camera. Science 2010;329:936–40.

[34] Watters TR, Robinson MS, Banks ME, Tran T, Denevi BW. Recent extensional tectonics on the Moon revealed by the Lunar Reconnaissance Orbiter Camera. Nat Geosci 2012;5:181–5.

[35] van der Bogert CH, Hiesinger H, Banks ME, Watters TR, Robinson MS. Derivation of absolute model ages for lunar lobate scarps. Lunar Planet Sci 2012;43: (#1847).

[36] Binder AB. Post-imbrian global lunar tectonism: evidence for an initially totally molten Moon. Moon Planets 1982;26:117–33.

[37] Brunner B. Mond: Die Geschichte einer Faszination. München: Antje Kunstmann Verlag; 2011.

[38] Gortner RA. The water content of medusae. Science 1994;77:282–3.

[39] de Leeuw NH, Catlow CRA, King HE, Putnis A, Muralidharan K, Deymier P, et al. Where on Earth has our water come from? Chem Commun 2010;46:8923–5.

[40] Schmidt-Nielsen K. Animal physiology. 5th ed. Cambridge University Press; 1997.

[41] Brasier M, McLoughlin N, Green O, Wacey D. A fresh look at the fossil evidence for early Archaean cellular life. Philos Trans R Soc B 2006;361:887–902.

[42] Stromatolites. The oldest fossils. http://www.fossilmuseum.net/Tree_of_Life/Stromatolites.htm; 2012.

[43] Oró J, Kimball AP. Synthesis of purines under possible primitive Earth conditions. I. Adenine from hydrogen cyanide. Arch Biochem Biophys 1961;94:217–27.

[44] Hartogh P, Lis DC, Bockelée-Morvan D, de Val-Borro M, Biver N, Küppers M, et al. Ocean-like water in the Jupiter-family comet 103P/Hartley 2. Nature 2011;478:218–20.

[45] Schadwinkel A. Quelle im All. Zeit Online, 2011, October 6.

[46] European Space Agency. Exobiology in the solar system & the search for life on Mars. Report from the ESA exobiology team study 1997-1998. European Space Agency; 1999

[47] Mora C, Tittensor DP, Adl S, Simpson AGB, Worm B. How many species are there on Earth and in the ocean? PLoS Biol 2011;9:http://dx.doi.org/10.1371/journal.pbio.1001127.

[48] Glaw F, Köhler J, Townsend TM, Vences M. Rivaling the world's smallest reptiles: discovery of miniaturized and microendemic new species of leaf chameleons (*Brookesia*) from Northern Madagascar. PLoS ONE 2012;7(2):e31314.

[49] Bhanoo SN. A deep-sea ecosystem unlike any other. The New York Times 2012; (January 9, 2012).

[50] Cowan DA. The upper temperature for life – where do we draw the line? Trends Microbiol 2004;12:58–62.

[51] Stetter KO. History of discovery of the first hyperthermophiles. Extremophiles 2006;10: 357–62.

[52] Rogers AD, Tyler PA, Connelly DP, Copley JT, James R, Larter RD, et al. The Discovery of new deep-sea hydrothermal vent communities in the Southern Ocean and implications for biogeography. PLoS Biol 2012;10:http://dx.doi.org/10.1371/journal.pbio.1001234.

[53] de Duve C. Ursprung des Lebens. Präbiotische Evolution und die Entstehung der Zelle. Heidelberg, Berlin, Oxford: Spektrum; 1994.

[54] Joyce GF. The antiquity of RNA-based evolution. Nature 2002;418:214–21.

[55] Schopf JW. Major events in the history of life. Jones & Bartlett Learning; 1992.

[56] Schopf JW. Microfossils of the Early Archean Apex chert: new evidence of the antiquity of life. Science 1993;260:640–6.

[57] Schopf JW. Disparate rates, differing fates: tempo and mode of evolution changed from the Precambrian to the Phanerozoic. Proc Natl Acad Sci 1994;91:6735–42.

[58] Schopf JW. Cradle of life: the discovery of Earth's earliest fossils. Princeton University Press; 1999.

[59] Stetter KO. Hyperthermophiles in the history of life. In: Bock GR, Goode JA, editors. Evolution of hydrothermal ecosystems on Earth (and Mars?). Chichester: Wiley; 1996.

[60] Rasmussen B, Bengtson S, Fletcher IR, McNaughton NJ. Discoidal impressions and trace-like fossils more than 1200 million years old. Science 2002;296:1112–5.

[61] Tice MM, Lowe DR. Photosynthetic microbial mats in the 3,416-Myr-old ocean. Nature 2004;431:549–52.

[62] Westall F. Life on the early Earth: a sedimentary view. Science 2005;308:366–7.

[63] Briggs EG, Crowther PR. Palaeobiology, a synthesis. Blackwell Science; 1990.

[64] Strickberger MW. Evolution. 3rd ed. Sudbury, MA: Jones and Bartlett; 2000.

[65] Zhang X, Liu W, Zhao Y. Cambrian Burgess Shale-type Lagerstätten in South China: distribution and significance. Gondwana Res 2008;14:255–62.

[66] Liem KF. Form and function of lungs: the evolution of air breathing mechanisms. Am Zool 1988;28:739–59.

[67] Perry SF. Lungs: comparative anatomy, functional morphology, and evolution. In: Gans C, Gaunt AS, editors. Biology of the reptilia, vol. 19, morphology G, visceral organs. Ithaca, N.Y., Society for the Study of Amphibians and Reptiles; 1998. p. 1–92.

[68] Perry SF, Sander PM. Reconstructing the evolution of the respiratory apparatus in tetrapods. Respir Physiol Neurobiol 2004;144:125–39.

[69] Janvier P, Clément G. Muddy tretrapod origins. Nature 2010;463:40–1.

[70] Smithson TR, Wood SP, Marshall JEA, Clack JA. Earliest carboniferous tetrapod and arthropod faunas from Scotland populate Romer's Gap. Proc Natl Acad Sci 2012; [published ahead of print].

[71] Gunga H-C, Suthau T, Bellmann A, Stoinski S, Friedrich A, Trippel T, et al. A new body mass estimation of Brachiosaurus brancai Janensch, 1914 mounted and exhibited at the Museum of Natural History (Berlin, Germany). Fossil Rec 2008;11:33–8.

[72] Economos AC. The largest land mammal. J Theor Biol 1981;89:211–5.

[73] Günther B, Morgado E, Kirsch K, Gunga H-C. Gravitational tolerance and size of *Brachiosaurus brancai*. Mitteilungen aus dem Museum für Naturkunde in Berlin. Geowiss Reihe 2002;5:263–7.

[74] Rauhut OWM, Fechner R, Remes K, Reis K. How to get big in the mesozoic: the evolution of the sauropodomorph body plan. In: Klein N, Remes K, Gee CT, Sander PM, editors. Biology of the sauropod dinosaurs. Understanding the life of giants. Bloomington and Indianapolis: Indiana University Press; 2011. p. 119–49.

[75] Gunga H-C, Kirsch K, Baartz F, Röcker L, Heinrich W-D, Lisowski W, et al. New data on the dimensions of Brachiosaurus brancai and their physiological implications. Naturwissenschaften 1995;82:190–2.

[76] Sander PM. The pachypleurosaurids (reptilia: nothosauria) from the Middle Triassic of Monte San Giorgio (Switzerland) with the description of a new species. Philos Trans R Soc Lond B 1989;30:561–666.

[77] Klein N, Remes K, Gee CT, Sander PM. Biology of the sauropod dinosaurs. Understanding the life of giants. Bloomington, Indianapolis: Indiana University Press; 2012.

[78] Gunga H-C, Suthau T, Bellmann A, Friedrich A, Schwanebeck T, Stoinski S, et al. Body mass estimations for *Plateosaurus engelhardti* using laser scanning and 3D reconstruction methods. Naturwissenschaften 2007;94:623–30.

[79] Preuschoft H, Hohn B, Stoinski S, Witzel U. Why so huge? Biomechanical reasons for the acquisition of large size in sauropod and theropod dinosaurs. In: Klein N, Remes K, Gee CT, Sander PM, editors. Biology of the sauropod dinosaurs. Understanding the life of giants. Bloomington and Indianapolis: Indiana University Press; 2012. p. 197–218.

[80] Stoinski S, Suthau T, Gunga H-C. Reconstructing body volume and surface area of dinosaurs using laser scanning and photogrammetry. In: Klein N, Remes K, Gee CT, Sander PM, editors. Biology of the sauropod dinosaurs. Understanding the life of giants. Bloomington and Indianapolis: Indiana University Press; 2011. p. 94–104.

[81] Smith FA, Boyer AG, Brown JH, Costa DP, Dayan T, Ernest SKM, et al. The evolution of maximum body size of terrestrial mammals. Science 2010;330:1216–9.

[82] Luo Z-X, Yuan C-X, Meng Q-J, Ji Q. A Jurassic eutherian mammal and divergence of marsupials and placentals. Nature 2011;476:442–5.

[83] Zalmout IS, Sanders WJ, MacLatchy LM, Gunnell GF, Al-Mufarreh YA, Ali MA, et al. New oligocene primate from Saudi Arabia and the divergence of apes and Old World monkeys. Nature 2010;466:360–4.

[84] Lambert D. The diagram group. The field guide to early man. New York: The Diagram Visual Information Ltd.; 1987.

[85] Henke W, Rothe H. Paläoanthropologie. Berlin, Heidelberg, New York: Springer; 1994.

[86] Henke W, Rothe H. Stammesgeschichte des Menschen. Springer; 1999.

[87] Fitzsimons JT. The physiology of thirst and sodium appetite. Cambridge University Press; 1979.

[88] Denton D. The hunger for salt. Berlin, Heidelberg, New York: Springer-Verlag; 1982.

[89] Grossman SP. Thirst and sodium appetite. Academic Press; 1990.

[90] Eaton SB, Konner M. Paleolithic nutrition – a consideration of its nature and current implications. N Engl J Med 1985;312:283–9.

[91] Watanabe S, Kang D-H, Feng L, Nakagawa T, Kanellis J, Lan H, et al. Uric acid, hominoid evolution, and the pathogenesis of salt-sensitivity. Hypertension 2002;40:355–60.

[92] Wu X, Muzny DM, Lee CC, Caskey CT. Two independent mutational events in the loss of urate oxidase during hominoid evolution. J Mol Evol 1992;34:78–84.

[93] Johnson RJ, Kang D-H, Feig D, Kivlighn S, Kanellis J, Watanabe S, et al. Is there a pathogenetic role for uric acid in hypertension and cardiovascular and renal disease? Hypertension 2003;41:1183–90.

[94] Ames BN, Cathcart R, Schwiers E, Hochstein P. Uric acid provides an antioxidant defense in humans against oxidant- and radical-caused aging and cancer: a hypothesis. Proc Natl Acad Sci USA 1981;78:6858–62.

[95] Barriel V. Der genetische Ursprung des modernen Menschen. Spektrum der Wissenschaft 2000;3:80–7.

[96] Wood BA. Evolution of australopithecines. In: Jones S, Martin R, Pilbeam D, editors. The Cambridge encyclopedia of human evolution. Cambridge: Cambridge University Press; 1995. p. 231–40.

[97] Facchini F. Der Mensch. Ursprung und Entwicklung. Augsburg: Natur Verlag; 1991.

[98] Hochachka PW, Gunga HC, Kirsch K. Our ancestral physiological phenotype: an adaptation for hypoxia tolerance and for endurance performance. Proc Natl Acad Sci 1998;95:1915–20.

[99] Hochachka PW. Muscles as molecular and metabolic machines. Boca Raton: CRC Press; 1994.

[100] Vrba ES, Denton GH, Partridge TC, Burckle LH. Paleoclimate and evolution with emphasis on human origins. New Haven, CT: Yale University Press; 1994 p. 1–547.

[101] Brooks GA, Fahey TD, White TP. Exercise physiology – human bioenergetics and its applications. London: Mayfield; 1996.

[102] Levine BD, Stray-Gundersen J. A practical approach to altitude training: where to live and train for optimal performance enhancement. Int J Sports Med 1992;13:S209–12.

[103] Hochachka PW, Mottishaw PD. Evolution and adaptation of the diving response: phocids and otariids. In: Pörtner HO, Playle RC, editors. Cold Ocean Physiology. Cambridge, U.K.: Cambridge University Press; 1998. p. 391–431.

[104] Hochachka PW, Somero GN. Biochemical adaptation. Princeton University Press; 1984.

[105] Lahiri S. Environmental physiology. Handbook of physiology. New York: Oxford University Press; 1996 p. 1183–206.

[106] Heath D, Williams DR. Man at high altitude: pathophysiology of acclimatization and adaptation. Churchill Livingstone; 1981.

[107] Richardson RS, Wagner H, Mudaliar SR, Henry R, Noyszewski EA, Wagner PD. Human VEGF gene expression in skeletal muscle: effect of acute normoxic and hypoxic exercise. Am J Physiol 1999;277:H2247–52.

[108] Jelkmann W. Erythropoietin: molecular biology and clinical use. F.P. Graham; 2003.

[109] Harik N, Harik SI, Kuo N-T, Sakai K, Przybylski RJ, LaManna JC. Time-course and reversibility of the hypoxia-induced alterations in cerebral vascularity and cerebral capillary glucose transporter density. Brain Res 1996;737:335–8.

[110] Bauer C, Kurtz A. Oxygen sensing in the kidney and its relation to erythropoietin production. Annu Rev Physiol 1989;51:845–56.

[111] Goldberg MA, Dunning SP, Bunn HF. Regulation of the erythropoietin gene: evidence that the oxygen sensor is a heme protein. Science 1988;242:1412–5.

[112] Jelkmann W. Erythropoietin: structure, control of production, and function. Physiol Rev 1992;72:449–89.

[113] Wenger RH. Cellular adaptation to hypoxia: O_2-sensing protein hydroxylases, hypoxia-inducible transcription factors, and O_2-regulated gene expression. FASEB J 2002;16:1151–62.

[114] Wenger RH, Gassmann M. Oxygen(es) and the hypoxia-inducible factor-1. Biol Chem 1997;378:609–16.

[115] Gunga H-C, Kirsch KA, Röcker L, Kohlberg E, Tiedemann J, Steinach M, et al. Erythropoietin regulations in humans under different environmental and experimental conditions. Resp Physiol Neurobiol 2007;158:287–97.

[116] Hochachka PW, Buck LT, Doll C, Land SC. Unifying theory of hypoxia tolerance: molecular/metabolic defense and rescue mechanisms for surviving oxygen lack. Proc Natl Acad Sci 1996;93:9493–8.

[117] Bunn HF, Poyton RO. Oxygen sensing and molecular adaptation to hypoxia. Physiol Rev 1996;76:839–85.

[118] West JB. High altitude physiology. Hutchinson Ross; 2006.

[119] Winslow RM, Monge C. Hypoxia, polycythemia, and chronic mountain sickness. Baltimore: Johns Hopkins University Press; 1987.

[120] Brooks GA. Mammalian fuel utilization during sustained exercise. Comp Biochem Physiol Part B 1998;120:89–107.

[121] Tattersall I, Matternes JH. Wir waren nicht die Einzigen. Spektrum der Wissenschaft 2000;3:40–7.

[122] Jones S, Martin R, Pilbeam D. The Cambridge encyclopedia of human evolution. New York: Cambridge University Press; 1992.

[123] Leakey MD, Hay RL, Curtis GH, Drake RE, Jackes MK, White TD. Fossil hominids from the Laetoli Beds. Nature 1976;262:460–6.

[124] Boyd R, Silk JB. How humans evolved. W. W. Norton & Company; 1997.

[125] Brandt M. Der Ursprung des aufrechten Ganges. Neuhausen-Stuttgart. Hänssler; 1995.

[126] Beall CM, Cavalleri GL, Deng L, Elston RC, Gao Y, Knight J, et al. Natural selection on *EPAS1* (*HIF2α*) associated with low hemoglobin concentration in Tibetan highlanders. Proc Natl Acad Sci 2010;107:11459–64.

[127] Beall CM. Andean, Tibetan, and Ethiopian patterns of adaptation to high-altitude hypoxia. Integr Comp Biol 2006;46:18–24.

[128] Reich D, Patterson N, Kircher M, Delfin F, Nandineni MR, Pugach I, et al. Denisova admixture and the first modern human dispersals into Southeast Asia and Oceania. Am J Hum Genet 2011;89:516–28.

[129] Haynes G. The early settlement of North America: the Clovis era. Cambridge: Cambridge University Press; 2002.

[130] Diamond J. Arm und Reich. Frankfurt/Main: S. Fischer; 1998.

[131] Rule S, Brook BW, Haberle SG, Turney CSM, Kershaw AP, Johnson CN. The aftermath of megafaunal extinction: ecosystem transformation in Pleistocene Australia. Science 2012;335:1483–6.

[132] Järvelä I, Torniainen S, Kolho K-L. Molecular genetics of human lactase deficiencies. Ann Med 2009;41:568–75.

[133] Laktoseintoleranz. http://de.wikipedia.org/wiki/Laktoseintoleranz; 2012.

[134] Mellaart J. Çatal Hüyük: a Neolithic town in Anatolia. New York: McGraw-Hill; 1967.

[135] Belcaro G. Once we were hunters. Imperial College Press; 2001.

[136] Andrews P, Martin L. Hominoid dietary evolution. Philos Trans R Soc Lond B 1991;334:199–209.

[137] Boström P, Wu J, Jedrychowski MP, Korde A, Ye L, Lo JC, et al. A PGC1-α-dependent myokine that drives brown-fat-like development of white fat and thermogenesis. Nature 2012;481:463–8.

[138] Preedy VR. Handbook of anthropometry. Physical measures of human form in health and disease. Berlin: Springer; 2012.

[139] Cunnane SC, Crawford MA. Survival of the fattest: fat babies were the key to evolution of the large human brain. Comp Biochem Physiol A Mol Integr Physiol Elsevier 2003;136(1):17–26.

[140] Higham T, Douka K, Wood R, Ramsey CB, Brock F, Basell L, et al. The timing and spatio-temporal patterning of Neanderthal disappearance. Nature 2014;512:306–9.

Chapter 2

Methodology

Alexander Christoph Stahn* and Hanns-Christian Gunga†

*Postdoctoral Research Associate, Center for Space Medicine and Extreme Environments, Institute of Physiology, CharitéCrossOver (CCO), Charité University Medicine Berlin, Berlin, Germany

†Professor, Center for Space Medicine and Extreme Environments, Institute of Physiology, CharitéCrossOver (CCO), Charité University Medicine Berlin, Berlin, Germany

2.1 INTRODUCTION

A cornerstone of natural sciences is measurement. Technically, measurement refers to the process of assigning numbers to objects or events. These values are turned into meaningful data by quantifying them into specific units, based on the International System of Units (abbreviated SI for Le Système International d'Unités, the French name for the system). The process of measurement requires the use of tools, ranging from simple measurement tapes and questionnaires to the most sophisticated technologies, such as confocal laser microscopy, protein microarray, genetic analysis, and cutting-edge brain imaging. The acceleration of technology is continuously leading to medical instruments that are smaller, faster, and cheaper, support web-based monitoring and have the potential to significantly change medical health care and research. This chapter provides an overview of some present technologies that could have promise for physiological and psychophysiological monitoring in extreme environmental conditions and their potential benefits in the clinical setting. To fully understand the strengths and limitations of these technological advances, one must appreciate the historical background underpinning their development.

2.2 HISTORICAL ASPECTS

The development of adequate methodologies to study physiological phenomena started approximately 100 years ago. The German physiologist Nathan Zuntz at the Königlich Landwirtschaftliche Hochschule in Berlin played a decisive role in the design and construction of such field methodologies [1]. At the time, research was dominated by progress in basic sciences and technological breakthroughs. Therefore, the main focus was on equations instead of philosophical elaborations [2]. By pursuing experimental research in the laboratory, Zuntz

Human Physiology in Extreme Environments. http://dx.doi.org/10.1016/B978-0-12-386947-0.00002-2

39

was a trailblazer in investigating individual physiological issues in a controlled manner under defined environmental conditions. Field research required the development of a separate set of physiological instruments, which had to be light, sturdy, and multipurpose. In 1906, Zuntz already hinted at the concepts of "laboratory physiology" and "field physiology" in his book "Höhenklima und Bergwanderungen in ihrer Wirkung auf den Menschen" (The Effect of High-Altitude Climate and Mountain Hiking on Human Beings) [3]. This was a first sign of how scientific research began to leave the confines of laboratories [4], as well as the starting point of new experimental systems and experimental cultures [5].

However, from the start, Zuntz recognized that any field physiological approach required the implementation of sound reference techniques, similar to the "Gold Standards" used in the laboratory. He believed that later phases of research could then be performed under field conditions. This consistent two-pronged, methodical approach surely contributed to Zuntz's scientific success. After moving from Bonn to Berlin in 1880, Zuntz first focused on issues of respiratory physiology, working with Geppert to develop the Zuntz-Geppert respiration apparatus for determining metabolic cost (Figure 2.1).

The Zuntz-Geppert respiration apparatus was comprised of a mouthpiece, a valve mechanism for the separation of expired or inspired gas, a gas meter to quantify breath volume, and the actual analysis apparatus, with the last component determining the carbon dioxide and oxygen volumes of expired air. Given that Zuntz's respiration apparatus could be used to determine both carbon dioxide release and oxygen consumption, it also became possible to record changes in metabolism quite accurately for relatively short intervals of time (direct determination of the respiratory quotients (RQ) or respiratory coefficients). This

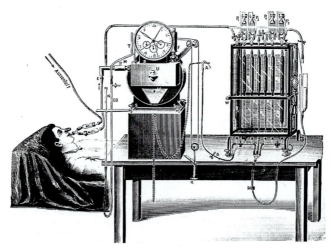

FIGURE 2.1 The "Zuntz-Geppert respiratory apparatus" in a typical experimental setup under laboratory conditions [3, p. 159].

was a decisive advantage of the Zuntz-Geppert apparatus over Pettenkofer's respiration chamber procedure, which continued to be widely used at the time and which only permitted a direct measurement of CO_2 release [1]. However, the Zuntz-Geppert method of determining metabolism without using the respiration chamber initially fell under sharp criticism from several professional colleagues [6]. Reviewing his own methods, Zuntz performed control trials in 1891 in Göttingen, at a facility that had been built according to the Pettenkofer principle. These experiments confirmed that concerns regarding the Zuntz-Geppert methods were widely unfounded [6]. Nonetheless, Zuntz regarded the optimal solution to be a combination of both methods. At the Institute for Veterinary Physiology on Chausseestrasse 103 in Berlin, Zuntz was able to realize his plans and establish the most advanced respiration chamber worldwide (Figure 2.2).

This chamber was later equipped with a treadmill and further complemented by an X-ray facility to gain better insights into the cardiovascular changes observed during different levels of exercise. Using this most advanced and spacious respiration chamber, along with the treadmill, Zuntz could address individual physiological issues in the laboratory with the most suitable measurement methods. Moreover, this set up allowed the validation of portable, small, light, mostly noninvasive methodologies that could be used in the field—the starting point for the incredible development of smart medical devices used in field conditions. For metabolic measurements Zuntz developed a mobile "dry gas meter." The "dry gas meter" was a hexagonal box made from sheet steel, which weighed only a few kilograms and could be easily carried like a backpack (Figure 2.3). After the subject occluded his nostrils with a clip and was connected to the gas meter via a tube with a mouthpiece, he would breathe into the gas meter. This movement of air was conveyed via an axis on cogwheels to the upper side of the gas meter, which, in turn, was connected to a dial on the

FIGURE 2.2 Schematic side view of the largest respiration chamber in the world, combined with a treadmill in Zuntz's Institute of Veterinary Physiology in Berlin in 1913 [136].

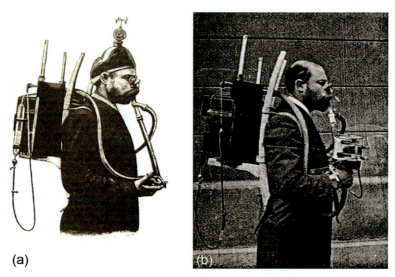

(a)   (b)

FIGURE 2.3 (a) Zuntz's transportable dry gas meter used during a marching experiment. On his head the subject is wearing an anemometer cap to measure wind speeds [3, p. 165]. (b) In addition to the dry gas meter, Zuntz added a sphygmograph (right hand) to measure pulse rate, pulse wave velocity, and blood pressure during rest and exercise [3, p. 249]. Zuntz had a limited financial budget, as demonstrated by the fact that frame of the transportable dry gas meter was made of a shortened wooden chair.

back of the gas meter. In this manner, the volume entering the gas meter could be determined and read immediately. Zuntz and his staff used this device for determining metabolic costs during marching, bicycling, swimming, and occupational health activities, such as typewriting or playing musical instruments. To better understand the cardiovascular responses during such conditions, he also added a handheld sphygmograph to measure blood pressure and pulse rate to the system (Figure 2.3b).

2.3 ACCELERATING GROWTH OF TECHNOLOGY

This previous section serves as a vivid illustration of the remarkable acceleration of technological developments in the twentieth century, which has been exponentially increasing (Figure 2.4). In the past 600 years, inventions such as the printing press, telescope, and telephone were centuries apart. In contrast, in the last few decades, the time between major technological achievements has tremendously decreased. Some experts have hypothesized that these rapid innovations, and particularly the acceleration of computing power, could even lead to a point at which artificial intelligence exceeds human intelligence, radically changing civilization and human nature. This point at which computing power surpasses all human brainpower combined has been coined the technological singularity by [137], a mathematician, computer scientist, and futurist. Beyond this point, Vinge argues, no current models of reality are sufficient for making

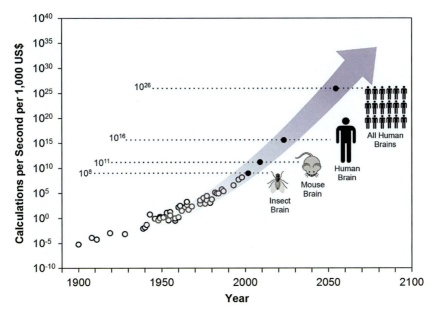

FIGURE 2.4   Exponential growth in computing power over time. While it took 90 years to achieve the first million instructions per second (MIPS), now about 1 MIPS per 1000 dollars are added every hour. Accordingly, processing power is exponentially getting cheaper and fast. Futurists have suggested that this ever-quickening rate of technological development could lead to a point at which computing power outperforms human brain power, making the future of mankind unpredictable. Data based on http://www.singularity.com/charts/page67.html. *Figure adapted from http://www. bit-tech.net/bits/2009/04/29/the-future-of-artificial-intelligence/7.*

predictions about the future of mankind. Irrespective of these speculations, technological progress has undoubtedly had an invaluable impact on human health care and research, with the International Space Station (ISS) representing one of the most complex technological systems created by humankind. Further cutting-edge examples of human achievements are advances in genetics, nanomedicine (e.g., drug delivery and visualization, lab-on-chip devices, neuroelectronic interfaces, and nanorobots), surgical telemetry, and new imaging techniques.

Mankind's drive for innovation has left remarkable footprints on the development of measurement approaches and technologies in health and life sciences, as exemplified by anthropometry. For instance, while measurement tapes are still very important and precise laboratory tools routinely used for assessing indices of body composition, such as waist-to-hip ratio or height, new 3D scanning techniques have revolutionized research in anthropometry and various other research fields such as textile and medical applications as well as palaeobiological studies (see Chapter 1) [7,8]. These completely noninvasive and harmless methods allow the acquisition of nearly any anthropometric measurement within a few seconds. Basically, a laser or light source is projected onto the body, and various cameras positioned around the body record how the body's shape deforms the light. From these data a 3D image of the human body can be

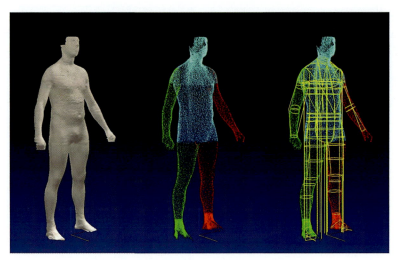

FIGURE 2.5   3D body scan from our Anthropometry and Body Composition Lab at the Center for Space Medicine Berlin. The 3D model can be used to generate nearly any length, circumference, or volume measurement of a body segment. The measurement does not involve any irradiation and is completed in less than 10 s.

created, allowing virtual measurement of the length, circumference, and volume of distinct body segments. Although laser scanning is considered to be most accurate, patterned light projection (typically white, red, or near-infrared) has the advantage of speed. A system that operates with good accuracy and relatively low cost is shown Figure 2.5, operated in our labs [9].

The advantage of such a system is that it rapidly speeds up measurement time, reduces interrater variability, and promotes the development and application of new measurement methodologies. For instance, combining the technique with bioelectrical impedance analysis might lead to improvements in noninvasive and fast approaches for body composition assessment. In addition, commercially available systems for whole-body scanning require substantial space and are typically limited to the laboratory. Yet, the introduction of the Kinect™ camera, an add-on for Microsoft's X-box™, a few years ago has opened further opportunities for 3D body scanning. The developers now provide open access to their software (http://www.microsoft.com/en-us/kinectforwindows/Develop/developerdownloads.asp), and specific software packages for 3D scanning using the Kinect™ are commercially available (www.kscan3d.com). In spite of a relatively low depth map resolution, the system can achieve a precision of about 5 mm using the latest hardware [10]. Such developments could have considerable implications for research in extreme environments. This has recently been demonstrated by a research group led by Carmelo Velardo and Marco Paleari. They were able to develop an algorithm for vision-based body-mass estimation with reasonable accuracy by using the Kinect™ camera system [11]. This development could either replace more expensive, energy- and space-consuming systems presently used in spaceflight, or it could be combined with larger systems to obtain detailed data on the change of posture and body shape during space missions. Moreover, these approaches could provide an alternative to the presently established body mass index (BMI) in the near future. Whole-body scanning allows to quickly determine a so-called body volume index (BVI), which is much more sophisticated index for health risks associated with increased fat mass as it takes into consideration body fat distribution.

2.4   CHARACTERISTICS OF MOBILE TECHNOLOGIES FOR FIELD USE

Importantly, research on humans in extreme environments is performed out of the lab. Such field conditions typically place strict limits on the technologies used to record human biosignals. For example, recent advances in imaging technologies, such as a functional magnetic resonance imaging of the brain, cannot be performed at the ISS or during extreme mountaineering expeditions or scuba diving. Similarly, even established and invaluable recording devices that are highly portable, such as polysomnographs, can be of limited use under field conditions. This can be due to time constraints, low compliance, uncontrolled electromagnetic noise, and lack of professional personnel or crew trained to carry out the measurements properly. Therefore, research on human physiology in extreme environments requires human-monitoring technologies that are innovative and unique, but also simple and straightforward, both technically and methodologically. Moreover, as pointed out by Hoyt and Friedl [12], these new methodologies, including high-efficiency network systems, are of particular interest because individuals in extreme environments commonly face mentally and physically demanding missions, often involving sleep deprivation, exhaustion, and dehydration. In these situations real-time physiological and psychophysiological status-monitoring systems can be used to (1) assess baseline physiological characteristics (physical fitness) prior to a mission, (2) decisively support mission planning (i.e., logistic requirements), (3) provide mission support (i.e., planning of rest-work cycles), (4) facilitate casualty evacuation in the case of an accident, and (5) improve the quality of after-action reviews [12]. For instance, such data monitoring could be used to estimate energy balance, thermal and hydration status, activity, work-and-rest schedules, sleep, mental workload, situational awareness, vigilance, and cognitive performance. Specifically, at high altitude the severity of acute mountain sickness could be assessed by combining questionnaires, such as the Lake Louise Consensus Score (for details see Chapter 4), with blood oxygen saturation measurements so that effective countermeasures could be initiated early, avoiding further risks. As later described in this chapter, the combination of actigraphy, mobile electroencephalogram (EEG), and cognitive performance measures can provide invaluable information about an individual's fatigue and reserves. Moreover, vital-sign monitoring could be used to assess remote life signs (body position, body motion, heart beats, respiration, skin temperatures). Finally, with unstoppable advances in telecommunications, these technologies could also be of value in the clinical setting for remote patient monitoring. For instance, a constantly worn, miniaturized ECG, perhaps even with electrodes integrated in fabrics, could be immediately activated when a patient began experiencing chest pain. The signal from this device could then be transmitted to the hospital or practitioner. In addition, such devices might also be used to identify arrhythmia and trigger an emergency alarm. Irrespective of whether they are used for remote patient monitoring or research in extreme environments, these technologies should generally be characterized by the following features: (i) noninvasive; (ii) easy-to-operate; (iii) fast; (iv) energy efficient; (v) miniaturized, light weight, and portable (ideally mobile); (vi) compatible with online monitoring; (vii) relatively inexpensive (particularly if multiple devices are needed for online monitoring); (viii) well protected and robust; and (ix) capable of local data storage and wireless data transfer.

As a result of the power and speed of technological acceleration outlined at the beginning of the chapter memory chips, batteries, and electric components continue to shrink in scale and price and allow the development of new continuously decreasing and more capable devices. Moreover, improvements of web-based communications allow to immediately access and even monitor subjects. Such instant data accessibility can change substantially scientific work and perspectives.

Certainly, devices with these characteristics occur on a continuum, ranging from so-called e-textiles (fabrics or intelligent clothing with electronics integrated into the material) to wearable technologies that can sense vital signs such as heart rate, respiration rate, temperature, activity, and posture. Portable devices, not attached to the body, can then be set up at field sites and used to remotely monitor data [13,14]. It is noteworthy that this continuum is not fixed. Rather, newer technologies and materials, increasing computer power, increasing internet accessibility, and the availability of smart phones and tablets allow a miniaturization of devices, so that measurements formerly bound to the lab can be obtained via highly mobile equipment. A classic example is indirect calorimetry, which can now be employed under nearly any condition. A more recent example is bioelectrical impedance devices. The bioimpedance technique determines body fluids and body composition by applying an indirect alternating current below the threshold of perception and measuring the voltage drop. Due to the lipid bilayer of the cell membrane, low-frequency currents (<5 kHz) rarely penetrate the cell and can be considered to be an indication of the extracellular compartment. In contrast, high-frequency alternating currents (>500 kHz) can polarize and repolarize the cell membrane so that the current also penetrates the entire cell. In turn, these high frequency currents are assumed to be a reflection of both the extracellular and intracellular compartments, or total body water. Although the bioimpedance approach has also been highly criticized, it is still the most widely used technology for body composition assessment. Moreover, it has the advantage of fulfilling all of the above-mentioned characteristics for field devices, except being mobile in the sense of being worn on the body and allowing for continuous monitoring. However, this final challenge has also been overcome recently by developing a minituarized bioimpedance device capable of continuous monitoring at various frequencies and channels [12]. They developed a miniaturized bioimpedance device capable of continuous monitoring at various frequencies and channels [15]. In addition to its compact size, the system has the unique ability to simultaneously track several channels and frequencies (Figure 2.6).

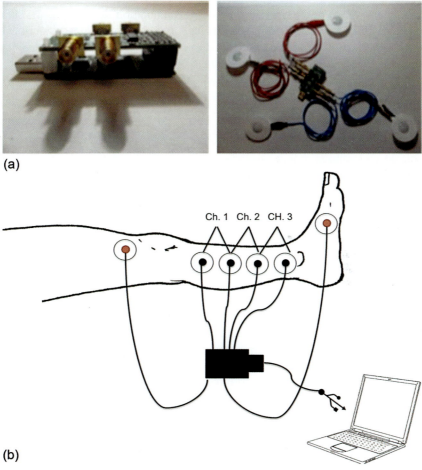

(a)

(b)

FIGURE 2.6 Real (a) and schematic (b) displays of a new segmental, multichannel, multifrequency bioelectrical impedance device for mobile, noninvasive, and continuous monitoring of fluid shifts. A current at 10 frequencies (1, 2, 4, 8, 17, 48, 128, 234, 488, and 769 kHz) is introduced between two outer electrodes and picked up by four inner electrodes. Data can be stored in the device's internal memory or on a laptop for extended measurements at high sampling rates. Future developments might lead to an increase in the number of channels. *Courtesy of Federica Villa, Politecnico di Milano, Italy.*

Simultaneous tracking is relevant for two reasons: (1) The acquisition of a frequency range between 1 kHz and approximately 800 kHz allows one to extrapolate resistance at zero and infinite frequency and, in turn, to derive extra- and intracellular resistance and their respective fluid compartments. This overcomes an often-criticized flaw of classic single-frequency bio-impedance analysis (BIA). Specifically, these approaches employ a frequency of 50 kHz, at which the extra- and intracellular spaces are partially measured, but their exact proportions can vary substantially both within and between subjects [16,17]. (2) The multiple channels allow the application of sound biophysical models for segmental body composition measurements as compared to statistically derived regression equations [18] and tracking fluid accumulation along segments. Such a device could be promising in all situations in which fluid balance is critical, such as in the clinical setting, with athletes and astronauts, and in fundamental research tracking fluid shifts during human centrifugation. This example represents a wide range of methodological and technological advances in the field of mobile human physiological monitoring.

2.5   SMART DEVICES

The development of mobile physiological monitoring technologies is presently skyrocketing, and the present summary provides an example of this rapid development. The summary focuses on devices in the field of sleep, chronobiology, and cognitive function. It should be noted that these examples reflect the authors' research experience, and the authors have no affiliations with or involvement in any organization or entity with any financial or nonfinancial interest in the devices presented.

2.5.1   Circadian Rhythm in Extreme Environments

A growing body of research indicates that desynchronization of the circadian system can be detrimental to mental and physical health [19–23]. Given the increasing duration of space missions, along with a concomitant increase in confinement and isolation, problems associated with changes of the circadian rhythm might be increasingly prominent during future exploratory missions [24]. Previous research has shown that sleep in space is often restricted, more disturbed, structurally altered, and shallower than it is on Earth [25–28]. In fact, it has been reported that 50% of crew-members on dual-shift shuttle flights and 19.4% on single-shift shuttle flights have used sleep medications at least once during their flights. A qualitative and quantitative lack of sleep might impact daytime sleepiness and, consequently, negatively affect mental and physical performance. The deleterious effects of sleep deprivation during aviation have been previously reviewed by Caldwell and colleagues, who found that short-term sleep deprivation can negatively affect working memory, reduce alertness, and slow down information processing [29]. In turn, the risk for aviation errors and accidents are increased. These reductions in alertness and performance have been observed in laboratory settings [30], and they have been confirmed by data from Russian space missions, during which crew errors were closely associated with unusual sleep-wake cycles [31]. Furthermore, on long-duration flights, disruptions of regular sleep-work cycles may lead to higher incidences of gastrointestinal problems, menstrual irregularities, colds, flu, weight gain, and cardiovascular problems [29]. The underlying mechanisms for these effects during space flight are still unknown. In particular, structural alterations of sleep and suppression of delta sleep remain areas of speculation and are unlikely to be explained by environmental factors such as noise, air temperature, sleeping compartments, and unusual working hours [24]. A possible explanation might be related to changes in the circadian rhythm. Only a few years ago, researchers discovered a new physiological system involved in the maladaptation of day-night cycle processes. This system comprises a new class of retinal photoreceptors distinct from rods and cones. These intrinsically photosensitive retinal ganglion cells (ipRGCs) express melanopsin as their light-sensitive photopigment, and melanopsin seems to play an essential role in the regulation of the circadian rhythm [32,33]. Depending on the intensity and the spectral composition of the incident light, the ganglion cells project their information onto the circadian master clock located in the suprachiasmatic nucleus (SCN) [34]. Synchronized by exogenous light-dark cycles, the SCN regulates various parameters of the circadian timing, such as body temperature and the synthesis of several hormones, including melatonin. Darkness is a potent stimulus for the pineal gland's synthesis and release of melatonin into the blood stream, and brightness suppresses melatonin production [35]. Daytime illumination during a space shuttle mission can be very low in various compartments of the spacecraft (<80lx [36]), and phototransduction in ipRGCs is relatively insensitive and cannot drive physiological responses at low light intensities [37]. As a result, melatonin might not be suppressed during "daytime," and thus it may be associated with mood and sleep disturbances. In addition, it is well documented that sudden changes in external cues (zeitgeber) are associated with sleep deprivation, mood disorders, reduced alertness, and fatigue, resulting in decreases of physical and mental performance [29]. During space missions, important external cues, such as usual work-rest cycles and regular changes of daylight and darkness, are lacking, and the light spectrum itself is attenuated. So, astronauts and cosmonauts might be prone to suffer from symptoms related to a "free run" of circadian rhythms or a shift in their circadian phase relationships [24]. These observations are also supported by studies in analog environments (e.g., expeditions to Antarctica), which suggest that desynchronization might be related to weaker diurnal cues (photic pacemakers) and social routines [24,38–42]. The main effects arising from such a misalignment include insomnia, daytime sleepiness, and performance deficits. Because space and Antarctic habitats share weak (or absent) photic zeitgebers, similar effects might be expected to arise during space flight. The substantial interindividual variation has also been noted in a previous short-term mission investigating the effect of space flight on the circadian rhythm of core body temperature (CBT), showing a phase advance, a phase delay, and no changes in acrophase [25]. So far, only a few studies have investigated these mechanisms in humans [25,26,28,43] and animals [44–46]. Although it has been documented that external pacemakers, such as room illumination, regular meals, and strict wake-up schedules, are useful sources for promoting adaptation of the human circadian system to a 24-h rhythm in space [26,28], these studies occurred during single-shift orbital spaceflights with limited duration (covering the first 8-30 days of space flight except for a few anecdotal reports from the cosmonauts). Moreover, a crucial aspect that might have confounded previous research is that no prior study separated the exogenous and endogenous components of the circadian rhythm of CBT. As the following subsections discuss this point in detail, we now only mention the endogenous circadian rhythm of CBT, known as the circadian timing system, most often revealed using so-called "constant routine" protocols, which standardize activity, meal composition and timing, posture, ambient temperature, and sleep-wake periods. Although it is apparent that such conditions cannot be replicated during space missions, previous data on the circadian rhythm of CBT during space missions must therefore be interpreted with care. Moreover, there is need for future studies to apply cutting edge approaches for controlling the external effects on CBT, or to simulate near "constant routine" protocols. Taking these needs into account, future research can address important issues such as whether the available zeitgeber in space are strong enough to keep the circadian rhythms stable and synchronized even given short-term disturbances such as workload-related shifts of sleep-wake cycle. Moreover, researchers must elucidate if and to what extent circadian disruptions might also explain changes in sleep duration and architecture and, eventually, directly or indirectly affect mental workload, vigilance, situational awareness, and cognitive performance.

2.5.1.1 Recording Core Body Temperature

A straightforward global indicator for monitoring circadian rhythms in humans is the continuous recording of CBT. In research settings CBT is typically measured by inserting a thermosensor into the esophagus, nasopharynx, rectum, or tympanum/auditory meatus. The relative advantages and disadvantages of these and other measurement sites are well understood [47–59] and have been reviewed in depth [60–66]. Clearly, rectal recording is the most widely used and accepted measurement of temperature used by physiologists for CBT. This approach has also been successfully employed in previous short-duration space missions [25,26,28] and ground-based analog studies [42]. However, these methods are neither applicable during daily routines on the ground nor in space, and they are particularly unsuitable for repeated measurements during long-duration missions. This unsuitability is due to the demanding requirements of the CBT-measuring process: the thermosensor must be (1) noninvasive, (2) easy to operate, (3) compatible with basic hygiene standards, (4) unbiased towards various environmental conditions, (5) sensitive enough to quantitatively reflect minor changes in arterial blood temperature, and (6) fast-responding [67,68].

2.5.1.2 Ingestible Temperature Capsule

Gut temperature, as determined by a telemetric, miniaturized, and ingestible sensor, which is known as the "temperature pill," has been shown to be an acceptable alternative to rectal and esophageal temperatures (bias <0.1 °C and 95% limits of agreement within ±0.4 °C for esophageal temperature) for long-term studies and shows excellent utility for ambulatory field-based applications [64]. The ingestible sensor probe has been successfully applied in numerous sport and occupational settings, such as the continuous measurement of CBT in deep-sea saturation divers, distance runners, and soldiers undertaking sustained military training exercises. Compared to rectal CBT recordings, the temperature capsule is better tolerated by subjects [58] and is characterized by a faster response time during exercise and heat stress [64]. In addition, it has also been successfully employed for studying the circadian rhythm of CBT [53,54,58]. Interestingly, in a collaborative effort with NASA, Johns Hopkins University developed one of the first commercially available ingestible temperature capsules, the Cor-Temp™ pill (HQInc., formerly Human Technologies, Inc., Palmetto, FL) (http://www.nasa.gov/vision/earth/technologies/thermometer_pill.html). The 1.9-cm-long capsule is powered by a silver oxide hearing aid battery that holds enough energy for 9 days. It also contains a quartz sensor that vibrates at different frequencies relative to the body's temperature, transmitting a harmless, low-frequency signal through the body. A recorder outside of the body can read this signal and display CBT. In 1998, astronaut and United States Senator John Glenn swallowed and tested a temperature capsule in space as part of the Space Shuttle Discovery medical experiments.

More recently, another ingestible temperature capsule (VitalSense® core body temperature capsule, Equivital™, Hidalgo Ltd, Cambridge, England) has become commercially available. The capsule passes easily through the GI tract without affecting bodily functions. Transit time varies between individuals, ranging from 12 to 136h, with an average of 48h [58]. The individual variations in gastrointestinal motility depend on the vegetative system, but also substantially on the volume of food ingested directly (decrease in transit time with larger meals). The observed level of agreement, compared to the rectal probe, has recently been reported to be 0.04 ± 0.03 °C [58]. The capsule weighs 1.6g and has a diameter of 8.7 mm and a length of 23 mm (Figure 2.7).

The VitalSense® capsule is also equipped with a silver oxide battery (1.55 V nominal voltage and a total capacity of 22.5 mAh), and it is activated by an external handheld device (Figure 2.7e), after which time the capsule is swallowed with liquid. Data are then transmitted and stored (four times per minute with an accuracy of ±0.1 °C) on a separate device (Equivital™ EQ02 SEM, Hidalgo Ltd, Cambridge, England), which is integrated into a chest belt (Equivital™ EQ02 Series Sensor Belt, Hidalgo Ltd, Cambridge, England)

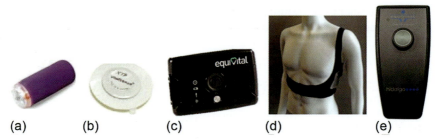

(a)   (b)   (c)   (d)   (e)

FIGURE 2.7   (a) Ingestible CBT capsule (VitalSense® core body temperature capsule, Equivital™, Hidalgo Ltd, Cambridge, England). (b) Disposable sensor patch for wireless skin surface recordings (VitalSense® Dermal Temperature Patch, Equivital™, Hidalgo Ltd, Cambridge, England). (c) Data-recording module (EQ02, Equivital™, Hidalgo Ltd, Cambridge, England) for CBT by ingestible temperature capsule, ECG (heart rate), and skin surface temperature. (d) Chest belt that houses recording module under left arm and includes three integrated fabric ECG electrodes for heart-rate monitoring. (e) Activation switch for starting temperature recordings.

(Figure 2.7c and d). The EQ02 SEM is equipped with an 8 GB memory, and is powered by a 3.7 V (300-mAh) Li-Po rechargeable cell. The device conforms to EN60601-1 Classification, and is EU, FDA, and CE marked (FDA Device Classification Class II, EU Device Classification Class IIb). It weighs 38 g (without the external battery pack) and has the following length, height, and width: 78 mm × 53 mm × 10 mm. To allow continuous 36-h recordings, the chest belt houses an external battery pack (60 mm × 35 mm × 14 mm), including three AAA batteries. Once data collection is completed, the data can be downloaded via a USB interface to a laptop computer, using the software package Equivital Manager (Equivital™, Hidalgo Ltd, Cambridge, England). A unique aspect of the system is that it can integrate a variety of additional vital signals, such as skin surface temperature, body position, and ECG. Skin surface temperature can be determined by disposable telemetric sensor patches (57.2 mm diameter × 5.3 mm thickness, 7.5 g) (Figure 2.7b). The hypoallergenic, nonirritating, self-adhesive patches consist of a thermistor (±0.01 °C accuracy), which wirelessly transmits the data to the SEM sensor module integrated in a chest belt. The sensor patches are waterproof and have a battery life of 10 days following activation. The chest belt also monitors body position by detecting the angle between acceleration sensors integrated in the belt and the horizontal plane. Heart rate can be continuously monitored by a single-lead ECG. The ECG is integrated in the Equivital™ chest belt, which contains three fabric ECG electrodes. Data are recorded at 256 Hz and stored in the sensor module integrated in the chest belt.

2.5.1.3   A Noninvasive Heatflux Sensor

Although ingestible temperature capsules have major advantages compared to rectal recordings, these devices can be expensive (about 80 US$ per capsule), and their transit time can be as fast as 12 h, requiring up to three capsules per

36-h recording. In addition, due to the substantial interindividual variation of transit time (12-136h), the exact position of the capsule in the gastrointestinal tract is not known. Moreover, the capsule might be prone to the same challenges associated with rectal temperature recordings, such as tendency to quantitatively reflect minor changes in arterial blood temperature and fast response time (see also Section 2.5.1.1). Taking these circumstances into account, we recently developed a combined skin temperature and heatflux sensor (Double Sensor) together with Draegerwerk AG Luebeck. This patented sensor concept works with only two temperature probes and has been miniaturized (Figure 2.8b). The recording hardware consists of a core piece called Healthlab-Master (HFM-Master), having dimensions similar to those of a smart phone (6cm×10cm×2cm; 100g) as shown in Figure 2.8a (Koralewski Industrie-Elektronik, Hambühren, Germany). In addition to supporting CBT using the Double Sensor, the device is also equipped with a 3-axial accelerometer for activity recording, as well as a sensor for ambient temperature, pressure, and humidity. Notably, the system can also be extended by additional amplifiers, increasing the number of vital signs that are recorded to include ECG, rectal or skin temperature, respiration, oxygenation, and blood pressure. The device is powered by two AAA batteries, allowing recordings up to 10h, but it can also be powered by an external battery pack for extended acquisition times. Data are stored on replaceable SD cards, which can be downloaded via a Bluetooth connection or USB interface. Either connection can also be employed for online monitoring.

Only recently, we have confirmed the validity of the sensor for determining circadian rhythm under constant ambient conditions [55]. Briefly, as part of the Berlin Bed Rest Study 2 (BBR2-2), we measured CBT using a rectal probe and the *Double Sensor* continuously for 24h. Figure 2.9 shows the raw data and the data fitted by Cosinor Analysis for an individual subject for both rectal and Double Sensor recordings.

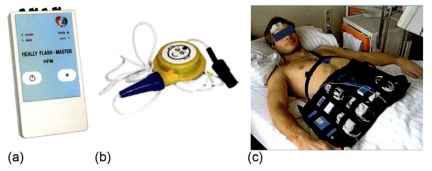

(a)   **(b)**   **(c)**

FIGURE 2.8 (a) HFM-Master (Korelewski Industrie Elektronik, Hambühren, Germany). (b) Double Sensor for data amplification and non-invasive recording of CBT at the forehead. (c) The Double Sensor can be combined with various additional biosignals, such as skin temperature, rectal CBT, GSR, respiration, oxygenation, pulse wave, blood pressure, heart rate, and actigraphy. The picture shows a typical measurement setup during the Berlin Bed Rest Study 2 (BBR2-2).

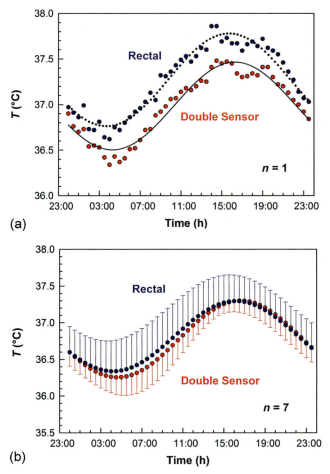

FIGURE 2.9 Raw and fitted data from Cosinor analysis for rectal (red) and Double Sensor (blue) recordings, as a function of time. The data represents a single subject (a) and the average for seven subjects (b) during head-down-tilt bed rest in the Berlin Bed Rest Study 2 (BBR2-2).

The high degree of visual agreement between measurements indicated in Figure 2.9b is also confirmed by the parameters of the fitted cosine curves ($y(t) = 36.63 + 0.53\cos(t + 20.68)$; $R = 0.979$ and $y(t) = 36.63 + 0.50\cos(t + 20.46)$; $R = 0.962$ for rectal and Double Sensor data, respectively). When circadian temperature profiles were quantified by mesor, acrophase, and amplitude, no significant differences were found between rectal and Double Sensor temperature recordings ($P = 0.310-0.866$). Further information about the degree of agreement between methods is provided by the limits of agreement. Of particular interest is the random error component for acrophase and amplitude. As indicated in Table 2.1, individual amplitude differences range from -0.21 to $0.19\,°C$.

TABLE 2.1 Cosinor Analysis for Rectal and Double Sensor Temperature Recordings (mean ± SD)

	Rectal	Double Sensor	Bias	Limits of Agreement	P
Mesor (°C)	36.79 ± 0.13	36.90 ± 0.24	−0.11	−0.67 to 0.45	0.310
Acrophase (rad)	5.55 ± 0.21	5.51 ± 0.22	0.04	−0.31 to 0.39	0.735
Amplitude (°C)	0.48 ± 0.07	0.49 ± 0.08	−0.01	−0.21 to 0.19	0.866

Although the differences in acrophase could be considered rather large (corresponding to 1.18-1.49 h), the sensor is not intended to serve as a surrogate for rectal recordings, and differences in technical and physiological aspects such as sensitivity to changes in arterial temperature changes, response times, and variations in "hot spots" in the body (several different cores that a single core could account for). In fact, we recently showed that, under controlled induction of hypothermia and rewarming during heart surgery, CBTs exhibit spatial and temporal dispersions, depending on the measurement site and probe [69]. Finally, Double Sensor recordings were characterized by higher variation than was found in rectal temperature measurements, yielding a poorer model fit. This was also statistically confirmed by a significantly greater amount of total variance accounted for by the rectal cosine curve model (0.93 vs. 0.79, $P < 0.05$) and significantly lower residual summed squares (0.48 vs. 1.69, $P < 0.05$) compared to the Double Sensor fitted cosine curve. In summary, these results confirm findings from our previous studies, through which we validated the Double Sensor during treadmill exercise [56] with limits of agreement ranging from −0.90 to +1.06 °C. Thus, in spite of the relative marked variability for single "spot checks," we conclude that the Double Sensor, placed at the forehead, seems to be a valuable approach for long-term monitoring of CBT and for determining circadian rhythm [55]. Presently, the sensor is used as part of the experiment Deutsches Zentrum für Luft- und Raumfahrt (DLR)/European Space Agency (ESA) "Circadian Rhythms" on the ISS. A typical example of a recording using the heatflux technology during long-duration spaceflight is shown below, confirming its feasibility for use in space (Figure 2.10).

2.5.1.4   Core Body Temperature—Role of Masking Effects and How to Account for Them

As already noted in Section 2.5.1, CBT recordings completed in field conditions are masked by a variety of factors. When CBT is measured in humans following a conventional lifestyle, the circadian rhythm of CBT is characterized by a plateau between 14:00 and 20:00 and a minimum around 05:00 in the morning. However,

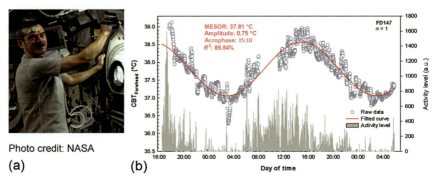

Photo credit: NASA

(a)   (b)

FIGURE 2.10   (a) Chris Hadfield performing the experiment circadian rhythms on the ISS. (b) Typical 36-h CBT (Double Sensor) and simultaneous activity profile (Actiwatch Spectrum, Respironics, Inc. Murrysville, PA, USA) recorded during long-term spaceflight on the ISS.

this phase reflects both an endogenous component (the circadian timing system) and a superimposed exogenous component (direct effects of sleep, physical and mental activity, changes in posture, light exposure, and the composition and timing of meals). Aschoff [70] pointed out the masking of such "lifestyle" factors in the 1960s. Typically, these effects have been removed by employing constant routine protocols, which are considered the "gold standard" for determining circadian rhythm. Those protocols keep as many exogenous factors constant as possible. Ideally, subjects remain awake and sedentary for at least 24 h, while light intensity, ambient temperature, humidity, meals, and meal times are kept constant. Standardizing the exogenous components reduces the amplitude by approximately 50% and is expected to lead to a pure indication of the circadian timing system [19], with a trough at 05:00, but a more stable peak at 17:00 [71]. Apparently, the major disadvantage of constant routine protocols is that they are limited to laboratory settings. The protocols also create substantial stress on the subjects and are simply not feasible for certain experiments, particularly for recordings under extreme environmental conditions, including space flight. In addition, constant routines are unsuitable for assessing the circadian rhythm of CBT during extended periods of time, such as consecutive days after time-zone transitions or during night work [19]. To reveal the endogenous component of circadian rhythm under field conditions (e.g., measurements in occupational health care, athletes, jet-lagged travellers, and humans exposed to extreme environmental conditions), an alternative approach is required. Waterhouse and colleagues have suggested a mathematical approach for separating the endogenous and exogenous components to yield a purified data set that is as close as possible to a sinusoid. Essentially, this approach corrects raw CBT for effects due to sleep and differing amounts of activity using diary entries or activity monitoring [19,72–74]. More recently, heart rate has been found to provide even better estimates of the true sinusoid, compared to data obtained from accelerometry and diaries [75]. The additional advantage of heart rate monitoring is likely to be explained by the fact that changes in CBT are evoked both by stressor-induced thermogenesis and physical activity. Hence, for some activities, accelerometry might yield a false increase in whole-body thermogenesis [75]. Basically, two purification approaches can be distinguished. The first method is referred to as *Purification by Categories* [76,77], and it has mainly been applied to activity and heart rate. This approach classifies CBT into categories by activity levels. The idea is that selecting temperature values that correspond to the lowest category can simulate a "constant routine" approach by compensating for the masking factors [72–75,78–83]. The second method is called *Purification by Intercepts* or more recently *Purification by ANCOVA* (analysis of covariance) [19,79,83,84]. Basically, this approach employs regression analysis to calculate the CBT that would correspond to zero activity (i.e., low heart rate) for each time bin, again to yield estimates of CBT profiles corresponding to a "constant routine" temperature curve. These models have been shown to be superior to earlier approaches and were very effective in describing thermal response to mild exercise taken at different times of day [84]. Despite some criticisms of the purification approaches, as compared to constant routine protocols [85], they have been shown to be valid and reliable alternatives for determining the endogenous rhythm of CBT [74]. First, results from different methodologies yield similar results [74,86]. Second, purified data clearly show a significantly smaller exogenous component when compared to the raw data [74]. More impressively, however, Waterhouse and colleagues also demonstrated that the purified data approach yields results similar to those of the constant routine protocol [79]. The authors therefore concluded that purification methods (i) achieve an accuracy similar to that of constant routine protocols for estimating the circadian rhythm phase of CBT; (ii) are presently the only solution for unmasking the effects of external influences and revealing the endogenous component of the CTS, when laboratory conditions are too challenging or simply not feasible; (iii) can be used in the field to estimate the size of masking effects caused by sleep and activity, as well as the shift of the endogenous component of the circadian rhythm [74,79,87]. Finally, an extension of purification techniques using only heart rate or activity monitoring has recently been proposed by Martinez-Nicolas [83]. In addition to physical activity, this approach accounts for ambient temperature, posture, sleep-wake period, and light exposure, each of which are divided into different categories (e.g., for posture and sleep-wake period, standing vs. lying down and sleep vs. wake are distinguished, respectively). This approach allows researchers to only select data that nearly matches conditions typically arranged by constant routine protocols.

2.5.1.5   Activity, Sleep and Light Spectrum Recordings

A straightforward surrogate for determining sleep and circadian rhythm is actigraphy [88–92]. Typically, actigraphy is based on the use of accelerometry to derive data about total sleep time, sleep efficiency, sleep-onset latency, and wake-after-sleep onset. Some concerns have been raised about the accuracy of such systems [93], particularly in populations with fragmented sleep or in situations where the sleep-wake cycle is challenged, such as jet lag and shift work [94]. Yet, actigraphy remains an excellent alternative for providing indications of circadian disorders and estimates of sleep-wake cycles, when polysomnography (the combination of electroencephalography, electromyography, and electrooculography, with possible monitoring of airflow, breathing movements, oxygen saturation, limb movements, and electrocardiography) is simply not available or feasible. Particularly, when longitudinal time series data are acquired, actigraphy can reveal its full potential, as has been recently demonstrated in the Mars500 study by Basner et al. [95].

In addition to acceleration-based sleep monitoring, some recently introduced technologies do not require the attachment of equipment to the body, instead using so-called contactless methods. One such commercially available technology (http://www.bed-dit.com) employs a flexible piezoelectric force sensor ($4\,cm \times 70\,cm \times 0.04\,cm$), which is positioned under the bed sheet or topper [96]. The force signal is sampled at 140 Hz, while additional sensors periodically record temperature, ambient noise level, and brightness to assess the quality of the sleeping environment. Based on the principle of ballistocardiography [97,98], the system detects tiny movements caused by respiration and heartbeats and turns them into measures of sleep time, sleep latency, snoring, and resting heart rate [96]. Interestingly, sleep onset time can lack accuracy when determined by actigraphy, but it has been shown to improve through the addition of ballistocardiography [98]. Ongoing work by Choi and colleagues is attempting to determine sleep architecture based on heart rate variation [99], respiratory variation [100], and activity information [101,102]. Presently, differentiating sleep stages (i.e., REM vs. non-REM sleep) using the technique have reportedly achieved accuracies between 75% and 90% [99,100,102,103]. Another fully noncontact and wireless approach uses short-range radio-frequency sensing. This type of device transmits a low-power pulse of radiofrequency energy to the subject and records the echo. An algorithm can then be used to determine respiration, including sleep-disordered breathing; movement; and sleep-wake times. The technique has shown promise in both healthy people [93] and people with breathing disorders [104]. In fact, O'Hare and colleagues observed a small advantage of the radiofrequency systems over wrist-worn actigraphy monitors, although they conceded that this difference might also be related to the software settings of the actigraphy system [93]. Yet, radiofrequency sensing certainly has the advantage of allowing biosignal monitoring without any constraints or sensors attached to the body. Future research is necessary, however, in order to determine the full potential of the technique and evaluate its limitations.

A clear benefit of acceleration-based activity monitoring devices is that they can be worn for 24 h, thus recording daily activity and rest patterns. This is highly advantageous for unmasking the underlying circadian rhythm of CBT from activity, as pointed out in the previous section. Compared to some more obtrusive monitoring systems for heart rate and/or body posture, wrist-worn activity monitors have only minimal constraints. Importantly, although originally only intended for activity monitoring using accelerometry, some wrist-worn devices

FIGURE 2.11   Activity monitor including a sensor for light spectrum measurement (Actiwatch Spectrum, Respironics Inc., Murrysville, PA, USA).

have already been equipped with light sensors, providing an important advantage for applying state-of-the-art purification techniques for measuring CBT (see Section 2.5.1.4). These accelerometers are small (4.8 cm × 3.7 cm × 1.5 cm), light-weight (16 g), rugged, water-resistant, actigraphy-based data loggers that record a digitally integrated measure of gross motor activity, as well as sleep schedule variability, and sleep quantity and quality statistics. In addition, some of these systems (e.g., Actilight Spectrum, Respironics, Inc. Murrysville, PA, USA) are equipped with three-color light sensors that provide irradiance and luminous flux recordings in three color bands of the visible spectrum: red, green, and blue. They also feature LCD displays for time, date, and device status indicators, as well as off-wrist detection. Data are sampled at a frequency of 32 Hz and can be recorded every 15 s for several days (up to 60 days at 1-min epochs). The devices can also be configured to automatically start and stop the data recording. Once data collection is completed, the data can be downloaded via a USB interface to a laptop computer, using Actiware Software (Figure 2.11).

2.5.2   Fatigue and Cognitive Performance

The determination of the circadian rhythm in space and other extreme environments is related to sleep, fatigue, and workload, which could ultimately impact mental performance and increase the risk of errors and failure. Specifically, disruptions of the circadian timing system, as a result of travel across different time zones, changes in rest-work cycles, sleep disruptions, and stress, may lead to increased mental fatigue and impaired mental performance, which are characterized by slower reaction times, reduced vigilance, and deficits in information processing [105–109]. Moreover, dehydration [110–112] and hypo- or hyperthermia [113–115] can further impair cognitive skills. Finally, emerging evidence suggests that performance impairments due to sleep deprivation are substantially explained by trait individual differences [116]. Thus, any measure characterizing workload and fatigue could be extremely useful in better understanding performance and scheduling of work-rest cycles. In fact, online monitoring of such biosignals could allow the monitoring party to alert crew members about their attentional resources at an early stage. In addition, such a technology could most certainly have applications for professionals such as truck drivers, pilots, air traffic controllers, and military personnel. Workload, vigilance, and cognitive performance must be clearly distinguished, however. Workload is the mental capacity required for task demand [117]. Accordingly, although two individuals may achieve similar cognitive performance, their respective workloads can be substantially different because they posses different skills in problem solving and coping strategies, have different experiences, or are not comparable in terms of sleep and work-rest cycles and/or mood [118]. Vigilance refers to sustained attention or tonic alertness and implies both the degree of arousal on the sleep-wake axis and the level of cognitive performance, with arousal referring to nonspecific activation of the cerebral cortex in relation to sleep-wake states [119]. Finally, mental fatigue is the gradual and cumulative impairment of alertness and mental performance.

2.5.2.1   Mental Workload, Vigilance and Cognitive Performance

Workload, vigilance, and cognitive performance can be assessed by behavioral or physiological measures or a combination of both. A classic test for sustained attention is the psychomotor vigilance task (PVT), a computer-based reaction-time task, pioneered by Dinges and Powell [120]. It objectively assesses fatigue-related changes in alertness associated with sleep loss, extended wakefulness, circadian misalignment, and time on task [108] for review. The standard version of the test has a length of 10 min, but it has recently been modified to a 3-min version (PVT-B) for contexts in which the original version is impractical [108]. Together with Pulsar Informatics Inc. (Philadelphia, USA) Dinges and Basner recently developed a test battery that combines the modified PVT with nine additional tasks assessing a variety of cognitive domains (Table 2.2), including sensory motor speed, visual object learning and memory, attention and working memory, abstraction and mental flexibility, abstract reasoning, spatial orientation, emotion recognition, complex scanning, tracking and atttention, and risk taking behavior [118].

This so-called Cognition Battery (also formerly called NeuroCATS) presents a substantial change in terms of both test paradigms and mobility compared to previous long-established test batteries such as WINSCAT [119] and the Automated Neuropsychological Assessment Metrics (ANAM) [120]. This new battery is innovative, mobile, can be flexibly adjusted and includes a well-refined vigilance test based on nearly 30 years of research. It is a promising tool for efficiently assessing a variety of cognitive domains that can be affected by

TABLE 2.2 Cognition (originally NeuroCATS) test battery. It should be noted though that the outlined tests are just summary of the core battery, which be adjusted and extended by additional paradigms and functions *(adapted from Basner et al. 2012)*.

Number	Test	Task	Cognitive Domain	Primary brain regions
1	Motor Praxis Test (MPT)	Reaction time task: continuously shrinking boxes are displayed on different locations on the screen, that need to be hit as quickly as possible.	sensory Motor Speed	motor, pre-motor, sand sensory motor cortex
2	Visual Object Learning Task (VOLT)	During a brief learning period several different 3-dimensional geometrical figures are shown. Subsequently, these figures are again presented, but mixed with distractor shapes which need to be distinguished from the original figures.	visual learning and spatial working memory	lateral occipital complex, medial temporal cortex, hippocampus
3	n-back	Classical n-back task. However, instead of number or letters images are shown and it needs to be determined whether a subsequent images appeared "n" trials ago or not. working memory		lateral premotor cortex; dorsal cingulate and medial premotor cortex; dorsolateral and ventrolateral prefrontal cortex; frontal poles; and medial and lateral posterior parietal cortex, hippocampus
4	Abstract Matching (AM)	Geometrical figures have to be matched to other pairs of shapes by detecting certain underlying rules.	abstraction, mental flexibility	prefrontal cortex
5	Line Orientation Task (LOT	Two lines at different angles and length have to be arranged in parallel by increment left/right rotations.	spatial orientation	right temporo-parietal cortex, visual cortex

(Continued)

TABLE 2.2 Cognition (originally NeuroCATS) test battery. It should be noted though that the outlined tests are just summary of the core battery, which be adjusted and extended by additional paradigms and functions *(adapted from Basner et al. 2012).—Cont'd*

Number	Test	Task	Cognitive Domain	Primary brain regions
6	Emotion Recognition Task (ERT)	List of emotions have to be matched to photographs of human faces.	emotion recognition	temporo-limbic regions including the fusiform gyrus, amygdala and insular cortex
7	Matrix Reasoning Task (MRT)	A series of patterns has to be completed by another pattern from a list of patterns.	abstract reasoning	dorsolateral prefrontal cortex, presupplementary motor area, posterior parietal lobe
8	Digital Symbol Substitution Task (DSST)	Classcial digit symbol substitution: a number of symbols paired with a number are constantly displayed. In addition, a single symbol is presented and its needs to determined which number matches the target symbol.	complex scanning, visual tracking and attention	fronto-parietal network, inferior frontal sulcus; MFG, middle frontal gyrus; SPL, superior parietal lobule
9	Balloon Analog Risk Task (BART)	A balloon is required to be inflated incrementally. Each additional pump leads to an increased reward, but at the same time increases the risk of popping the ballong, after which all gains are lost.	risk decision making	Dorsolateral prefrontal cortex, parietal and motor regions, anterior cingulate cortex, bilateral insula, and bilateral lateral orbitofrontal cortex
10	Psychomotor Vigilance Test (PVT)	Within a red rectangle constantly displayed on the screen a number will appear that changes by counting every milisecond. As soon as the first number appears a target key needs to be pressed.	vigilance and motor speed	fronto-parietal network

Basner, M., Gur, R., Mollicone, D., & Dinges, D. (2012). Neurocats: Individualized real-time neurocognitive assessment toolkit for space flight fatigue. In NASA human research program investigators' workshop, 14 February 2012, Houston: NASA.

fatigue, sleep deprivation, circadian misalignment, changing work-rest schedules as well as environmental stressors including dehydration, hypo- and hyperthermia and hypoxic conditions. Presently, it is used to assess behavioral alertness and fatigues in a variety of space analog settings (e.g., NEEMO, HIGH-SEAS) as well as onboard ISS. Very recently, a tablet version of the software has been developed that is particularly useful for field physiological conditions. We have recently tested the feasibility of the tablet version for assessing fatigue and cognitive performance in managers during short-time vacation. The hardware and the battery were very well accepted by the subjects and allowed reliable recordings with immediate online access of the acquired data. The mobility of the approach, short administration time, the diversity of cognitive domains assessed, the use of well validated paradigms, its potential for repeated administrations, and instant online access to acquired data are important features compared to classical laboratory based behavioral testing systems and batteries, which make the Cognition Battery a promising platform for straightforwardly, quickly, and reliably assessing cognitive function and fatigue in extreme environmental conditions as well as various occupational health settings [121].

2.5.2.2 Mobile EEG—Present Challenges and Promises

Certain experimental conditions, such as behavioral testing platforms, might be complemented by acquiring additional biosignals. Typical measures for this purpose include the EEG and the electrooculogram. The EEG is sensitive to fluctuations in vigilance and has been shown to predict performance degradation due to sustained mental work (see Borghini et al. [118] for review). Generally, decreases in vigilance and deterioration in performance have been shown to be associated with increased EEG power spectra in the theta band (4-8 Hz) and a change in EEG alpha power (8-12 Hz), whereas increased alertness might correlate with higher beta power in the range of 12-18 Hz [124,125]. Notably, during tasks requiring sustained attention, multitasking, and situations characterized by emerging fatigue, the strongest effects seem to be associated with increases in theta power and decreases in alpha power over the frontal and parietal cortex (see Borghini et al. [118] for review). The sensitivity of EEG spectral power measures as indicators of task complexity and fatigue has been controversial. In part this controversy may be due to the fact that recent concepts of EEG spectral power have been too simple, both in regard to fixed frequency ranges and their underlying neurophysiological interpretation [126]. For instance, although low alpha power may indeed be related to good cognitive performance during actual task demands, the opposite is true for the resting condition [126]. Yet, this only seems to hold true for tasks related to memory performance, but not visual discrimination tasks. In other words, perception performance seems to be enhanced if the cortex is already activated (low alpha), whereas memory performance is enhanced if the cortex is deactivated (high alpha) before a task is performed [127]. Complicating matters further, the current predefined ranges of frequency bands might be too general to reflect inter- and intraindividual differences [126]. A promising approach might also be to train the acquisition system using support vector machines, artificial neural networks, or self-organizing maps based on predefined conditions, before applying the individual classification systems for the assessment of future task demands [128,129]. The downside of individual classification systems is that they can be highly specific with regard to conditions and therefore lack generalizability. This drawback might be overcome by hierarchical Bayes models. Wang et al. [130] demonstrated that, using this approach, group-based models yielded a robust classification accuracy of individual workloads similar to that observed in individually trained models. In addition to spectral power, different feature-extraction approaches might also show promise for assessing specific neurophysiological states. For instance, the feature extraction of alpha EEG spindles might be superior to approaches determining EEG in the frequency domain for detecting fatigue [131]. Finally, it might be promising to combine EEG data with additional biosignals, such as heart rate, eye blink duration and frequency, and indices of the autonomic nervous system using galvanic skin responses and/or heart rate variability, which could lead to a classification accuracy of 90% (see Borghini et al. [118] for review). Presently, such approaches are still limited to the research setting, and various challenges have to be tackled before these approaches can be applied to online monitoring and mental state identification. In addition to refining algorithms in terms of their speed, sensitivity, specificity, and robustness (validity under extreme environmental conditions), future technological improvements must produce devices that minimally distract and affect their users. If the technologies are not comfortable and easily worn or even integrated in garments, they will lack acceptability. Some commercially available devices have already been miniaturized to an extent that they can be stowed in a small backpack (http://www.brainproducts.com/productde-tails. php?id=15). These devices reportedly provide valid recordings of EEG in extreme conditions such as motorcycling, mountaineering expeditions, or rollercoaster rides. However, the technology is still too bulky, and the methodology too time-consuming (hardware instrumentation and gelling of electrodes) to consider them as standard mobile monitoring devices in the field. For various biosignals, such as ECG, actigraphy, skin and CBT, and GSR measurement, devices are readily available, but the equivalent EEG solutions are still emerging. One of the first devices employed a headset with 14 wet saline electrodes and a gyroscope, allowing 12 h of continuous use (http://emotiv.com/store/headset. php). However, although this headset has been successfully employed for human-machine interactions, such as controlling a car as part of the AutoNOMOS Labs project at the Freie Uninversität Berlin (http://www.autonomos.inf.fu-berlin. de), the fixed electrode position, the preparation of electrodes, and the rather loose fixation of the system on the head presently limit its functionality in field settings such as extreme environments or sleep studies. Neurosky has suggested a simplified approach, offering a system with a single prefrontal dry electrode (http:// neurosky.com/products-markets/eeg-biosensors/hardware/). Another forehead-electrode system has recently been developed by the Department of Applied Physics at the University of Eastern Finland and Kuopio University Hospital in cooperation with Screen- Tec LLC, Kuopio Academy of Design, and Mega Electronics Ltd. (Figure 2.12c). Dry electrodes, requiring no gel or wet solution, with a flexible electrode set-up would indeed push the frontiers of EEG monitoring in the field, because no special pretreatment of the skin and scalp would be required and set-up time would be minimized.

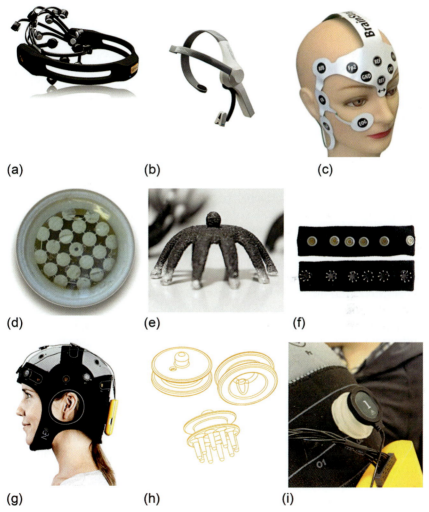

FIGURE 2.12   Examples of mobile EEG systems. (a) Fixed 14 saline electrode system (Emotiv, San Francisco, USA). (b) Single prefrontal dry electrode system (NeuroSky Inc., San Jose, USA). (c) Multiple hydrogel electrode system for frontal recordings (Mega Electronics Ltd, Kuopio, Finnland). (d) Dry pad electrode for use over skin (Cognionics Inc., San Diego, USA). (e) Dry, through-hair EEG sensor optimized for sliding through hair for high-quality scalp contact (Cognionics Inc., San Diego, USA). (f–i) Customized headset combining pad and through-hair electrodes for mobile recordings in the field, including sleep.

With no conductive gel, however, dry systems must fulfill the following requirements: (1) have direct contact with the scalp; (2) remain securely in place (there is no gel to buffer any gap between the sensor and the skin during measurement); (3) tolerate high-contact impedances, which could be 100- to 200-fold higher than those present for wet electrodes, while rejecting noise and interference. In addition, for use on hair-covered areas, the hair must be removed or the sensor designed to brush through hair with modest pressure, while retaining the ability to flatten for safety and comfort. Such a sensor system with up to 64 electrodes has recently been introduced for commercial use (http://www.cognionics.com). The system utilizes a proprietary combination of active shielding, high-input impedance amplifiers, and noise cancellation to support low-noise acquisition [132]. An example of a customized system that could show promise for field applications, including sleep, is presently being tested at the Center for Space Medicine Berlin. The device is shown in Figure 2.12f. It consists of a headband that can be flexibly equipped with dry-pad and through-hair electrodes. The miniaturized ($5\,cm \times 7\,cm \times 1.5\,cm$) amplifier is battery-powered (AAA battery, 1.5V), can be attached with snap connectors at nearly any position of the headband, and can be repositioned during the recording. Neuroelectrics (Barcelona, Spain) offers another promising EEG solution for field use. The so-called Enobio system has 8-32 EEG channels and allows the use of gel and dry electrodes. The system features a headband for frontal electrodes and a neoprene cap for individualized recording setups. The small high-input amplifier (65 g) includes a rechargeable 3.7V battery and is attached to the back of the cap, and it is equipped for both wireless recording to a laptop via a Bluetooth connection or to a micro SD card that can be inserted in the amplifier. Moreover, the system can be combined with synchronous recording of additional biosignals such as GSR or ECG. Finally, very recently, the company has been working with a "solid" gel electrode (Figure 2.12i) that is supposed to further improve signal quality and comfort.

Although sophisticated sensor technology can consistently detect more specific and accurate signals, the data processing of the acquired signals, via efficient and valid algorithms, allows the signals to be transformed into meaningful measures that are rapidly available. In addition to the contactless sleep-assessing technologies outlined in Section 2.5.1.5, a new mobile EEG device can reportedly differentiate between sleep stages. The device consists of a headband equipped with prefrontal fabric sensors located at approximately Fp1 and Fp2. The acquired signal (sample rate 128 Hz) is a combination of EEG, eye movements, and activity of the frontalis muscle. These data are bandpass-filtered (second-order bandpass 2-47 Hz) and transferred wirelessly to a base station, using an ultra-low-power propriety wireless protocol at 2.4 GHz. Sleep stages (determined every 2 s and smoothed using a 2 min moving average window) are then calculated in real-time, using a microprocessor within the base station. Classification of sleep stages is based on artificial neural networks using a combination of measures in the time and frequency domain. In a validation study of

18 healthy adults (6 women), a striking degree of agreement could be observed. Compared to standard polysomnographic recordings and evaluations of sleep stages by two independent observers, Shambroom and colleagues reported an agreement of 81% with a Cohen's kappa of 0.70, which was strikingly similar to the agreement between raters (83.2%, Cohen's kappa=0.74) [133]. The highest accuracy was achieved for REM and light sleep (86%), followed by deep sleep (71%) and wakefulness (64%). Notably, the system was also superior at evaluating sleep-onset latency compared to wrist-worn actigraphy, a measure that is known to yield unreliable estimates [93]. Some criticisms have been raised about the compliance of the headband. Yet, according to our experience, the compliance issue is simply due to overtightening of the headband.

Moreover, we have tested self-adhesive electrodes as a replacement for the headband in case the use of the headband might be restricted. For economic reasons the device is not yet commercially available, but as pointed out in Figure 2.12, we are testing a similar headband for general EEG recordings that could achieve similar accuracy. This approach might be particularly promising when combined with measurement of additional biosignals such as short-range radiofrequency sensing (Section 2.5.1.5) and/or actigraphy.

2.6 OUTLOOK

This chapter begins by considering the early development of biosignal acquisition more than a 100 years ago. The pioneers of the field, lead and inspired by Nathan Zuntz, already recognized and appreciated the value and impact of research outside the laboratory. The technological growth has only accelerated since then, with its most rapid advances within the last 30, and maybe even 10, years. Such developments largely result from the continuous increase in computing power, the miniaturization of electronics, and the expanded development of telecommunication technologies. Although the size and weight of mobile indirect calorimetry systems have decreased considerably since the era of Nathan Zuntz's "dry gas meter," these devices still need to be attached to the body. Thus, the next step in the miniaturization process has involved the creation of wireless sensors, such as fabric EEG sensors, and patches or capsules for the measurement of skin and CBT, as outlined in this chapter. Moreover, sensor probes requiring minimal preparation and cleaning time are further milestones in the widespread application of biosignal acquisition. Eventually, sensing systems without any need for attaching sensor probes to the body could promote the acceptance of and compliance with these technologies in research, the clinical setting, and particularly home-based monitoring for preventative healthcare and rehabilitation. In this regard, in addition to short-range radiofrequency sensing, unobtrusive signal monitoring is also attracting increasing interest by employing capacitive sensor technologies, and smart clothing or e-textiles could become increasingly important [13,14]. For instance, very recently, a so-called IMPACT Shirt (IMPedAnce Cardiography Textile) has been introduced—a shirt with integrated textile electrodes, textile wiring, and portable miniaturized hardware to record not only ECG, but also ICG (impedance cardiography), while transmitting these measures via Bluetooth to a mobile phone. Such a development might be very advantageous for monitoring high-risk patients at home [134]. Another exciting line of development involves biosensors, which are sometimes referred to as Tattoo sensors, because the sensors' electronic components are printed onto commercial tattoo transfer paper. Biosensors are characterized by very low power consumption and might be incorporated into lab-on-a-chip technologies. Thus, they can be placed on the skin like a thin plaster and easily removed [135]. Finally, nanosensors could also be employed for continuous and noninvasive real-time monitoring of biomarkers—applications of such technologies can be expected to exponentially rise in the coming years. This could not only provide groundbreaking technologies for physiological research in extreme environments and space, but also open completely new fields for applied research in the clinical setting, leading to breakthroughs in medical diagnostics and health care. Eventually, these advances could lead to the real-time monitoring of physiological parameters combined with genetic biomarkers, thus promoting a "translational real-time monitoring approach" that incorporates cellular, organ, and systemic levels. While undoubtedly such integrative approaches will promote a better understanding of human physiology, these technological advances will also raise new legal questions about personal data privacy and confidentiality, which need to be timely and appropriately addressed by medical, ethical and legal commissions before their implementation. Yet, to conclude with the Wayne Gretzky words, legendary Canadian hockey player, we should look where "the puck is going to be, not where it has been". The integration of approaches across molecular, cellular, organ and systemic levels combined with environmental and personalized data could promote a holistic understanding of human adaptation to extreme environments. Finally, in line with the systems biology and "P4 Medicine" approach coined by Lee Hood (http://p4mi.org), this could be a major step to predict, prevent, and personalize crew health during explanatory missions.

REFERENCES

[1] Gunga HC. Nathan Zuntz. His Life and Work in the Fields of High Altitude Physiology and Aviation Medicine. Burlington, London, New York: Elsevier; 2009.

[2] Dorn T, Wagner R. Die deutsche Seele. München: Knaus; 2011.

[3] Zuntz N, Müller F, Caspari W. Höhenklima und Bergwanderungen in ihrer Wirkung auf den Menschen…: Ergebnisse experimenteller Forschungen im Hochge-birge und Laboratorium. Berlin, Leipzig, Wien, Stuttgart: Deutsches Verlagshaus Bong & Co; 1906.

[4] Felsch P. Laborlandschaften: physiologische Alpenreisen im 19. Jahrhundert. Göttingen: Wallstein; 2007.

[5] Rheinberger H-J, Herrgott G. Experimentalsysteme und epistemische Dinge: eine Geschichte der Proteinsynthese im Reagenzglas. Frankfurt a.M: Suhrkamp; 2006.

[6] Lehmann F, Hagemann O, Zuntz N. Zur Kenntnis des Stoffwechsels beim Pferde. Landwirtsch Jahrb 1894;23:125.

[7] Gunga HC, Suthau T, Bellmann A, Friedrich A, Schwanebeck T, Stoinski S, et al. Body mass estimations for Plateosaurus engelhardti using laser scanning and 3D reconstruction methods. Naturwissenschaften 2007;94(8):623–30.

[8] Sander PM, Christian A, Clauss M, Fechner R, Gee CT, Griebeler EM, et al. Biology of the sauropod dinosaurs: the evolution of gigantism. Biol Rev Camb Philos Soc 2011;86(1):117–55.

[9] Daanen HAM, Ter Haar FB. 3D whole body scanners revisited. Displays 2013;34(4):270–5.

[10] Wang R, Choi J, Medioni G. Accurate full body scanning from a single fixed 3D camera. In: 2012 Second International Conference on 3D Imaging, Modeling, Processing, Visualization and Transmission (3DIMPVT). 2012. p. 432–9.

[11] Velardo C, Dugelay J, Paleari M, Ariano P. Building the space scale or how to weigh a person with no gravity. In: 2012 IEEE International Conference on Emerging Signal Processing Applications (ESPA). 2012. p. 67–70.

[12] Hoyt RW, Friedl KE. Metabolic Monitoring Technologies for Military Field Applications. Washington, DC: National Academy Press; 2004, p. 247–57.

[13] Zheng YL, Ding XR, Poon CC, Lo BP, Zhang H, Zhou XL, et al. Unobtrusive sensing and wearable devices for health informatics. IEEE Trans Biomed Eng 2014;61(5):1538–54.

[14] Zhang YT, Zheng YL, Lin WH, Zhang HY, Zhou XL. Challenges and opportunities in cardiovascular health informatics. IEEE Trans Biomed Eng 2013;60(3):633–42.

[15] Villa F, Magnani A, Merati G, Castiglioni P. Feasibility of long-term monitoring of multifrequency and multisegment body impedance by portable devices. IEEE Trans Biomed Eng 2014;61(6):1877–86.

[16] Stahn AC, Terblanche E, Gunga HC. Use of bioelectrical impedance: general principles and overview. In: Preedy VR, editor. Handbook of Anthropometry: Physical Measures of Human Form in Health and Disease. New York: Springer; 2012. p. 49–90.

[17] Stahn AC, Terblanche E, Gunga HC. Selected applications of bioelectrical impedance analysis: body fluids, blood volume, body cell mass and fat mass. In: Preedy VR, editor. Handbook of Anthropometry: Physical Measures of Human Form in Health and Disease. New York: Springer; 2012. p. 415–40.

[18] Stahn A, Terblanche E, Strobel G. Modeling upper and lower limb muscle volume by bioelectrical impedance analysis. J Appl Physiol 2007;103(4):1428–35.

[19] Waterhouse J, Drust B, Weinert D, Edwards B, Gregson W, Atkinson G, et al. The circadian rhythm of core temperature: origin and some implications for exercise performance. Chronobiol Int 2005;22(2):207–25.

[20] Santhi N, Aeschbach D, Horowitz TS, Czeisler CA. The impact of sleep timing and bright light exposure on attentional impairment during night work. J Biol Rhythms 2008;23(4):341–52.

[21] Reinberg AE, Ashkenazi I, Smolensky MH. Euchronism, allochronism, and dyschronism: is internal desynchronization of human circadian rhythms a sign of illness? Chronobiol Int 2007;24(4):553–88.

[22] Scheer FA, Hilton MF, Mantzoros CS, Shea SA. Adverse metabolic and cardiovascular consequences of circadian misalignment. Proc Natl Acad Sci USA 2009;106(11):4453–8.

[23] Schmidt C, Collette F, Cajochen C, Peigneux P. A time to think: circadian rhythms in human cognition. Cogn Neuropsychol 2007;24(7):755–89.

[24] Manzey D. Limiting factor for human health and performance: psychological issues. In: Comet B, Facius R, Horneck G, editors. Study on the Survivability and Adaptation of Humans to Long-Duration Interplanetary and Planetary Environments—HUMEX-TN-002: Critical Assessments of the Limiting Factors for Human Health and Performance and Recommendation of Countermeasures. 2001. p. 1–45.

[25] Gundel A, Polyakov VV, Zulley J. The alteration of human sleep and circadian rhythms during spaceflight. J Sleep Res 1997;6(1):1–8.

[26] Monk TH, Kennedy KS, Rose LR, Linenger JM. Decreased human circadian pacemaker influence after 100 days in space: a case study. Psychosom Med 2001;63(6):881–5.

[27] Santy PA, Kapanka H, Davis JR, Stewart DF. Analysis of sleep on Shuttle missions. Aviat Space Environ Med 1988;59(11 Pt 1):1094–7.

[28] Monk TH, Buysse DJ, Billy BD, Kennedy KS, Willrich LM. Sleep and circadian rhythms in four orbiting astronauts. J Biol Rhythms 1998;13(3):188–201.

[29] Caldwell JA, Mallis MM, Caldwell JL, Paul MA, Miller JC, Neri DF. Fatigue countermeasures in aviation. Aviat Space Environ Med 2009;80(1):29–59.

[30] Dinges DF, Pack F, Williams K, Gillen KA, Powell JW, Ott GE, et al. Cumulative sleepiness, mood disturbance, and psychomotor vigilance performance decrements during a week of sleep restricted to 4-5 hours per night. Sleep 1997;20(4):267–77.

[31] Nechaev AP. Work and rest planning as a way of crew member error management. Acta Astronaut 2001;49(3–10):271–8.

[32] Guler AD, Ecker JL, Lall GS, Haq S, Altimus CM, Liao HW, et al. Melanopsin cells are the principal conduits for rod-cone input to non-image-forming vision. Nature 2008;453(7191):102–5.

[33] Hattar S, Liao HW, Takao M, Berson DM, Yau KW. Melanopsin-containing retinal ganglion cells: architecture, projections, and intrinsic photosensitivity. Science 2002;295(5557):1065–70.

[34] Altimus CM, Guler AD, Alam NM, Arman AC, Prusky GT, Sampath AP, et al. Rod photoreceptors drive circadian photoentrainment across a wide range of light intensities. Nat Neurosci 2010;13(9):1107–12.

[35] Simonneaux V, Ribelayga C. Generation of the melatonin endocrine message in mammals: a review of the complex regulation of melatonin synthesis by norepinephrine, peptides, and other pineal transmitters. Pharmacol Rev 2003;55(2):325–95.

[36] Dijk DJ, Neri DF, Wyatt JK, Ronda JM, Riel E, Ritz-De CA, et al. Sleep, performance, circadian rhythms, and light-dark cycles during two space shuttle flights. Am J Physiol Regul Integr Comp Physiol 2001;281(5):R1647–64.

[37] Wong KY, Dunn FA, Graham DM, Berson DM. Synaptic influences on rat ganglion-cell photoreceptors. J Physiol 2007;582(Pt 1):279–96.

[38] Boivin DB, Czeisler CA, Dijk DJ, Duffy JF, Folkard S, Minors DS, et al. Complex interaction of the sleep-wake cycle and circadian phase modulates mood in healthy subjects. Arch Gen Psychiatry 1997;54(2):145–52.

[39] Gander PH, Macdonald JA, Montgomery JC, Paulin MG. Adaptation of sleep and circadian rhythms to the Antarctic summer: a question of zeitgeber strength. Aviat Space Environ Med 1991;62(11):1019–25.

[40] Stuster JW. Bold Endeavors: Lessons from Polar and Space Exploration. Annapolis, MD: Naval Institute Press; 1996.

[41] Palinkas LA. Sociocultural influences on psychosocial adjustment in Antarctica. Med Anthropol 1989;10(4):235–46.

[42] Yoneyama S, Hashimoto S, Honma K. Seasonal changes of human circadian rhythms in Antarctica. Am J Physiol 1999;277(4 Pt 2):R1091–7.

[43] Gundel A, Nalishiti V, Reucher E, Vejvoda M, Zulley J. Sleep and circadian rhythm during a short space mission. Clin Investig 1993;71(9):718–24.

[44] Fuller CA, Hoban-Higgins TM, Griffin DW, Murakami DM. Influence of gravity on the circadian timing system. Adv Space Res 1994;14(8):399–408.

[45] Krotov VP, Iushin VA, Vatsek A, Korolkov VI, Shebela A, Truzhennikov AN, et al. Oxygenation of the frontal cerebral cortex in monkeys during a two-week space flight. Aviakosm Ekolog Med 1992;26(2):42–6.

[46] Fuller CA, Hoban-Higgins TM, Klimovitsky VY, Griffin DW, Alpatov AM. Primate circadian rhythms during spaceflight: results from Cosmos 2044 and 2229. J Appl Physiol 1996;81(1):188–93.

[47] Aikas E, Karvonen MJ, Piironen P, Ruostennoja R. Intramuscular, rectal and oesophageal temperature during exercise. Acta Physiol Scand 1962;54:36–70.

[48] Cooper KE, Kenyon JR. A comparison of temperatures measured in the rectum, oesophagus, and on the surface of the aorta during hypothermia in man. Br J Surg 1957;44(188):616–9.

[49] Cranston WI, Gerbrandy J, Snell ES. Oral, rectal and oesophageal temperatures and some factors affecting them in man. J Physiol 1954;126(2):347–58.

[50] Gerbrandy J, Snell ES, Cranston WI. Oral, rectal, and oesophageal temperatures in relation to central temperature control in man. Clin Sci (Lond) 1954;13(4):615–24.

[51] Mairiaux P, Sagot JC, Candas V. Oral temperature as an index of core temperature during heat transients. Eur J Appl Physiol Occup Physiol 1983;50(3):331–41.

[52] Saltin B, Hermansen L. Esophageal, rectal, and muscle temperature during exercise. J Appl Physiol 1966;21(6):1757–62.

[53] Edwards B, Waterhouse J, Reilly T, Atkinson G. A comparison of the suitabilities of rectal, gut, and insulated axilla temperatures for measurement of the circadian rhythm of core temperature in field studies. Chronobiol Int 2002;19(3):579–97.

[54] Darwent D, Zhou X, van den Heuvel C, Sargent C, Roach GD. The validity of temperature-sensitive ingestible capsules for measuring core body temperature in laboratory protocols. Chronobiol Int 2011;28(8):719–26.

[55] Gunga HC, Werner A, Stahn A, Steinach M, Schlabs T, Koralewski E, et al. The Double Sensor—a non-invasive device to continuously monitor core temperature in humans on earth and in space. Respir Physiol Neurobiol 2009;169(Suppl. 1):S63–8.

[56] Gunga H-C, Sandsund M, Reinertsen RE, Sattler F, Koch J. A non-invasive device to continuously determine heat strain in humans. J Therm Biol 2008;33(5):297–307.

[57] Eichna LW. Thermal gradients in man; comparison of temperatures in the femoral artery and femoral vein with rectal temperatures. Arch Phys Med Rehabil 1949;30(9):584–93.

[58] McKenzie JE, Osgood DW. Validation of a new telemetric core temperature monitor. J Therm Biol 2004;29(7):605–11.

[59] Domitrovich JW, Cuddy JS, Ruby BC. Core-temperature sensor ingestion timing and measurement variability. J Athl Train 2010;45(6):594–600.

[60] Sawka MN, Gonzalez RR, Young AJ, Muza SR, Pandolf KB, Latzka WA, et al. Polycythemia and hydration: effects on thermoregulation and blood volume during exercise-heat stress. Am J Physiol 1988;255(3 Pt 2):R456–63.

[61] Lim CL, Byrne C, Lee JK. Human thermoregulation and measurement of body temperature in exercise and clinical settings. Ann Acad Med Singapore 2008;37(4):347–53.

[62] Mazerolle SM, Ganio MS, Casa DJ, Vingren J, Klau J. Is oral temperature an accurate measurement of deep body temperature? A systematic review. J Athl Train 2011;46(5):566–73.

[63] Insler SR, Sessler DI. Perioperative thermoregulation and temperature monitoring. Anesthesiol Clin 2006;24(4):823–37.

[64] Byrne C, Lim CL. The ingestible telemetric body core temperature sensor: a review of validity and exercise applications. Br J Sports Med 2007;41(3):126–33.

[65] Mcilvoy L. Comparison of brain temperature to core temperature: a review of the literature. J Neurosci Nurs 2004;36(1):23–31.

[66] Moran DS, Mendal L. Core temperature measurement: methods and current insights. Sports Med 2002;32(14):879–85.

[67] Cooper KE, Cranston WI, Snell ES. Temperature in the external auditory meatus as an index of central temperature changes. J Appl Physiol 1964;19(5):1032–4.

[68] Shiraki K, Konda N, Sagawa S. Esophageal and tympanic temperature responses to core blood temperature changes during hyperthermia. J Appl Physiol 1986;61(1):98–102.

[69] Opatz O, Trippel T, Lochner A, Werner A, Stahn A, Steinach M, et al. Temporal and spatial dispersion of human body temperature during deep hypothermia. Br J Anaesth 2013;111(5):768–75.

[70] Aschoff J. Exogenous and endogenous components in circadian rhythms. Cold Spring Harb Symp Quant Biol 1960;25:11–28.

[71] Kräuchi K, Wirz-Justice A. Circadian rhythm of heat production, heart rate, and skin and core temperature under unmasking conditions in men. Am J Physiol 1994;267(3 Pt 2):R819–29.

[72] Marques MD, Waterhouse JM. Masking and the evolution of circadian rhythmicity. Chronobiol Int 1994;11(3):146–55.

[73] Waterhouse J, Minors D, Akerstedt T, Hume K, Kerkhof G. Circadian rhythm adjustment: difficulties in assessment caused by masking. Pathol Biol (Paris) 1996;44(3):205–7.

[74] Waterhouse J, Weinert D, Minors D, Folkard S, Owens D, Atkinson G, et al. A comparison of some different methods for purifying core temperature data from humans. Chronobiol Int 2000;17(4):539–66.

[75] Waterhouse J, Edwards B, Mugarza J, Flemming R, Minors D, Calbraith D, et al. Purification of masked temperature data from humans: some preliminary observations on a comparison of the use of an activity diary, wrist actimetry, and heart rate monitoring. Chronobiol Int 1999;16(4):461–75.

[76] Wever RA. Internal interactions within the human circadian system: the masking effect. Experientia 1985;41(3):332–42.

[77] Folkard S. The pragmatic approach to masking. Chronobiol Int 1989;6(1):55–64.

[78] Minors DS, Waterhouse JM. Investigating the endogenous component of human circadian rhythms: a review of some simple alternatives to constant routines. Chronobiol Int 1992;9(1):55–78.

[79] Waterhouse J, Kao S, Weinert D, Edwards B, Atkinson G, Reilly T. Measuring phase shifts in humans following a simulated time-zone transition: agreement between constant routine and purification methods. Chronobiol Int 2005;22(5):829–58.

[80] Minors DS, Waterhouse JM. Removing masking factors from urinary rhythm data in humans. Chronobiol Int 1990;7(5–6):425–32.

[81] Minors DS, Folkard S, Waterhouse JM. The shape of the endogenous circadian rhythm of rectal temperature in humans. Chronobiol Int 1996;13(4):261–71.

[82] Rietveld WJ, Minors DS, Waterhouse JM. Circadian rhythms and masking: an overview. Chronobiol Int 1993;10(4):306–12.

[83] Martinez-Nicolas A, Ortiz-Tudela E, Rol MA, Madrid JA. Uncovering different masking factors on wrist skin temperature rhythm in free-living subjects. PLoS One 2013;8(4):e61142.

[84] Waterhouse J, Edwards B, Bedford P, Hughes A, Robinson K, Nevill A, et al. Thermoregulation during mild exercise at different circadian times. Chronobiol Int 2004;21(2):253–75.

[85] Klerman EB, Lee Y, Czeisler CA, Kronauer RE. Linear demasking techniques are unreliable for estimating the circadian phase of ambulatory temperature data. J Biol Rhythms 1999;14(4):260–74.

[86] Waterhouse J, Nevill A, Weinert D, Folkard S, Minors D, Atkinson G, et al. Modeling the effect of spontaneous activity on core temperature in healthy human subjects. Biol Rhythm Res 2001;32(5):511–28.

[87] Minors DS, Waterhouse JM. Separating the endogenous and exogenous components of the circadian rhythm of body temperature during night work using some 'purification' models. Ergonomics 1993;36(5):497–507.

[88] Blood ML, Sack RL, Percy DC, Pen JC. A comparison of sleep detection by wrist actigraphy, behavioral response, and polysomnography. Sleep 1997;20(6):388–95.

[89] Chesson Jr. MD, Coleman MD, Lee-Chiong MD, Pancer DDS. Practice parameters for the use of actigraphy in the assessment of sleep and sleep disorders: an update for 2007. Sleep 2007;30(4):519.

[90] Kosmadopoulos A, Sargent C, Darwent D, Zhou X, Roach GD. Alternatives to polysomnography (PSG): a validation of wrist actigraphy and a partial-PSG system. Behav Res Methods 2014; http://dx/doi.org/10.3758/s13428-013-0438-7.

[91] Marino M, Li Y, Rueschman MN, Winkelman JW, Ellenbogen JM, Solet JM, et al. Measuring sleep: accuracy, sensitivity, and specificity of wrist actigraphy compared to polysomnography. Sleep 2013;36(11):1747–55.

[92] de Souza L, Benedito-Silva AA, Pires MLN, Poyares D, Tufik S, Calil HM. Further validation of actigraphy for sleep studies. Sleep 2003;26(1):81–5.

[93] O'Hare E, Flanagan D, Penzel T, Garcia C, Frohberg D, Heneghan C. A comparison of radiofrequency biomotion sensors and actigraphy versus polysomnography for the assessment of sleep in normal subjects. Sleep Breath 2014; http://dx/doi.org/10.1007/s11325-014-0967-z.

[94] Paquet J, Kawinska A, Carrier J. Wake detection capacity of actigraphy during sleep. Sleep 2007;30(10):1362–9.

[95] Basner M, Dinges DF, Mollicone D, Ecker A, Jones CW, Hyder EC, et al. Mars 520-d mission simulation reveals protracted crew hypokinesis and alterations of sleep duration and timing. Proc Natl Acad Sci USA 2013;110(7):2635–40.

[96] Paalasmaa J, Waris M, Toivonen H, Leppakorpi L, Partinen M. Unobtrusive online monitoring of sleep at home. In: 2012 Annual International Conference of the IEEE Engineering in Medicine and Biology Society (EMBC). 2012. p. 3784–8.

[97] Karlen W, Mattiussi C, Floreano D. Improving actigraph sleep/wake classification with cardio-respiratory signals. Conf Proc IEEE Eng Med Biol Soc 2008;2008:5262–5.

[98] Mack DC, Patrie JT, Felder RA, Suratt PM, Alwan M. Sleep assessment using a passive ballistocardiography-based system: preliminary validation. Conf Proc IEEE Eng Med Biol Soc 2009;2009:4319–22.

[99] Choi BH, Chung GS, Lee JS, Jeong DU, Park KS. Slow-wave sleep estimation on a load-cell-installed bed: a non-constrained method. Physiol Meas 2009;30(11):1163–70.

[100] Chung GS, Choi BH, Lee JS, Lee JS, Jeong DU, Park KS. REM sleep estimation only using respiratory dynamics. Physiol Meas 2009;30(12):1327–40.

[101] Devot S, Dratwa R, Naujokat E. Sleep/wake detection based on cardiorespiratory signals and actigraphy. Conf Proc IEEE Eng Med Biol Soc 2010;2010:5089–92.

[102] Kortelainen JM, Mendez MO, Bianchi AM, Matteucci M, Cerutti S. Sleep staging based on signals acquired through bed sensor. IEEE Trans Inf Technol Biomed 2010;14(3):776–85.

[103] Migliorini M, Bianchi AM, Nisticó D, Kortelainen J, Arce-Santana E, Cerutti S, et al. Automatic sleep staging based on ballistocardiographic signals recorded through bed sensors. In: 2010 Annual International Conference of the IEEE Engineering in Medicine and Biology Society (EMBC). 2010. p. 3273–6.

[104] De Chazal P, Fox N, O'Hare E, Heneghan C, Zaffaroni A, Boyle P, et al. Sleep/wake measurement using a non-contact biomotion sensor. J Sleep Res 2011;20(2):356–66.

[105] Goel N, Basner M, Rao H, Dinges DF. Circadian rhythms, sleep deprivation, and human performance. Prog Mol Biol Transl Sci 2013;119:155–90.

[106] Basner M, Rao H, Goel N, Dinges DF. Sleep deprivation and neurobehavioral dynamics. Curr Opin Neurobiol 2013;23(5):854–63.

[107] Lim J, Dinges DF. A meta-analysis of the impact of short-term sleep deprivation on cognitive variables. Psychol Bull 2010;136(3):375–89.

[108] Basner M, Mollicone D, Dinges DF. Validity and sensitivity of a brief psychomotor vigilance test (PVT-B) to total and partial sleep deprivation. Acta Astronaut 2011;69(11–12):949–59.

[109] Minkel J, Moreta M, Muto J, Htaik O, Jones C, Basner M, et al. Sleep deprivation potentiates hpa axis stress reactivity in healthy adults. Health Psychol 2014; May 12, epub ahead.

[110] Adan A. Cognitive performance and dehydration. J Am Coll Nutr 2012;31(2):71–8.

[111] Lieberman HR. Hydration and cognition: a critical review and recommendations for future research. J Am Coll Nutr 2007;26(Suppl. 5):555S–61S.

[112] Masento NA, Golightly M, Field DT, Butler LT, van Reekum CM. Effects of hydration status on cognitive performance and mood. Br J Nutr 2014;111(10):1841–52.

[113] Racinais S, Gaoua N, Grantham J. Hyperthermia impairs short-term memory and peripheral motor drive transmission. J Physiol 2008;586(Pt 19):4751–62.

[114] Mäkinen TM, Palinkas LA, Reeves DL, Pääkkönen T, Rintamäki H, Leppäluoto J, et al. Effect of repeated exposures to cold on cognitive performance in humans. Physiol Behav 2006;87(1):166–76.

[115] Simmons SE, Saxby BK, McGlone FP, Jones DA. The effect of passive heating and head cooling on perception, cardiovascular function and cognitive performance in the heat. Eur J Appl Physiol 2008;104(2):271–80.

[116] Van Dongen HP, Bender AM, Dinges DF. Systematic individual differences in sleep homeostatic and circadian rhythm contributions to neurobehavioral impairment during sleep deprivation. Accid Anal Prev 2012;45(Suppl.):11–6.

[117] O'Donnell RD, Eggemeier FT. Workload Assessment Methodology. New York: John Wiley & Sons; 1986.

[118] Borghini G, Astolfi L, Vecchiato G, Mattia D, Babiloni F. Measuring neuro-physiological signals in aircraft pilots and car drivers for the assessment of mental workload, fatigue and drowsiness. Neurosci Biobehav Rev 2012;44C:58–75.

[119] Oken BS, Salinsky MC, Elsas SM. Vigilance, alertness, or sustained attention: physiological basis and measurement. Clin Neurophysiol 2006;117(9):1885–901.

[120] Dinges DF, Powell JW. Microcomputer analyses of performance on a portable, simple visual RT task during sustained operations. Behav Res Methods Instrum Comput 1985;17(6):652–5.

[121] Basner M, Gur R, Mollicone D, Dinges D. Neurocats: individualized real-time neurocognitive assessment toolkit for space flight fatigue. In: NASA Human Research Program Investigators' Workshop. Houston: NASA; 2012.

[122] Kane RL, Short P, Sipes W, Flynn CF. Development and validation of the spaceflight cognitive assessment tool for windows (WinSCAT). Aviat Space Environ Med 2005;76(Suppl. 6):B183–91.

[123] Reeves DL, Winter KP, Bleiberg J, Kane RL. ANAM® genogram: historical perspectives, description, and current endeavors. Arch Clin Neuropsychol 2007;22:15–37.

[124] Okogbaa OG, Shell RL, Filipusic D. On the investigation of the neurophysiological correlates of knowledge worker mental fatigue using the EEG signal. Appl Ergon 1994;25(6):355–65.

[125] Zhao C, Zhao M, Liu J, Zheng C. Electroencephalogram and electrocardiograph assessment of mental fatigue in a driving simulator. Accid Anal Prev 2012;45:83–90.

[126] Klimesch W. EEG alpha and theta oscillations reflect cognitive and memory performance: a review and analysis. Brain Res Brain Res Rev 1999;29(2–3):169–95.

[127] Hanslmayr S, Klimesch W, Sauseng P, Gruber W, Doppelmayr M, Freunberger R, et al. Visual discrimination performance is related to decreased alpha amplitude but increased phase locking. Neurosci Lett 2005;375(1):64–8.

[128] Honal M, Schultz T. Determine task demand from brain activity. In: Biosignals, Springer-Verlag: Berlin, Heidelberg; vol. 1. 2008. p. 100–7.

[129] Shen KQ, Li XP, Ong CJ, Shao SY, Wilder-Smith EP. EEG-based mental fatigue measurement using multi-class support vector machines with confidence estimate. Clin Neurophysiol 2008;119(7):1524–33.

[130] Wang Z, Hope RM, Wang Z, Ji Q, Gray WD. Cross-subject workload classification with a hierarchical Bayes model. Neuroimage 2012;59(1):64–9.

[131] Simon M, Schmidt EA, Kincses WE, Fritzsche M, Bruns A, Aufmuth C, et al. EEG alpha spindle measures as indicators of driver fatigue under real traffic conditions. Clin Neurophysiol 2011;122(6):1168–78.

[132] Chi YM, Maier C, Cauwenberghs G. Ultra-high input impedance, low noise integrated amplifier for noncontact biopotential sensing. IEEE J Emerg Sel Top Circuits Syst 2011;1(4):526–35.

[133] Shambroom JR, Fábregas SE, Johnstone J. Validation of an automated wireless system to monitor sleep in healthy adults. J Sleep Res 2012;21(2):221–30.

[134] Ulbrich M, Mühlsteff J, Sipilä A, Kamppi M, Koskela A, Myry M, et al. The IMPACT shirt: textile integrated and portable impedance cardiography. Physiol Meas 2014;35(6):1181–96.

[135] Jia W, Bandodkar AJ, Valdés-Ramírez G, Windmiller JR, Yang Z, Ramírez J, et al. Electrochemical tattoo biosensors for real-time noninvasive lactate monitoring in human perspiration. Anal Chem 2013;85(14):6553–60.

[136] von der Heide R, Klein W, Zuntz N. Untersuchungen über den Nährwert der Kartoffelschlempe und ihres Ausgangsmaterials. Respirations- und Stoffwechselversuche am Rinde. Landwirtschaftliches Jahrbuch 1913;44:765–832.

[137] Vinge V. The coming technological singularity. Whole Earth Review 1993;81:88–95.

Chapter 3

Exercise Physiology

Mathias Steinach* and Hanns-Christian Gunga[†]

*Postdoctoral Research Associate, Center for Space Medicine and Extreme Environments, Institute of Physiology, CharitéCrossOver (CCO), Charité University Medicine Berlin, Berlin, Germany
[†]Professor, Center for Space Medicine and Extreme Environments, Institute of Physiology, CharitéCrossOver (CCO), Charité University Medicine Berlin, Berlin, Germany

3.1 INTRODUCTION

Exercise physiology is the science of human performance under physical stress, foremost the muscular system that, if regarded as a whole, is the largest organ system of the human body. From an evolutionary standpoint, as outlined in one of the previous chapters, humans have an ancestral phenotype that has developed in the East African Rift Valley system under challenging environmental conditions that got progressively higher, dryer, and colder [1,2]. These environmental conditions, when combined with an upright posture, put special evolutionary stresses on the thermoregulatory, cardiovascular, respiratory and muscular systems. This led, with respect to the thermoregulatory system, to extraordinary i) sweating capacities in humans and ii) a cardiovascular and muscular system designed for extraordinary high-endurance capacities as compared to other mammals. Today the muscular mass constitutes between 36% and 44% of human body mass and is subject to changes due to training, gender, and age. By means of ergometry, performance can be quantitatively measured and recorded by comparing the physical performance (given in watts) with a biological equivalent (i.e., oxygen consumption, heart rate, blood pressure, blood lactate concentration). Given that human performance, especially in extreme environments, depends crucially on the functioning of these systems—that is, the thermoregulatory, the cardiovascular, respiratory and the muscular systems, including nutritional and metabolic aspects—in this chapter on exercise physiology, these subsystems and their integration will be given a broader space in order to understand the limits of human performance under challenging physical and environmental conditions.

Naturally, it follows that exercise physiology contributes to other fields of science and medicine such as sports, occupational and environmental

Human Physiology in Extreme Environments. http://dx.doi.org/10.1016/B978-0-12-386947-0.00003-4

medicine, as well as high altitude and space medicine. For the most part, endurance and strength loads are being differentiated from each other when quantifying a person's physical capabilities. In addition, physiological strains are often augmented if humans exercise under special conditions such as extreme environments, for example, very hot (desert climate) and warm-humid (tropical climate) or very cold and dry environments (arctic and/or high altitude climate [3–9]). Therefore, field studies in such environments as well as under hypo- and hyperbaric conditions (at high altitude and diving) and even under microgravity pose interesting possibilities for studying the adaptability of human physiology in general. Furthermore, it should always be recognized that the observed momentary performance of a human has to be seen in relation to the actual possible maximum performance, and the different capacities of males and females as well [10,11]. Humans are usually not able to mobilize all their performance reserves deliberately. This is the so-called autonomic protected range. Below that there is a range that can be mobilized in exceptional cases (life-threatening fear) or under maximal will (operational resources). Last but not least, the range we deal with in our everyday life and the individual capacities of mental motivation are also subject to marked circadian variations [12].

3.2   PHYSICAL PRINCIPLES

In order to measure and understand several aspects encountered when dealing with exercise physiology, some basic physical principles need to be addressed. Force, work, power, and energy are physical values and the relationship between them enables the measurement and description of human performance capabilities.

Force (F, unit: Newton [N] = [kg $\cdot$ m $\cdot$ s^{-2}]) resembles accelerated mass; in other words, force is necessary to accelerate or decelerate a mass of otherwise unimpeded continuous velocity (e.g., such as an object floating in space).

Work (W, unit: Newtonmeter [Nm]) resembles a force that acts along a way. The unit Newtonmeter is interchangeable with the unit Joule [J].

Power (P, unit: Watt [W] = [J $\cdot$ s^{-1}]) resembles a given amount of work per time. One Watt equals one Newtonmeter—or Joule—per second. Two examples can represent the same amount of work; however, if one is performed in half the time of the other, the first example yields twice the power than the second one.

Energy (E, unit: Joule [J] = [Nm]) resembles the ability of a system to perform work. The convertibility of Nm and J expresses this fact mathematically. Energy for movement in the human body is converted via metabolic processes from the chemical energy of the nutrients. The efficiency of these processes in humans is only about 25-30% (see below); the major proportion of the energy is always lost as heat, which means the higher the force, work, and power are, the higher is the energy demand needed to fuel the physiological processes.

3.3   ERGOMETRY

Ergometry (from Greek *ergos*=work and *metry*=to measure) is the measurement and quantification of human physical performance. Traditionally, ergometric measurements evaluate the endurance capabilities of a person, although strength capabilities can also be quantified using appropriate tests.

Ergometric measurements are specific to the type of sports, for example, a cyclist should be tested on a cycle ergometer and a runner on a treadmill. Applying tests not specific to the type of sport—and thus not addressing the patterns of movements, used musculature, and energy supply—will yield false data regarding the athlete's capabilities, in most cases an underestimation of the true values [13,14]. Therefore, a vast array of testing devices exist today to closely emulate the physical load that the athlete experiences when exercising in each specific sport, such as in cycle ergometers, treadmills, rowing ergometers, and others.

3.4   ENERGY EXPENDITURE

Each cell of the 10^{13} to 10^{14} cells in the human body (see chapter one) requires energy to maintain its structure and functional capabilities. This energy requirement to maintain the structure of all cells of an organism is called Basal Metabolic Rate (BMR) and is sometimes used synonymously with Resting Metabolic Rate (RMR). If the BMR is to be measured, the human should be dressed in light clothing, in supine position under complete rest at a neutral ambient temperature (28 °C), with no forced convection, and abstinence from eating for 10 h. Eating and the subsequent processes of digestion, absorption, and metabolism will produce a further energy requirement, called Thermic Effects of Feeding (TEF). This energy requirement can be pictured as the energy necessary to access the energy of the nutrients. All other energy requirements such as movement, thermal stress, or psychological stress are subsumed as the Thermic Effects of Activity (TEA). It is obvious that both the degree of physical load (e.g., given in Watts on a cycle ergometer or as speed and grade on a treadmill) as well as the amount of time spent under physical strain will consequently affect the amount of the TEA, both in its absolute value as well as its relative value with regard to the daily energy expenditure (DEE) [15]. The DEE combines the values of the BMR, TEF, and TEA. A person participating in an endurance event such as a triathlon will have high values in the TEA and therefore also in the DEE. A person lying supine and being fed intravenously such as an ICU patient will have a DEE that closely resembles the BMR (although medication and the type of illness or injury can alter the DEE by increasing energy requirements). Typically, a DEE consists of 60-75% BMR, 15-30% TEA, and around 10% TEF. Figure 3.1 illustrates the typical DEE composition [15]. Given that the type and mass of different tissues and cells affect the energy requirements as well as hormone levels, BMR and DEE can vary greatly between the sexes, different ages, and individuals in general [16].

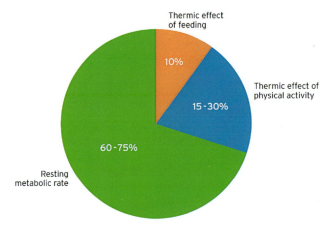

FIGURE 3.1   Composition of an average DEE, *adapted from [15]*.

A person's BMR is closely correlated to the amount and proportion of the fat free mass (FFM). FFM was shown to be seven times more metabolically active than fat mass (FM) [17]. This explains why males exhibit a 5-10% higher BMR than females because of their somewhat higher FFM percentage. The higher rate of metabolism of infants and adolescents due to the demands of growth also leads to an increased BMR. The FFM is not a homogenous mass but consists of different organs and tissues exhibiting different metabolic rates as Table 3.1 summarizes [18,19].

Table 3.1 shows that during rest only a few internal organs making up approximately 5% of body mass nevertheless account for 60% of the BMR [20]. At rest, the musculature is metabolically relatively inactive, but still more active than the FM. However, during intense exercise, the $\dot{V}O_2$ and thus the energy expenditure of working musculature can increase 100- to 200-fold, reaching oxygen consumptions of up to 400 ml/kg musculature per minute with the result that exercise can greatly increase the overall energy expenditure [21–23].

TABLE 3.1 Mass, $\dot{V}O_2$ and % BMR of different organs and tissues

	Organ Mass (kg)	% Body Mass	$\dot{V}O_2$ (ml/min)	% BMR
Heart	0.3	0.5	30	10
Brain	1.4	2	45	15
Liver	1.5	2	65	25
Kidneys	0.3	0.5	25	10
Muscle	30	40	65	25

The table shows the mass, the percentage of body mass, the oxygen consumption and the approximated percentage of basal metabolic rate of different organs and tissues [18,19].

Energy expenditure can be determined using different technical approaches; some of them have been discussed already in chapter 2 on methodology. The current gold standard is Doubly Labeled Water method (DLW) [24,25]:

- When this method is applied, a test subject ingests a volume of water containing a known concentration of $^2H_2^{18}O$. The isotopes eventually leave the body in form of $C^{18}O_2$, $H_2^{18}O$, and ^2H_2O through respiration, sweat, and urine. Using a spectroscopic analysis of these samples, it is possible to determine the CO_2 production and thus the energy expenditure.

Although highly accurate, the DLW method is expensive and mostly used to validate other methods [26,27]. Therefore, for practical reasons, much more frequently actimetry and indirect calorimetry are applied to determine energy expenditure in humans under rest and exercise.

- An actimeter—as already discussed in the previous chapter—is a usually wrist-worn mobile device that continuously measures acceleration and deceleration along with other parameters such as body temperature and galvanic skin reaction, in some cases. From these data an analysis of daily activity and energy expenditure is possible [28–30]. A disadvantage of the actigraphy is the need to adjust the algorithm used for calculation of the energy expenditures according to the population that is being measured (e.g., age, sex, weight, training status [31,32]). Advantages include the high mobility and relative low price of the method.
- Indirect calorimetry: For most applications, indirect calorimetry (IC), which measures oxygen consumption ($\dot{V}O_2$) and thus subsequently allows the calculation of the energy expenditure, yields accurate results and is both relatively easy to handle and acceptable for the person being measured. The IC will be described in more detail below because several parameters derived from the IC are used in exercise physiology to describe the fitness level of an organism.

3.4.1   Indirect Calorimetry

The method of IC is based on the concept that eventually all energy-yielding processes of an organism require oxygen [33]. Therefore measuring the amount of oxygen consumed allows determination of the amount of energy used. From a technical point, at least the minute ventilation (V_E), and the gas fractions of oxygen (F_EO_2) and carbon dioxide (F_ECO_2) have to be measured. In most cases also the heart rate, blood pressure, perceived exertion by the subject (e.g., Borg Scale [34]) and blood lactate levels as well as an electrocardiogram (ECG) are usually obtained—these concepts were combined first in a mobile device in the 1940s by German scientists Koffranyi and Michaelis [35].

Combustion of nutrients (carbohydrates, fats, and proteins) yields an average energy amount of 20 kJ/l oxygen (the caloric equivalent). Consequently, when measuring 1 l consumed oxygen during IC, an energy release of 20 kJ has taken place for that period of time.

More detailed scrutiny reveals that the amount of energy released per liter oxygen depends on the type of nutrient combusted. Carbohydrates release approximately 21 kJ/l oxygen whereas fats release approximately 19 kJ/l oxygen. Because less fat is required to combust with 1 l oxygen (0.5 g fat versus 1.23 g carbohydrates per liter oxygen), the caloric value of fats is much higher (approximately 39 kJ/g fat) than the caloric value of carbohydrates (approximately 17 kJ/g carbohydrate). In order to differentiate between combusted carbohydrates and combusted fats, the Respiratory Quotient (RQ, also synonymously Respiratory Exchange Ratio, RER) can be measured and analyzed [36]. The RQ is the amount of carbon dioxide output divided by the amount of oxygen consumption. The following chemical reactions will demonstrate the use of the RQ:

$$\text{Carbohydrates (glucose)} : C_6H_{12}O_6 + 6O_2 \rightarrow 6CO_2 + 6H_2O \quad RQ = 6/6 = 1$$

$$\text{Fats (palmitic acid)} : C_{16}H_{32}O_2 + 23O_2 \rightarrow 16CO_2 + 16H_2O \quad RQ = 16/23 \approx 0.7$$

It is obvious that for carbohydrates the amount of carbon dioxide produced is equal to the amount of oxygen consumed, so the RQ = 1. For fats, a relatively greater amount of oxygen is required compared to the amount of carbon dioxide produced, so the RQ ≈ 0.7. Thus, measuring the RQ enables the user to determine if carbohydrates or fats (or a mixture of both) are combusted in the metabolism, which eventually makes it possible to determine the exact amount of energy released per liter oxygen. It was Nathan Zuntz, the German pioneer in nutrition, exercise, and high altitude physiology, who first realized this relationship at the end of the nineteenth century [37,38]. Proteins take a special position with regard to IC. In order to determine the RQ of a person, it is beneficial to calculate with just two unknown variables (the RQ of carbohydrates and the RQ of fats) so that the contribution of these nutrients to the overall RQ can be determined. It is therefore useful to ignore the energy metabolism of proteins. In addition, the organism is conserving proteins, because many structural elements can be made out of proteins; only about 10-15% are being used for energy metabolism [39]. Also, the RQ of proteins of 0.8 lies between the 1.0 of carbohydrates and 0.7 of fats, so that their overall influence to the RQ can be assumed to be relatively small. The physiological caloric value of proteins is approximately 17 kJ/g, which is somewhat lower than the physical value of approximately 23 kJ/g, because not all chemical energy can be used in metabolism; some chemical energy is lost by excreting urea. This leads to the information in Table 3.2, which illustrates several mixtures of carbohydrates and fats and the resulting RQ and energy output per liter oxygen [37].

If, for example, one reads an RQ of 0.88, then, while neglecting the amount of energy derived from proteins for ease of calculation, the assumption can be

TABLE 3.2 Ratio of metabolized carbohydrates and fats depending on the respiratory quotient (RQ)

Amount Carb (%)	Amount Fat (%)	RQ (CO_2/O_2)	kJ/lO_2	g Carb/l O_2	g Fat/l O_2
100	0	1.00	21.11	1.23	0.00
80	20	0.94	20.78	0.96	0.11
60	40	0.88	20.44	0.70	0.21
40	60	0.82	20.15	0.45	0.31
20	80	0.76	19.86	0.21	0.41
0	100	0.70	19.56	0.00	0.50

The table shows the ratio of carbohydrates and fats, the energy yield and the masses of carbohydrates and fats that burn with one liter oxygen, depending on the respiratory quotient (RQ) [37].

made that 60% carbohydrates and 40% fats have contributed to the total energy released while measuring that RQ. Because the amount of energy released per liter oxygen fluctuates around 20 kJ, this value has also been labeled the Caloric Equivalent of Oxygen. If, for technical or other reasons, the measurement of carbon dioxide output is not possible and therefore the RQ cannot be determined, then this average value of 20 kJ/l oxygen should be used in calculation of energy expenditure.

3.4.2 Metabolic Equivalent and Oxygen Consumption

The Metabolic Equivalent (MET) is the amount of oxygen required during rest calculated per mass of 1 kg body weight and time [33]. An average of 3.5 ml O_2/kg/min has been determined as the oxygen consumption at the BMR; however, age, gender, and illness can have considerable effects on this value [40,41]. Because this value is given per kilogram body weight, it is also called the Relative Oxygen Consumption (relative $\dot{V}O_2$). Multiplication with the bodyweight yields the Absolute Oxygen Consumption (absolute $\dot{V}O_2$). For example, a person weighing 80 kg would have an absolute $\dot{V}O_2$ of 280 ml/min at rest. However, the relative value is of greater interest, because it easily allows comparison of the values of different persons with different body weights by eliminating the influence of the aforementioned body weight. The MET has been introduced to allow for quantification of different amounts of oxygen consumption and also to reduce the necessity of calculating with the somewhat cumbersome value of 3.5. For example, a 10-fold increase of oxygen consumption (and thus energy expenditure) can either be expressed as 35 ml O_2/kg/min or simply as 10 MET and so on.

3.4.3   Energy Efficiency

Knowledge of one's oxygen consumption during rest or during exercise allows for the calculation of the amount of energy per time (which in essence is power) that is being released by metabolism. For example, a person consuming 3 l oxygen per minute would have an average energy expenditure of 60 kJ/min (3 l×20 kJ/l). Considering W = J/s, 60 kJ/min = 1000 W (=J/s) of released chemical energy. If it is known how much mechanical power is being generated (e.g., by calculating this value or by simply reading the value given by the ergometer display), one can calculate the mechanical efficiency of a person.

Efficiency (Greek letter η read "eta", no unit) is calculated by the released mechanical power divided by the released chemical power measured through IC as mentioned above. If, for example, an athlete consumes 3 l oxygen, yielding a released power of 1000 W while cycling on an ergometer, yielding a mechanical power of 200 W, then the efficiency would be 200 W/1000 W = 0.2 or 20%. The efficiency of humans' physical work rarely increases to values over 25-30%—the rest, about 75-70%, is lost as heat and needs to be released to the environment in order to keep the body core temperature within physiological range. Biomechanical factors and muscle fiber distribution, as well as movement economy, influence mechanical efficiency [42–46]. It should be noted that this determination of efficiency does not include energy generated by anaerobic processes. Thus it should be obvious that efficiency increases as an athlete performs under increasing loads (e.g., a graded exercise test) in which the number of anaerobic processes increases with increasing loads in order to fuel the energy demands of high load strain (see also later sections). Other approaches for the calculation of the efficiency exist, for example, by subtracting the amount of energy of the BMR from the amount measured first through IC [47]. The latter approach yields the energy expenditure and efficiency of only the TEA, while the former also includes the BMR and thus yields the total energy expenditure and efficiency while exercising; one should make sure to note the method used and apply only one approach when conducting several measurements at different points of time.

3.5 $\dot{V}O_2$ MAX

The previous sections have explained that energy expenditure and oxygen consumption are closely tied, to a point that allows the calculation from one to the other. This process reveals that the capability to consume and use more oxygen in metabolism also yields more energy, which can be used to fuel processes to maintain physical exertion. The higher the amount of oxygen an athlete can consume, the higher the amount of energy released will be. Therefore, the Maximum Oxygen Consumption (or $\dot{V}O_2$max) is determined as a predictor for the maximum aerobic capacity [48,49]. It is obtained usually by performing an all-out exercise test with successively increasing grades and by measuring the $\dot{V}O_2$max directly. Several other approaches exist to approximate the $\dot{V}O_2$max if a maximum test cannot be performed (e.g., if contraindications such as a heart condition of the test subject exist) or if the test equipment does not allow for direct $\dot{V}O_2max$ measurement. Accordingly, the $\dot{V}O_2max$ can be derived from the heart rate of a subject or from the amount of mechanical power released using preprepared reference charts. $\dot{V}O_2max$ is greatly influenced by the cardiovascular and respiratory capacity as well as factors on the local tissue level such as capillarization, aerobic enzymes, muscle fiber type distribution, and mitochondrial size and number [50–54]. The $\dot{V}O_2max$ of different athletes differs greatly depending on the type of sport the athlete engages in as well as the amount of training that has been applied and individual factors, gender and age. Figure 3.2 shows values of $\dot{V}O_2max$ for different male athletes as well as for a male sedentary person, values for women are about 15-20% lower [55].

It becomes obvious that endurance sports that activate large masses of musculature yield the highest values of $\dot{V}O_2max$. If expressed in MET, the highest values that can be achieved are a little above 20 MET for well-trained endurance athletes. Because the $\dot{V}O_2max$ is the maximum oxygen consumption—or peak oxygen consumption [56]—and because it is attained only for a short period of time, the load achieved during its measurement is not the one of a steady state exercise. As a practical figure, the running speed at the $\dot{V}O_2max$ can be

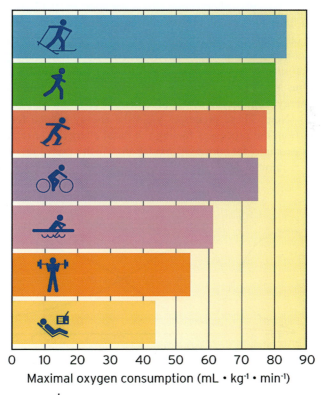

FIGURE 3.2 Values of $\dot{V}O_2max$ in male athletes and in male sedentary persons, *adapted from [55].*

compared to the one maintained during an all-out 3000 m run [57]; at longer distances the athlete will have to run somewhat slower than at the $\dot{V}O_2max$ to complete the run; at shorter distances the athlete will instead not exceed his or her oxygen consumption but will only increase the energy released through anaerobic processes to fuel the energy requirements resulting from the higher running speed.

3.6   ENERGY SOURCES AND STORAGE

Eventually all energy necessary to fuel the various processes in an organism comes from adenosine triphosphate (ATP), which has been called the universal energy currency in the body, necessary to fuel the various energy requirements such as movement, chemical activation in metabolism, and ion transport work work [58,59]. ATP is usually formed through aerobic and anaerobic glycolysis, β-oxydation of fats, and introduction of desaminated amino acids into the citric acid cycle [60]. A small amount of ATP is available within the muscle cells at the start of an exercise to fuel the energy requirements for the first few seconds. Phosphocreatine (PCr) is a compound that can regenerate ATP from previously formed ADP; this allows for ATP-formation without the need for ATP de-novo synthesis [61]. Thus, phosphocreatine allows for another 10-15 s fueling energy requirements. Further exertion and energy requirement results in the formation of ADP and other compounds such as AMP, adenosine, which will eventually increase the conversion rate —through allosteric stimulation—of glycolytic enzymes and subsequently the enzymes and transport mechanisms of the citric acid cycle, thus allowing for de-novo ATP synthesis [62]. Figure 3.3 illustrates the use of energy from these various processes at different times of a physical exertion [63] (Note: These processes yield the energy needed for fueling the physical exertion; the processes fueling the basal energy requirements are not included, for example, fat burning/β-oxidation, which ensures up to 30% of the basal energy requirements.)

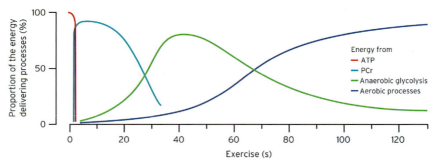

FIGURE 3.3   Energy contribution from various processes depending on the time during the exertion, *adapted from [63].*

Aside from these short-term energy stores, energy can also be stored for longer periods of time, especially when food is not readily available. Carbohydrates are stored as glycogen in the musculature and the liver [64]. While glucose from glycogen from the liver can be released to be utilized outside the liver (e.g., by the musculature or the brain), glycogen in the musculature is available as an energy source only in these respective musculature cells due to the absence of the enzyme glucose-6-phosphatase in musculature cells, which is necessary to export glucose outside the cell. Approximately 400 g glycogen can be stored on the musculature, while about 100 g can be stored in the liver for a total sum of approximately 500 g. However, these values can differ greatly among different individuals and also depend on training status and nutritional condition. (Note: Only a very limited amount of carbohydrates is found as plasma glucose, about 3 g in total, reflecting an energy of not more than about 12 kcal or 50 kJ. Plasma glucose, however, is of crucial importance because it distributes the glucose to all other cells and therefore has to be continuously refilled to keep the glucose concentrations in the plasma constant.) Applying the relationships described in the previous sections, 500 g glycogen yields an energy of 8500 kJ. Exercising with a continuous oxygen consumption of 3 l/min would require an average of 60 kJ/min and 3600 kJ/h. Thus 500 g glycogen could fuel such an exercise for a little over 2 h and 20 min [65,66]. The importance of carbohydrates for exercise capabilities and the limited glycogen stores in the muscle and liver will be discussed in later sections.

Fat can be stored in adipose tissue in large amounts. Gender, age, as well as nutritional status and the amount of physical exertion, determine the amount of adipose tissue. Young male adults have about 10-20% body fat, while young adult females have about 20-30% body fat, which sums to a total of about 10-20 kg FM. Again, applying the relationships described in the previous sections, this would yield an energy amount of approximately 400,000-800,000 kJ, which in theory is enough to supply the basal metabolic energy requirements of 50-100 days. A small part of triacylglycerol can be found in the musculature (about 300 g) as well as in the plasma (plasma triacylglycerols and free fatty acids). Some of the adipose tissue has a structural purpose and is therefore also called "essential fat" (males: approximately 3%, females approximately 12%). It can be found in the nervous and the skeletal system as well as around internal organs such as the kidneys and the heart [67]. In addition, although large amounts of energy can be released from fats, this metabolic pathway is relatively slow compared to the one of carbohydrates [68] and therefore has its limitations with regard to fueling energy requirements during high-intensity exercise as shall be described in later sections.

Proteins are of high importance for an organism due to their versatile use as structural building blocks for all kinds of cells and tissues. A high number of proteins amass in the musculature. Because this tissue is well hydrated (approximately 73% of which is water), a young male adult carries only about 5-10 kg dry muscle (protein). In theory, this would yield an energy amount of approximately 85,000-170,000 kJ. However, the musculature would cannibalize itself if the muscle protein were used as main energy source in the metabolism, so these values are somehow of theoretical nature. Nevertheless, during extreme metabolic demands (starvation, malnutrition), these resources will be also metabolized step by step until the individual dies.

3.7 MUSCULATURE

As mentioned before, the skeletal musculature can be construed as the largest organ of the human body. It comprises about 40% of the body mass of a young male adult. Females tend to have somewhat less musculature (about 30%). Age as well as nutritional status and exercise regimen greatly influence this number. The purposes of skeletal musculature range from the obvious tasks of generating movement to maintaining posture as well as generating heat if the body core temperature falls (known as "shivering"). Skeletal musculature also plays a certain role, as was more recently found, in the immune system [69–71].

Anatomically, the smallest functional unit is a motor unit comprised of a single α-motoneuron and all muscle fibers connected with this motoneuron. All muscle fibers will contract once an appropriate number of action potentials is commenced by the motoneuron. Graduation of the force generated by a muscle is achieved by gradually increasing the number of recruited numbers of motor units as well as by increasing the frequency by which the action potentials are commenced (also called "summation").

3.7.1 Fiber Types

Several different fiber types have been identified recently applying various biochemical and immunohistochemical identification methods. In a very simple general approach, two fiber types can be identified: the slow contracting but enduring type I fibers and the fast contracting but fast fatiguable type II fibers. Here the motor neuron and the myoneural junction govern the trophic and twitch characteristics of the connected fibers [72,73].

The characteristics of type I and II fibers (and their subtypes) lie within their genetic and structural properties. Type I fibers possess a plethora of mitochondria, oxydative enzymes and myoglobine, and are well vasculated; in other words, they can make very good use of oxygen through aerobic processes generating ATP. However, they have slow ATP-ase activity (the enzyme at the myosin head converting chemical energy of ATP into mechanical work), low PCr stores, and can therefore generate only a limited amount of force [74–76].

Type II fibers instead possess relatively few mitochondria, fewer oxydative enzymes and myoglobine, and are not as well vascularized, so they cannot make such good use of oxygen (regardless of whether oxygen is provided through the blood stream or not). However, they have a fast-acting ATP-ase activity and high PCr stores that quickly recycle ATP, enabling them to generate a high amount of force [76–78]. Figure 3.4 illustrates the two fiber types in terms of force generated and rate of fatigue [76].

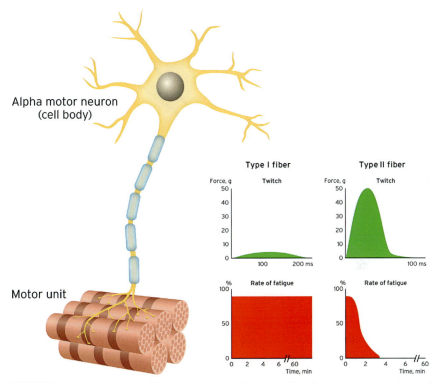

FIGURE 3.4 A motor unit and fiber type characteristics regarding force generation and rate of fatigue, *adapted from [76]*.

The distribution of fibers differs between individuals (which could be attributed to "genetic talent" depending on which fiber type dominates, ranging from 70:30 to 30:70 distributions) as well as within an individual where more type I fibers are found in postural muscle of the trunk and in deeper layers of the extremity musculature. Type II fibers are more abundant in superficial layers of the extremity musculature [79–80]. Training appropriate in length, frequency, specificity, and intensity can lead to fiber hypertrophy [81,82], an increase in functional fiber mass and diameter, and to fiber transformation [83–85], where fibers can switch their characteristics from one type to the other. Studies suggest that endurance training leading to low (around 10 Hz) but lengthy stimulation frequencies within the motor unit will switch a type II to a type I fiber. Cessation of these stimuli will transform the fiber back to type II [86–88]. However, type II fibers will require high-frequency (around 40 Hz) stimulation through appropriate resistive exercise in order to become effective strength fibers.

According to these characteristics, the motor units comprising different fiber compositions are recruited depending on the particular intensity of effort as Figure 3.5 illustrates [76].

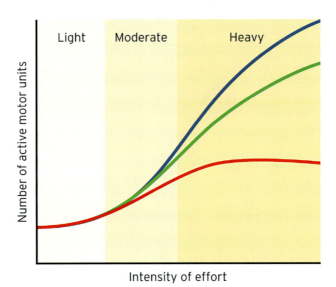

FIGURE 3.5   Fiber type recruitment depending on the intensity of effort (red, Type I; green, Type IIa; blue, Type IIx), *adapted from [76].*

Corresponding to the so-called "Size Principle", small motor units comprised of mostly type I fibers are recruited first if the intensity of effort is low; increasing intensity of effort will necessitate that additional, larger motor units are recruited, which are comprised of more type II fibers; finally, the highest intensities are necessary to recruit the largest motor units comprised of almost solely type II fibers [88,90]. An explanation is that small motor units with smaller motoneurons are excited easier than the larger ones. A teleological approach makes this relationship apparent: At low intensities, the type I fibers are able to handle the relatively low strain while generating the necessary energy requirements very efficiently through aerobic pathways. Once the intensity rises, the organism is constrained to recruit type II motor units, because only these are able to generate enough force to overcome the higher intensity. Even though the metabolism of type II fibers operates anaerobically (regardless of whether oxygen is present, due to their structural properties) and thus less efficiently generates less energy in form of ATP per glucose molecule, the next section should make it obvious why the switch from type I to type II fibers takes place.

3.8   ENERGY AND EXERCISE INTENSITY

The rate—not the amount but the speed—of ATP generation by the various metabolic processes varies considerably. As Table 3.3 shows, the anaerobic processes are up to 10 times faster in making ATP available for muscle contraction than aerobic processes [91–93].

TABLE 3.3 Velocity of different ATP-yielding processes

Process	Rate of ATP Production (mM/kg dry muscle/s) (Approximated Values)
ATP and PCr system	10
Anaerobic glycolysis	5
Aerobic glycolysis	2
β-Oxydation	1

The various ATP-yielding processes in a muscle cell exhibit different velocities. Thus aerobic processes are too slow to adequately deliver ATP for fast contracting type II fibers [91–93].

Aerobic processes are too slow in making ATP available for the fast-converting ATP-ase of the type II fibers to generate force because of the higher amount of involved enzymes and compartments, such as the mitochondria, as well as the overall greater distances within the cell that the substances have to travel. Only anaerobic processes (the already available ATP, the ATP recycling PCr and to a lesser degree the anaerobic glycolysis) are fast enough to supply the energy demands of the type II fibers' fast-converting ATP-ase necessary for generating large amounts of force due to their close proximity to the contractile proteins.

Aerobic processes, however, are not as fast as the anaerobic ones, but they are efficient at producing great amounts of ATP from aerobic glycolysis and β-oxydation, which are needed to fuel lengthy energy demands of everyday movements, maintaining posture and low intensity endurance movements generated by the type I fibers. Figure 3.6 illustrates the switch from the use of fats to carbohydrates with increasing intensity of effort in accordance with the need for fast availability of ATP for type II fibers. Note that near 100% of $\dot{V}O_2max$, almost all energy generated, arises from the use of carbohydrates; the RQ or RER changes accordingly [63].

It should become clear why tasks of great intensity such as weightlifting, shotput, or running at sprinting speed cannot be maintained for more than a few seconds, because the high speed by which the ATP-ase of type II fibers convert the chemical energy of ATP into mechanical work quickly consumes the small ATP and PCr stores [94,95]. For submaximal efforts (e.g., a 400 m run) the anaerobic glycolysis is fast enough to resupply the energy demands with ATP de-novo synthesis. These facts explain why carbohydrates are the important nutrient for fuelling exertion of higher intensity and why, therefore, an endurance athlete must also have the glycogen stores filled up prior to competition. Gradually, lower intensities require less quick regeneration of ATP and thus allow the efficient aerobic de-novo synthesis of ATP in the type I fibers. Given that the adaptation of the metabolism within the muscle cells takes some time,

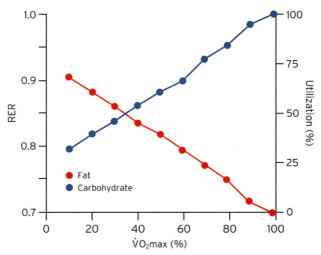

FIGURE 3.6   Change in RER and carbohydrate utilization (blue) and fat utilization (red), depending on the intensity of effort measured in % $\dot{V}O_2$max. As shown in table 3.3 the aerobic processes are slow with regard to their capability to quickly provide ATP for muscle contraction. Therefore carbohydrate utilization - and thereby RER - increase with increasing exercise load, *adapted from [63]*. The respiratory exchange ratio RER (or respiratory quotient RQ) approaches "1" once carbohydrates are used as a sole source for ATP.

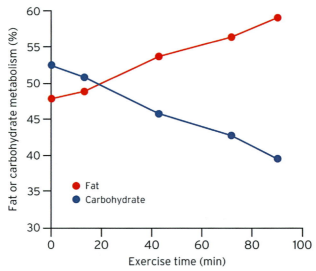

FIGURE 3.7   Change of utilization of fat and carbohydrate metabolism depending on exercise time, while the intensity of effort remains unchanged at 60% $\dot{V}O_2$max, *adapted from [63]*.

the amount of metabolism of carbohydrates is also greater for a given intensity at the beginning of the exercise than some time later (provided that the intensity remains the same throughout the exercise time), as Figure 3.7 illustrates [63].

In other words, during an exercise, the metabolism can become more efficient by gradually increasing the proportion of energy released from aerobic processes such as the β-oxydation of fats [96,97]; however, this is limited to low-to-moderate intensities [98]. At higher intensities the energy release greatly depends on carbohydrates [99] as mentioned before. It can therefore be stated that carbohydrates present an important energy source [52,64,100] both at the beginning of an exercise—when the aerobic metabolic processes have not yet reached their full functional capabilities (see Figure 3.3)—and when trying to maintain higher intensities (throughout the race and especially during the last few kilometers of a long distance run when the athletes increase their speed; see Figure 3.5)—because of the high rate of ATP production only attained by anaerobic processes in order to fulfill the higher energy requirements of type II fibers to maintain high intensity efforts. Therefore, it makes sense to ingest carbohydrates during an endurance event in order to fuel the exercise intensity of the race [101,102].

3.9   LACTATE THRESHOLD

As previously discussed, as the intensity of effort increases, the proportion of energy generated by anaerobic processes also increases because type II fibers rely on these pathways for the high rate of ATP production that can be achieved. Anaerobic de-novo synthesis of ATP involves the anaerobic glycolysis in which lactic acid is formed. The H^+-ions can impede muscle contraction by lowering the functionality of enzymes involved in metabolism [103–105] and by competing with the trigger-ion Ca^{2+} with its binding site at the troponin complex, thus hampering the cross bridge cycle of actin and myosin [63,106]. Furthermore, acidosis can impair central nervous system (CNS) capability to drive the muscle [107]. Depending on training status [108], a further increase of intensity of effort will therefore lead to a lactate accumulation, which can hamper muscle contraction and thus limit exercise capabilities. This intensity of effort is called the "Lactate Threshold" [109–111]. Endurance exercises with intensities above this threshold will become increasingly difficult to maintain. Reasons for the lactate threshold are:

- Increased recruitment of type II fibers that inherently rely, for the most part, on anaerobic metabolism for ATP generation [72,88,112].
- Decreased lactate clearance due to decreased blood flow through lactate-clearing organs such as the liver and kidneys [113–115].
- Hormonal stimulation of glycolysis by catecholamines as part of the overall activation of the organism during exercise [116–118].
- Exceeding the rate of the citric acid cycle by the rate of glycolysis [119–121].
- Decreased oxygen supply of the musculature, especially during static exercise [122,123].

The concept of limited oxygen supply to the musculature as a sole reason for the occurrence of the lactate threshold is obsolete. As previously described, type II muscle fibers rely on anaerobic metabolism due to their structural buildup, which is highly adjusted to their energy needs because aerobic processes are simply not fast enough to generate the amount of ATP needed by type II fibers to maintain high-intensity efforts. It is therefore the increase of intensity of efforts that leads to an increased recruitment of type II fibers, which will operate anaerobically—regardless of whether oxygen is present. As shall be described in later sections, the blood flow during exercise is shifted toward the working musculature and to some extent away from organs such as the liver and kidney, which are responsible for lactate clearance, thus leading to lactate accumulation at certain levels of intensity [124]. Therefore, the term *aerobic-anaerobic threshold* should be avoided as well, because only the blood level of lactate is evaluated, which is dependent on the rate of lactate buildup and clearance. Consider the following example: Two endurance athletes are identical except for their lactate clearance capabilities. The one with better clearance capabilities will have a lower blood lactate level and would be labeled "below an aerobic-anaerobic threshold" according to the old concept, while the other—whose musculature operates at the same aerobic and anaerobic level—would be labeled "above an aerobic-anaerobic threshold" simply due to a lower rate of lactate clearance. It is for this reason that the sole interpretation of blood lactate levels does not permit an evaluation regarding the amount or ratio of aerobic and anaerobic energy metabolism. While the athlete with better clearance abilities does have consequently higher exercise capabilities (e.g., running at a higher speed during an endurance race), the term *aerobic-anaerobic threshold* is a scientific misnomer, even though it is still in widespread use [125]. For a more accurate description, the terms *lactate threshold* (instead of aerobic threshold), occurring at 2 mM blood lactate level, and *Onset of Blood Lactate Accumulation (OBLA)* (instead of anaerobic threshold) occurring at 4 mM blood lactate level [126,127] provide better descriptions. Because these blood lactate levels are dependent on the training status of an athlete, they occur at different levels of intensity when comparing different athletes and can be expressed as running velocity, mechanical power (W), or in percent of $\dot{V}O_2max$. While the $\dot{V}O_2max$ is the maximum capacity of aerobic energy metabolism, expressed in the amount of oxygen that can be consumed, the lactate threshold is a more practical figure for an athlete, because intensities (or running speeds) above it will lead to lactate accumulation and eventually limit the athletes' endurance capabilities [128,129]. Average values of the lactate threshold are at 50-60% of $\dot{V}O_2max$ for untrained and 80-90% of $\dot{V}O_2max$ for endurance-trained persons [127], mostly due to a higher lactate clearance [114,130,131], which means they can perform in a steady state at higher intensities of effort than the untrained ones.

As mentioned earlier, blood lactate levels are obtained during a graded exercise test. If these levels are measured at several points during the exercise test (e.g., usually sometime after each stage of intensity increase), then a curve can be plotted showing the change of blood lactate level at the given intensity of effort (e.g., W, km/h), as Figure 3.8 illustrates [132].

The already mentioned blood lactate levels of 2 and 4 mM represent limiting points at which the lactate curve changes its kinetics and by which three zones can be isolated [133,134]. In the "green zone"—up to the lactate threshold (formerly known as aerobic threshold) of 2 mM—the lactate clearance is greater than the amount of lactate formation. Within this zone, the athlete can accelerate to

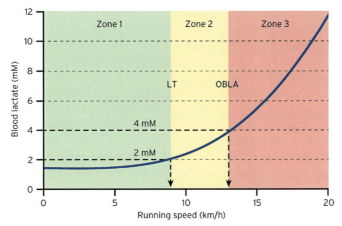

FIGURE 3.8 Change in blood lactate level for a given athlete with increase of blood lactate depending on intensity of effort (measured in running speed km/h). LT=Lactate Threshold, OBLA=Onset of Blood Lactate Accumulation, *adapted from [132].*

a higher speed or intensity and the lactate curve will remain almost the same. Between the lactate threshold and the point of OBLA (formerly known as anaerobic threshold) of 4 mM is the "yellow zone"—here the lactate clearance is adapted to the amount of formation. An increase of blood lactate level takes place; however, as long as the athlete does not increase the intensity of effort, the blood lactate level will remain stable at its new level. Above the point of OBLA, a "red zone" appears, representing blood lactate accumulation, even if the athlete does not increase the intensity of effort. It is this red zone that an endurance athlete wishes to avoid during endurance events such as a marathon or triathlon, so the athlete has to make sure to remain under a certain intensity, that is, running speed.

The lactate forming during anaerobic glycolysis is not a waste product. As was shown, several other compounds can be synthesized from lactate within the body. Also, highly oxidative tissues such as type I fiber musculature as well as heart muscle cells are able to use lactate as an energy source [135–137]. In addition, the lactate buildup during anaerobic glycolysis is necessary to allow for further glycolysis reaction by pyruvate accepting hydrogen from NADH (nicotinamide adenine dinucleotide and hydrogen) and thus forming lactate and NAD, which in turn is needed to allow anaerobic glycolysis to continue and thus further ATP generation.

During a graded exercise test, the RQ is measured as previously described. By definition, the RQ ranges between 0.7 (metabolism of fats) and 1.0 (metabolism of carbohydrates). At some point during an exercise test, if the intensity reaches a certain amount of strain, the RQ begins to rise above 1.0. This occurs because the hydrogen component of lactic acid is being buffered by the bicarbonate buffering system, leading to the formation of carbon dioxide, which is being eliminated by respiration:

$$H^+ + HCO_3^- \rightarrow H_2CO_3 \rightarrow H_2O + CO_2\uparrow$$

This carbon dioxide is being released by respiration in addition to the carbon dioxide forming because of energy metabolism, which is caused by the high intensity of effort now mostly involving carbohydrates. Therefore, the RQ rises above 1.0 and can attain values of 1.2, 1.3 or more, depending on how much strain the athlete is capable (and willing) to handle. Thus, a RQ which is clearly above 1 is an objective indicator for an all-out exercise [15]. In addition, these relationships also explain why exercise activities for health and recreational purposes should avoid the necessity of heavy breathing, thus making sure that no lactate buildup and subsequent additional carbon dioxide elimination leading to an RQ above 1.0 takes place. This has led to the term *conversational exercise*, where participants choose an intensity of effort (e.g., jogging speed) where they are still able to talk while exercising.

3.10   OXYGEN SUPPLY

Oxygen is necessary for survival: Without oxygen the processes of a living organism cease to exist within a matter of minutes. The neurons of the gray matter of the brain are especially sensitive to a decrease in oxygen supply. With regard to generating mechanical movement, only short-term exercises involving the highest intensities, such as weightlifting, rely solely on anaerobic energy metabolism. An adequate oxygen supply is therefore mandatory for an organism to maintain physical exertion. Because of this, the $\dot{V}O_2max$ is an important parameter to quantify the maximum aerobic capacities of an athlete, as previously described [48].

3.10.1   Respiration

A young male adult of average constitution has a minute ventilation of about 6 l/min, calculated by multiplying the breathing frequency (approximately 12 min^{-1}) by the tidal volume (approximately 0.5 l air). During intense exercise the value of minute ventilation can increase theoretically to 120 l or more (breathing frequency 30 min^{-1} and 4 l tidal volume). The respiratory center of the medulla oblongata integrates sensory information such as chemical state of the blood, core temperature, and input from proprioreceptors in joints and muscles to generate an increase in both breathing frequency and depth [138–140].

3.10.2   Blood Transport

The alveolar partial pressure of oxygen reaches approximately 100 mmHg both during rest and exercise—provided that no respiratory conditions such as restrictive or obstructive disorders are present—so that under most conditions the respiration poses no restriction for the oxygen supply and $\dot{V}O_2max$ [141]. However, in elite endurance athletes, the pulmonary system lags behind their highly trained cardiovascular and muscular capabilities [142], leading to exercise-induced arterial hypoxemia (EIH) [143,144]. After diffusion, oxygen is dissolved in the blood plasma and taken up to be carried by the red blood cells bound to the carrier molecule haemoglobin. Four oxygen molecules are transported by one molecule of haemoglobin. According to the binding curve of oxygen to haemoglobin, an arterial partial pressure of 100 mmHg leads to a saturation of 97-99% of the haemoglobin carriers. A young healthy male adult has about 150 g haemoglobin per liter blood. In order to calculate the oxygen content, the so called "Hüfner number" needs to be applied stating that 1.34 ml oxygen are carried by 1 g haemoglobin, full saturation provided. Calculating $150\,g\,Hb/l\,Blood \times 1.34\,ml\,O_2/g\,Hb = 200\,ml$ $O_2/l\,Blood$, so 200 ml oxygen are carried in 1 l arterial blood. This resembles a 70-fold increase compared to oxygen just being physically dissolved in the plasma. Assuming a total blood volume of 5 l in an adult human, this means that approximately 1 l O_2 is stored in the body, which theoretically—if oxygen supply to the body is completely interrupted— could last for 4 min, assuming an oxygen consumption of an adult (see above) under resting conditions of 250 ml/min. The real amount is lower however, since about 85% of the blood is of the venous type which is only 75% saturated with oxygen.

3.10.3   Cardiac Output and Blood Distribution

Cardiac output under resting conditions is approximately 5-6 l/min and can be increased to values of 25 to almost 40 l/min, depending on constitution and training status [48–51,145]. Figure 3.9 illustrates the distribution of these volumes to the various organs and tissues [146].

It is interesting to note that while receiving only 20% from 5 l/min (=1 l/min) of blood during rest, the musculature can increase its blood supply to 20 l/min or more by increasing its proportion to over 80% from the now increased cardiac output during intense endurance exercise [147]. As mentioned before, the blood supply to the lactate-clearing organs such as liver and kidneys reduces not only in proportion but also in the absolute values, thus explaining the decrease in blood lactate clearance during intense exercise [60,114,148].

The redistribution of blood toward the working musculature is achieved by autoregulation—the increase of blood flow caused by vasodilation results from

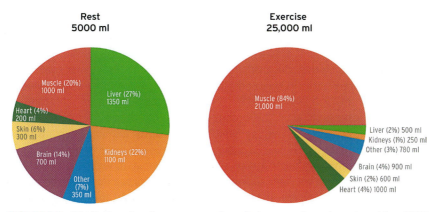

FIGURE 3.9   Distribution of cardiac output per minute during rest and exercise, *adapted from [146].*

increased amounts of vasoactive compounds such as adenosine, ADP, lactate, NO, and K^+—leading to a drop in systemic blood flow, which is compensated by an increased release of catecholamines and the activation of the renin-angiotensin system, leading to general vasoconstriction. The vasodilation caused by autoregulation predominates in the working musculature. The release of catecholamines causes positive chronotropy and inotropy of the heart (a higher heart rate and increased stroke volume) leading to an increased cardiac output, which is—due to the mechanisms described earlier—predominantly distributed toward the working musculature [149–151].

The oxygen provided via the blood stream needs to be taken up. The *arteriovenous oxygen difference* (*avO_2Diff*) resembles the oxygen takeup by the tissues [145]. During rest, only 25% of the oxygen is taken up (25% of 200 ml O_2/l Blood = 50 ml O_2/l Blood), thus leading to a venous saturation of still 75%. Light exercise causes about 50% to be taken up, while high intensities around the $\dot{V}O_2$max cause approximately 75% of the provided oxygen to be taken up. In highly trained endurance athletes, up to 85% can be taken up (85% of 200 ml O_2/l Blood = 170 ml O_2/l Blood).

3.10.4 Fick's Principle

Fick's Principle, named after German physiologist Adolf Fick (1829-1901), is a mathematical relationship intended to determine the cardiac output ($\dot{Q}$) using the values of oxygen consumption $\dot{V}O_2$ and the arterio-venous oxygen difference (avO_2Diff): $\dot{Q} = \dot{V}O_2/avO_2$Diff [49]. As an example, at rest 250 ml oxygen are consumed per minute while 50 ml oxygen are taken up per liter arterial blood; in turn, 5 liters of blood have to flow per minute in order to fulfill the equation. If the equation is transposed, it can be used to explain the relationships that limit maximum oxygen uptake: $\dot{V}O_2$max = $\dot{Q}$max × avO_2Diffmax. For an untrained person, the following average values would apply: 3000 ml = 20 l/min × 150 ml/l. For a highly trained endurance athlete, the values would be: 6000 ml = 35 l/min × 170 ml/l (rounded values), leading to a twofold increase of aerobic capacity, thus, together with a higher lactate clearance, allowing for greater speeds for a longer time as can be seen in world-class marathon runners completing the race at an average pace of 20 km/h. It becomes obvious that not the maximum oxygen uptake by the musculature (expressed as avO_2Diffmax) but the maximum cardiac output ($\dot{Q}$max) is the value that affects and thus most limits the maximum aerobic capacity expressed as $\dot{V}O_2$max [48]. Because the cardiac output is calculated by heart rate and stroke volume ($\dot{Q}$ = HR × SV with the maximum heart rate relatively close to the general value of "220 minus age (years)"), it becomes clear that the maximum stroke volume—ranging from 100 ml/beat for untrained persons to almost 200 ml/beat for highly trained endurance athletes—is the influential factor for the $\dot{Q}$max and ultimately the $\dot{V}O_2$max. It should be noted, though, that in addition to these relationships, neuronal factors also greatly influence the ability to sustain endurance exercises because they eventually induce central impulses leading to fatigue [49,102,152,153].

3.10.5   Excess Postexercise Oxygen Consumption

As described earlier and illustrated in Figure 3.3, during the first seconds to minutes of an exercise, no additional oxygen-consuming metabolic pathways are being used because the necessary oxidative enzymes in the mitochondria have not yet been activated. Therefore, an oxygen deficit arises during that first period of exercise [154,155]. The greater the intensity of effort, the greater is the arising oxygen deficit. Figure 3.10 illustrates this relationship with the green areas resembling the oxygen deficit during the beginning of an exercise [154–156].

Figure 3.10 also illustrates that during recovery oxygen consumption does not immediately drop to resting levels, but instead takes some time to return to

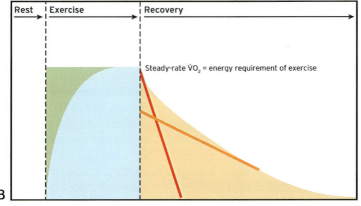

FIGURE 3.10   Oxygen deficit (green area) depending on the intensity (measured in $\dot{V}O_2 = y$-axis) and duration (time $= x$-axis) and the resulting excess postexercise oxygen consumption, EPOC (yellow areas), which resembles the oxygen uptake during recovery above resting levels. Note the fast-component of EPOC (red line) resulting from light exercise and the slow component (orange line) resulting from moderate to intense exercise, *adapted from [156].*

pre-exercise values [156–159]. During light intensity exercise, the area of this Excess Postexercise Oxygen Consumption (EPOC) closely resembles the area of the oxygen deficit, which is also called the "Fast Component" (red line in the figure). Exercises of higher intensities show an additional area that is greater than the original oxygen deficit, called "Slow Component" (orange line in the figure). Reasons for the fast component of EPOC include [156–159]:

- Reloading of the short-term oxygen storages (myoglobin).
- Resynthesis of the short-term energy storages (ATP pool and PCr).

During high-intensity exercises—especially if they are maintained over a longer period of time—the physical strain causes the following additional changes, leading to the additional slow component of EPOC [156,160,161]:

- Membrane leaks from the muscle fibers as result of the physical strain, necessitating Na^+-/K^+-ATPase to operate on a higher rate, thus requiring more energy and oxygen.
- Increased body core temperature causes an increase of the BMR, thus increasing the overall metabolic turnover, requiring more oxygen.
- Increased protein biosynthesis during recovery to repair and adapt the musculature to the previous strain of exercise, thus requiring more energy and oxygen.
- Increased metabolism of compounds such as lactate requiring more oxygen.

This slow component of EPOC can last several minutes to several hours after exercise has ended. This is also an argument for regular exercise, because energy expenditure is not only increased during exercise but also during recovery long after exercise has ended, thus burning more calories at rest than without previous exercise.

3.11   STRENGTH

Strength could be defined as the ability of a neuromuscular system to maintain or overcome an external resistance by exerting force. Furthermore, maximum strength (e.g., in weightlifting), springiness or power (e.g., in javelin or shotput), and finally strength-endurance (e.g., in sports climbing) can be differentiated [162–167]. A special form is plyometrics, which uses the elastic capabilities of the musculoskeletal system (e.g., in high jump or long jump) [168,169].

According to the size principle previously described, type II fibers are necessarily recruited in order to deliver the higher forces to overcome external resistance load.

Strength capabilities can be measured using dynamometers measuring the maximum exertible force—for example, knee extension or handgrip [170]. However, in most cases static isometric strength is measured, which is not necessarily convertible to strength capabilities applied during an actual athletic exercise. More expensive computer-aided tools allow for isokinetic (dynamic) testing [171,172]. Practical approaches involve the one-repetition maximum (1RM), in which an athlete lifts or pushes a maximum weight that can just be handled in one repetition [173,174]. Although of high practical value, if applied without proper instruction and warmup, the 1RM test can cause injuries [175].

Another approach trying to circumvent the danger of injury is the determination of the 10RM in which the weight is determined that can be lifted in 10 repetitions. Maximum strength capabilities are at or close to the weight of the 1RM. The maximum achievable force of a muscle is dependent on the Cross Sectional Area (CSA): The greater the CSA, the more sarcomere elements can exert force, or in other words "pull," side by side at the same time, thus generating more overall force [176,177]. Power strength requires a high Rate of Force Development (RFD) [178,179]. It usually takes 400 ms to achieve the maximum exertible force for a neuromuscular system. Efficient training (with loads around 40-50% of the 1RM) will increase the RFD, thus allowing faster contractions [180].

Strength endurance is necessary for continuous, static (isometric) contractions (e.g., in sports climbing when a crucial difficult point has been reached by the climber having to hold the grip). Here, the intermediate type IIa fiber types that are less fatigable than type IIx are used.

Plyometrics involves the Strength Shortening Cycle. Here 50% or more of the kinetic energy is stored in the elastic elements within and around the muscle as well as in tendons. This is beneficial for athletes in track and field where repetitive movements do not have to be newly generated but are, so to speak, recycled from the energy of the previous movement.

3.12   TRAINING

Training necessitates a challenge to the body's homeostasis which only then activates an adaptational response to the training stimulus. General training principles involve [132,181–183]:

- Individuality, in recognition of interindividual genetic differences regarding muscle fiber type distribution, ability for cellular repair, metabolism, hormonal and nervous systems.
- Specificity, in that adaptation to the training load is specific to the trained sport regarding the recruited muscle groups and fiber types as well as the energy systems used.
- Effective training load, meaning that musculature, cardiovascular system, and lactate clearance increase their capacities only if the training load is in excess of the current level of performance.
- Progressive overload, meaning that once an adaptation to a training load has taken place, these loads have to be increased through variation of frequency, length, and intensity of the training.

Supercompensation is the concept of an increase of performance level in adaptation to a previous training load. Originally supercompensation described the increase in glycogen stores after carbohydrate feeding subsequent to heavy exercise, which led to more rapid energy for the following training or competition [184,185]. However, performance capacity is not solely based on glycogen stores, which take about one to two days following various regimens of all-out exercise and subsequent carbohydrate loading. The buildup of contractile proteins as well as the replacement of damaged cell organelles and components take about three to seven days. Accordingly, there seems to be no single concept of supercompensation if one takes all different adaptational processes into account. Still, supercompensation efficiently describes the processes of adaptation to training loads and the necessity of altering training and recovery. This concept also explains *overtraining*, in which inadequate recovery is achieved when training loads follow each other too high in frequency [186–188]. Here, depletion of glycogen (energy deprivation), protein catabolism (functional deficit of the musculature), and imbalance in amino acids (leading to a deficit in immune function as well as hormone imbalances) can lead to a general decline in performance level and health.

Adequate endurance training will lead to [132,189–194]:

- An increase of $\dot{V}O_2max$ through an increase of $\dot{Q}max$ and to lesser degree of avO_2Diff.
- An increase of the lactate threshold through higher capillary density, higher proportion of type I fibers, an increase in oxidative enzymes, and greater lactate clearance.
- Improved biomechanics (e.g., running or cycling economy) requiring less energy and thus improving movement efficiency.

These in turn lead to an athlete capable of performing faster for a longer time. Training success depends on [132,195–198]:

- Current level of performance (a small amount of training can benefit untrained persons, while highly trained athletes require higher training loads and frequencies).
- Training intensity (measured in Watts or km/h, energy expenditure, in percent of HRmax or $\dot{V}O_2max$, METs, or lactate threshold and OBLA or as a rating of perceived exertion), with effective intensities ranging around 60-90% HRmax, depending on training regimen.
- Training length and frequency, again depending on the level of performance, where even only 5 min/day can have a benefit for the untrained and up to several hours per day for endurance athletes.

Because it is especially the cardiovascular system that is strained during endurance training, seeking qualified medical advice prior to a new endurance training is always recommended.

Adequate strength training will at first lead to an improved neuromuscular interaction with increased recruitment, innervation frequency, and inter- and intramuscular coordination; hypertrophy, the increase of CSA will occur later [199,200]. Here, most pathways about how hypertrophy takes place are not yet well understood. It would appear that signal molecules such as the Insulin-like Growth Factor-1 (IGF-1) and Mechano Growth Factor (MGF) are excreted by the muscle cell and act through an autocrine pathway, resulting in an increased protein biosynthesis of contractile proteins [201,202]. Also, satellite cells fuse with muscle cells—especially after microinjuries of the cell—resulting in additional nuclei again increasing protein biosynthesis [203].

The same concepts of training success apply as well regarding current level of performance, intensity, length, and frequency of training. It is important to note that while moderate strength training is beneficial for performance and health, some exercises can lead to high peaks in blood pressure [204]; in addition, tendons and ligaments adapt more slowly to resistive exercise, so caution and qualified instruction is recommended, especially when beginning strength training.

3.13   SPORT, HEALTH, AND EVOLUTIONARY ASPECTS

For ten thousands of years, mankind has strived to make life safer, less stressful, and more comfortable. Technological approaches allowed us to travel without the need for exertion, and agriculture, domestication and farming have led to an increase in the amount of food (see chapter one). Although a considerable proportion of the world population still lives in poverty and famine [205], the Western civilizations experience the downside of physical inactivity and overeating [206,207]. The prevalence of obesity reaches 50% in some areas of Western industrialized countries [208,209]. Obesity, coupled with physical inactivity leads to changes in metabolism and immunological reactions, which over time result in diabetes, hypertension, arteriosclerosis—which may eventually lead to stroke and myocardial infarction—as Figure 3.11 summarizes [206,210–212].

When evaluating the energy expenditures of our ancestors by means of anthropology and genetics, it becomes obvious that our ancestors had to be highly active throughout the day—hunting and gathering food as well as seeking protection [213,214]. It can be assumed that this was the case until 8000 years ago or less—a very small timescale with regard to evolution. This means that humans today are virtually not different from these ancestors who had to search for water and nutrients on a daily basis and had to cope with famine for most of the last 4 million years of evolution. As Figure 3.12 illustrates [214], today's "extreme environment" of our Western civilization, with no requirements for physical exertion, has led to extremely low values of DEEs (daily energy expenditures) as depicted for the Western office worker. Here also a lower BMR (basal metabolic rate) is detectable due to the lower proportion of musculature (fat-free mass) and the higher (hypometabolic) proportion of fat mass as a result of low exertion and overeating. It also becomes obvious that deliberate activity—as found in recreational athletes and even more so in highly trained elite endurance athletes—leads to energy expenditures in accordance with those of our ancestors.

It can be stated that although physical exertion seems to be no longer necessary for immediate survival, it is necessary to avoid obesity and the subsequent diseases associated with it. Consequently, mortality is increased in populations with low physical exertion and decreased in those with higher amounts of physical exertion. It could be shown that for every increased score of one metabolic equivalent (MET) in exercise capability, the risk of overall mortality

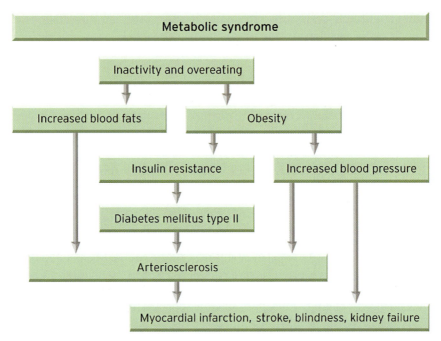

FIGURE 3.11   Pathomechanisms of the metabolic syndrome, *created from [206, 210–212].*

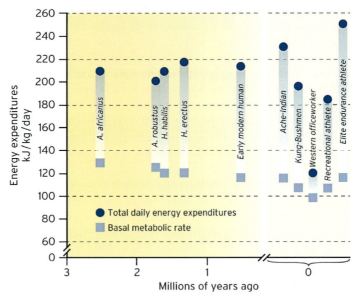

FIGURE 3.12   Energy expenditures of ancestral human populations (left panel) and human populations living at present time (right panel). Note the decreased amounts of basal metabolic rate and total daily energy expenditure of the "Western officeworker", *adapted from [214].*

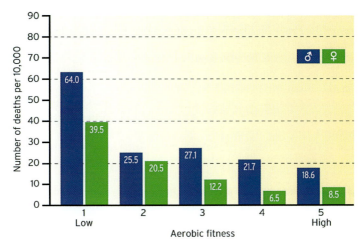

FIGURE 3.13 Mortality among males and females depending on aerobic fitness. Note that already a slight increase in aerobic fitness leads to a considerable decrease in mortality, *adapted from [216]*.

is decreased by 17% [215]. Figure 3.13 illustrates the relationship between mortality and aerobic fitness [216].

To sum up, we are required to respect our genes which are still virtually the same as those of our ancestors. In order to stay healthy and avoid critical illnesses today, we have to consider the circumstances that selected those genes, i.e. for example the necessity in former times for extended daily foraging to find enough food at all [214]. This understanding gradually finds its way into therapy, for example, in the treatment of obesity and diabetes. However, it would be more correct to note that the diseases mentioned above would not arise at all, or at least to a lesser extent, if people would engage in more regular physical exertion and a sound eating pattern [217,218].

REFERENCES

[1] Jones S, Martin RD, Pilbeam DR, editors. The Cambridge encyclopedia of human evolution. Cambridge: Cambridge University Press; 1992.

[2] Hochatchka PW, Gunga HC, Kirsch K. Our ancestral physiological phenotype: an adaptation for hypoxia tolerance and for endurance performance? Proc Natl Acad Sci USA 1998;.

[3] Taylor CR. Exercise and thermoregulation. In: Robertshaw D., editor. Environmental physiology, vol. 7. Baltimore: University Park Press; 1974. p. 163–84.

[4] Sawka MN, Wenger CB, Pandolf KB. Thermoregulatory response to acute exercise-heat stress and heat acclimation. In: Fregly M, Blatteis CM, editors. Handbook of physiology. Oxford: Oxford University Press; 1996. p. 157–86.

[5] Mack GW, Nadel ER. Body fluid balance during heat stress in humans. In: Fregly M, Blatteis CM, editors. Handbook of physiology. Oxford: Oxford University Press; 1996. p. 187–214.

[6] Johnson JM, Proppe DW. Cardiovascular adjustments to heat stress. In: Fregly M, Blatteis CM, editors. Handbook of physiology. Oxford: Oxford University Press; 1996. p. 215–44.

[7] Francesconi RP. Endocrinological and metabolic responses to acute and chronic heat exposures. In: Fregly M, Blatteis CM, editors. Handbook of physiology. Oxford: Oxford University Press; 1996. p. 245–60.

[8] Crawshaw LI, Wallace HL, Dasgupta S. Thermoregulation. In: Auerbach PS, editor. Wilderness medicine. 5th ed. Philadelphia: Mosby Elsevier; 2007. p. 110–24.

[9] Gaffin SL, Moran DS. Pathophysiology of heat-related illness. In: Auerbach PS, editor. Wilderness medicine. 5th ed. Philadelphia: Mosby Elsevier; 2007. p. 228–68.

[10] Astrand P-O. Experimental studies of physical working capacity in relation to sex and age. Copenhagen: Ejnar Munksgaard; 1952.

[11] Epstein Y, Yanovich R, Moran DS, Heled Y. Physiological employment standards IV: integration of women in combat units physiological and medical considerations. Eur J Appl Physiol 2012;.

[12] Jiricka MK. Activity tolerance and fatigue. In: Porth CM, Matfin G, editors. Pathophysiology concepts of altered health states. 8th ed. Wolters Kluwer Health/Lippincott Williams & Wilkins; 2009. p. 231–51.

[13] McArdle WD, Glaser RM, Magel JR. Metabolic and cardiorespiratory response during free swimming and treadmill walking. J Appl Physiol 1971 May;30(5):733–8.

[14] Magel JR, Faulkner JA. Maximum oxygen uptakes of college swimmers. J Appl Physiol 1967 May;22(5):929–33.

[15] McArdle WD, Katch FI, Katch VL. Human energy expenditure during rest and physical activity. In: McArdle WD, Katch FI, Katch VL, editors. Exercise physiology: energy, nutrition and human performance. Baltimore/Philadelphia: Lippincott Williams & Wilkins; 2007, p. 195–207.

[16] Poehlman ET, Arciero PJ, Goran MI. Endurance exercise in aging humans: effects on energy metabolism. Exerc Sport Sci Rev 1994;22:251–84.

[17] Illner K, Brinkmann G, Heller M, Bosy-Westphal A, Muller MJ. Metabolically active components of fat free mass and resting energy expenditure in nonobese adults. Am J Physiol Endocrinol Metab 2000 February;278(2):E308–15.

[18] Nelson KM, Weinsier RL, Long CL, Schutz Y. Prediction of resting energy expenditure from fat-free mass and fat mass. Am J Clin Nutr 1992 November;56(5):848–56.

[19] Wang Z, Heshka S, Gallagher D, Boozer CN, Kotler DP, Heymsfield SB. Resting energy expenditure-fat-free mass relationship: new insights provided by body composition modeling. Am J Physiol Endocrinol Metab 2000 September;279(3):E539–45.

[20] Muller MJ, Bosy-Westphal A, Kutzner D, Heller M. Metabolically active components of fat free mass (FFM) and resting energy expenditure (REE) in humans. Forum Nutr 2003;56:301–3.

[21] Richardson RS, Knight DR, Poole DC, Kurdak SS, Hogan MC, Grassi B, et al. Determinants of maximal exercise VO$_2$ during single leg knee-extensor exercise in humans. Am J Physiol 1995 April;268(4 Pt 2):H1453–61.

[22] Richardson RS. What governs skeletal muscle VO$_2$max? New evidence. Med Sci Sports Exerc 2000 January;32(1):100–7.

[23] Bangsbo J. Muscle oxygen uptake in humans at onset of and during intense exercise. Acta Physiol Scand 2000 April;168(4):457–64.

[24] Speakman JR. The history and theory of the doubly labeled water technique. Am J Clin Nutr 1998 October;68(4):932S–8S.

[25] Slinde F, Arvidsson D, Sjoberg A, Rossander-Hulthen L. Minnesota leisure time activity questionnaire and doubly labeled water in adolescents. Med Sci Sports Exerc 2003 November;35(11):1923–8.

[26] Schutz Y, Deurenberg P. Energy metabolism: overview of recent methods used in human studies. Ann Nutr Metab 1996;40(4):183–93.

[27] Conway JM, Seale JL, Jacobs Jr. DR, Irwin ML, Ainsworth BE. Comparison of energy expenditure estimates from doubly labeled water, a physical activity questionnaire, and physical activity records. Am J Clin Nutr 2002 March;75(3):519–25.

[28] Hart TL, McClain JJ, Tudor-Locke C. Controlled and free-living evaluation of objective measures of sedentary and active behaviors. J Phys Act Health 2011 August;8(6):848–57.

[29] Moran DS, Heled Y, Gonzalez RR. Metabolic rate monitoring and energy expenditure prediction using a novel actigraphy method. Med Sci Monit 2004 November;10(11):MT117–20.

[30] Lyden K, Kozey SL, Staudenmeyer JW, Freedson PS. A comprehensive evaluation of commonly used accelerometer energy expenditure and MET prediction equations. Eur J Appl Physiol 2011 February;111(2):187–201.

[31] Welk GJ. Principles of design and analyses for the calibration of accelerometry-based activity monitors. Med Sci Sports Exerc 2005 November;37(Suppl. 11):S501–11.

[32] Johannsen DL, Calabro MA, Stewart J, Franke W, Rood JC, Welk GJ. Accuracy of armband monitors for measuring daily energy expenditure in healthy adults. Med Sci Sports Exerc 2010 November;42(11):2134–40.

[33] Müller MJ, Bosy-Westphal A. Energieverbrauch. In: Speckmann E.-J., Hescheler J., Köhling R., editors. Physiologie, vol. 5. Auflage ed. München: Elsevier GmbH; 2008. p. 587–93.

[34] Borg GA. Perceived exertion. Exerc Sport Sci Rev 1974;2:131–53.

[35] Koffranyi E, Michaelis HF. Ein tragbarer Apparat zur Bestimmung des Gasstoffwechsels. Arbeitsphysiologie 1940;1940(11):148.

[36] Krogh A, Lindhard J. The relative value of fat and carbohydrate as sources of muscular energy: with appendices on the correlation between standard metabolism and the respiratory quotient during rest and work. Biochem J 1920 July;14(3–4):290–363.

[37] Zuntz N. Über die Bedeutung der verschiedenen Nährstoffe als Erzeuger der Muskelkraft. Pflügers Arch Physiol 1901;1901(83):557–71.

[38] Gunga HC. Nathan Zuntz his life and work in the fields of high altitude physiology and aviation medicine. Burlington, MA, USA: American Physiological Society; 2009.

[39] Pirnay F, Lacroix M, Mosora F, Luyckx A, Lefebvre P. Effect of glucose ingestion on energy substrate utilization during prolonged muscular exercise. Eur J Appl Physiol Occup Physiol 1977 May 10;36(4):247–54.

[40] Kwan M, Woo J, Kwok T. The standard oxygen consumption value equivalent to one metabolic equivalent (3.5 ml/min/kg) is not appropriate for elderly people. Int J Food Sci Nutr 2004 May;55(3):179–82.

[41] Byrne NM, Hills AP, Hunter GR, Weinsier RL, Schutz Y. Metabolic equivalent: one size does not fit all. J Appl Physiol 2005 September;99(3):1112–9.

[42] Morgan DW, Bransford DR, Costill DL, Daniels JT, Howley ET, Krahenbuhl GS. Variation in the aerobic demand of running among trained and untrained subjects. Med Sci Sports Exerc 1995 March;27(3):404–9.

[43] Saunders PU, Pyne DB, Telford RD, Hawley JA. Reliability and variability of running economy in elite distance runners. Med Sci Sports Exerc 2004 November;36(11):1972–6.

[44] Morgan D, Martin P, Craib M, Caruso C, Clifton R, Hopewell R. Effect of step length optimization on the aerobic demand of running. J Appl Physiol 1994 July;77(1):245–51.

[45] Coyle EF, Sidossis LS, Horowitz JF, Beltz JD. Cycling efficiency is related to the percentage of type I muscle fibers. Med Sci Sports Exerc 1992 July;24(7):782–8.

[46] Coyle EF, Feltner ME, Kautz SA, Hamilton MT, Montain SJ, Baylor AM, et al. Physiological and biomechanical factors associated with elite endurance cycling performance. Med Sci Sports Exerc 1991 January;23(1):93–107.

[47] Poole DC, Gaesser GA, Hogan MC, Knight DR, Wagner PD. Pulmonary and leg VO_2 during submaximal exercise: implications for muscular efficiency. J Appl Physiol 1992 February;72(2):805–10.

[48] Bassett Jr. DR, Howley ET. Limiting factors for maximum oxygen uptake and determinants of endurance performance. Med Sci Sports Exerc 2000 January;32(1):70–84.

[49] Levine BD. VO_2max: what do we know, and what do we still need to know? J Physiol 2008 January 1;586(1):25–34.

[50] Lepretre PM, Koralsztein JP, Billat VL. Effect of exercise intensity on relationship between VO_2max and cardiac output. Med Sci Sports Exerc 2004 August;36(8):1357–63.

[51] Wagner PD. Counterpoint: in health and in normoxic environment VO_2max is limited primarily by cardiac output and locomotor muscle blood flow. J Appl Physiol 2006 February;100(2):745–7.

[52] Holloszy JO, Coyle EF. Adaptations of skeletal muscle to endurance exercise and their metabolic consequences. J Appl Physiol 1984 April;56(4):831–8.

[53] Rowell LB. Human cardiovascular control. Oxford: Oxford University Press; 1993.

[54] Astrand P-O, Rodahl K. Textbook of work physiology. McGraw-Hill; 1977.

[55] Saltin B, Astrand PO. Maximal oxygen uptake in athletes. J Appl Physiol 1967 September;23(3):353–8.

[56] Smith TD, Thomas TR, Londeree BR, Zhang Q, Ziogas G. Peak oxygen consumption and ventilatory thresholds on six modes of exercise. Can J Appl Physiol 1996 April;21(2):79–89.

[57] Shave R, Franco A. The physiology of endurance training. In: Whyte G, editor. The physiology of training. Elsevier Limited; 2006. p. 61–85.

[58] Frascarelli R, Cesarini G. ATP and energy metabolism. Riforma Med 1956 January 14;70(2):36–8.

[59] McClare CW. How does ATP act as an energy source? Ciba Found Symp 1975;31:301–25.

[60] Brooks GA, Fahey TD, White TP. Exercise physiology. Mountain View: Mayfield Publishing Company; 1996.

[61] Trump ME, Heigenhauser GJ, Putman CT, Spriet LL. Importance of muscle phosphocreatine during intermittent maximal cycling. J Appl Physiol 1996 May;80(5):1574–80.

[62] Greenhaff PL, Timmons JA. Interaction between aerobic and anaerobic metabolism during intense muscle contraction. Exerc Sport Sci Rev 1998;26:1–30.

[63] Birch K, MacLaren D, George K. Sport and exercise physiology. Andover: Thomson Publishing Services; 2005.

[64] Shearer J, Graham TE. Novel aspects of skeletal muscle glycogen and its regulation during rest and exercise. Exerc Sport Sci Rev 2004 July;32(3):120–6.

[65] Bergstrom J, Hermansen L, Hultman E, Saltin B. Diet, muscle glycogen and physical performance. Acta Physiol Scand 1967 October;71(2):140–50.

[66] Ekelund LG, Ahlborg B, Bergstrom J, Hultman E. Muscle glycogen consumption during work with and without glucose administration. Nord Med 1970 January 15;83(3):89–90.

[67] Womack HC. The relationship between human body weight, subcutaneous fat, heart weight, and epicardial fat. Hum Biol 1983 September;55(3):667–76.

[68] van der Vusse GJ, Reneman RS. Lipid metabolism in muscle. In: Fregly MJ, Blatteis CM, editors. Handbook of physiology. New York: Oxford University Press; 1996.

[69] Mackinnon LT. Future directions in exercise and immunology: regulation and integration. Int J Sports Med 1998 July;19(Suppl. 3):S205–9.

[70] Jonsdottir IH, Hellstrand K, Thoren P, Hoffmann P. Enhancement of natural immunity seen after voluntary exercise in rats. Role of central opioid receptors. Life Sci 2000 February 18;66(13):1231–9.

[71] Lambert CP, Flynn MG, Braun WA, Mylona E. Influence of acute submaximal exercise on T-lymphocyte suppressor cell function in healthy young men. Eur J Appl Physiol 2000 May;82(1–2):151–4.

[72] Hakkinen K, Komi PV. Electromyographic changes during strength training and detraining. Med Sci Sports Exerc 1983;15(6):455–60.

[73] Pette D, Vrbova G. Neural control of phenotypic expression in mammalian muscle fibers. Muscle Nerve 1985 October;8(8):676–89.

[74] Karlsson J, Jacobs I. Onset of blood lactate accumulation during muscular exercise as a threshold concept. I. Theoretical considerations. Int J Sports Med 1982 November;3(4):190–201.

[75] Mackie BG, Terjung RL. Blood flow to different skeletal muscle fiber types during contraction. Am J Physiol 1983 August;245(2):H265–75.

[76] McArdle WD, Katch FI, Katch VL. Neural control of human movement. In: McArdle WD, Katch FI, Katch VL, editors. Exercise physiology: energy, nutrition and human performance. Baltimore/Philadelphia: Lippincott Williams & Wilkins; 2007, p. 391–415.

[77] Gollnick PD, Parsons D, Riedy M, Moore RL. Fiber number and size in overloaded chicken anterior latissimus dorsi muscle. J Appl Physiol 1983 May;54(5):1292–7.

[78] Kelso TB, Hodgson DR, Visscher AR, Gollnick PD. Some properties of different skeletal muscle fiber types: comparison of reference bases. J Appl Physiol 1987 April;62(4):1436–41.

[79] Aura O, Komi PV. Effects of muscle fiber distribution on the mechanical efficiency of human locomotion. Int J Sports Med 1987 March;8(Suppl. 1):30–7.

[80] Staron RS. Human skeletal muscle fiber types: delineation, development, and distribution. Can J Appl Physiol 1997 August;22(4):307–27.

[81] Carson JA. The regulation of gene expression in hypertrophying skeletal muscle. Exerc Sport Sci Rev 1997;25:301–20.

[82] Kimball SR, Farrell PA, Jefferson LS. Invited review: role of insulin in translational control of protein synthesis in skeletal muscle by amino acids or exercise. J Appl Physiol 2002 September;93(3):1168–80.

[83] Adams GR, Hather BM, Baldwin KM, Dudley GA. Skeletal muscle myosin heavy chain composition and resistance training. J Appl Physiol 1993 February;74(2):911–5.

[84] Howald H. Training-induced morphological and functional changes in skeletal muscle. Int J Sports Med 1982 February;3(1):1–12.

[85] Schantz P, Billeter R, Henriksson J, Jansson E. Training-induced increase in myofibrillar ATPase intermediate fibers in human skeletal muscle. Muscle Nerve 1982 October;5(8):628–36.

[86] Beermann DH, Cassens RG, Couch CC, Nagle FJ. The effects of experimental denervation and reinnervation on skeletal muscle fiber type and intramuscular innervation. J Neurol Sci 1977 March;31(2):207–21.

[87] Grow WA, Kendall-Wassmuth E, Grober MS, Ulibarri C, Laskowski MB. Muscle fiber type correlates with innervation topography in the rat serratus anterior muscle. Muscle Nerve 1996 May;19(5):605–13.

[88] Wickham JB, Brown JM. Muscles within muscles: the neuromotor control of intra-muscular segments. Eur J Appl Physiol Occup Physiol 1998 August;78(3):219–25.

[89] Senn W, Wyler K, Clamann HP, Kleinle J, Luscher HR, Muller L. Size principle and information theory. Biol Cybern 1997 January;76(1):11–22.

[90] Studer LM, Ruegg DG, Gabriel JP. A model for steady isometric muscle activation. Biol Cybern 1999 May;80(5):339–55.

[91] Maughan RJ, Gleeson M, Greenhaff PL. Biochemistry of exercise and training. Oxford: Oxford University Press; 1997.

[92] Maughan RJ, Gleeson M. The biochemical basis of sports performance. Oxford: Oxford University Press; 2004.

[93] Romijn JA, Coyle EF, Sidossis LS, Gastaldelli A, Horowitz JF, Endert E, et al. Regulation of endogenous fat and carbohydrate metabolism in relation to exercise intensity and duration. Am J Physiol 1993 September;265(3 Pt 1):E380–91.

[94] Korhonen MT, Cristea A, Alen M, Hakkinen K, Sipila S, Mero A, et al. Aging, muscle fiber type, and contractile function in sprint-trained athletes. J Appl Physiol 2006 September;101(3):906–17.

[95] Billeter R, Jostarndt-Fogen K, Gunthor W, Hoppeler H. Fiber type characteristics and myosin light chain expression in a world champion shot putter. Int J Sports Med 2003 April;24(3):203–7.

[96] Edwards HT. Metabolic rate, blood sugar and utilization of carbohydrate. Am J Physiol 1934;108(203).

[97] Askew EW. Role of fat metabolism in exercise. Clin Sports Med 1984 July;3(3):605–21.

[98] De FP, Di LC, Lucidi P, Murdolo G, Parlanti N, De CA, et al. Metabolic response to exercise. J Endocrinol Invest 2003 September;26(9):851–4.

[99] Jeukendrup AE. Carbohydrate and exercise performance: the role of multiple transportable carbohydrates. Curr Opin Clin Nutr Metab Care 2010 July;13(4):452–7.

[100] Murakami T, Shimomura Y, Fujitsuka N, Sokabe M, Okamura K, Sakamoto S. Enlargement glycogen store in rat liver and muscle by fructose-diet intake and exercise training. J Appl Physiol 1997 March;82(3):772–5.

[101] Rennie MJ. How to avoid running on empty. J Physiol 2000 October 1;528(Pt 1):3.

[102] Nybo L. CNS fatigue and prolonged exercise: effect of glucose supplementation. Med Sci Sports Exerc 2003 April;35(4):589–94.

[103] Hogan MC, Gladden LB, Kurdak SS, Poole DC. Increased [lactate] in working dog muscle reduces tension development independent of pH. Med Sci Sports Exerc 1995 March;27(3):371–7.

[104] Sahlin K. Intracellular pH and energy metabolism in skeletal muscle of man. With special reference to exercise. Acta Physiol Scand Suppl 1978;455:1–56.

[105] Sahlin K, Harris RC, Nylind B, Hultman E. Lactate content and pH in muscle obtained after dynamic exercise. Pflugers Arch 1976 December 28;367(2):143–9.

[106] Vestergaard-Poulsen P, Thomsen C, Sinkjaer T, Henriksen O. Simultaneous 31P-NMR spectroscopy and EMG in exercising and recovering human skeletal muscle: a correlation study. J Appl Physiol 1995 November;79(5):1469–78.

[107] Cairns SP. Lactic acid and exercise performance: culprit or friend? Sports Med 2006;36(4):279–91.

[108] Wilber RL, Zawadzki KM, Kearney JT, Shannon MP, Disalvo D. Physiological profiles of elite off-road and road cyclists. Med Sci Sports Exerc 1997 August;29(8):1090–4.

[109] Morton RH. Detection of a lactate threshold during incremental exercise? J Appl Physiol 1989 August;67(2):885–8.

[110] Weltman A, Seip R, Bogardus AJ, Snead D, Dowling E, Levine S, et al. Prediction of lactate threshold (LT) and fixed blood lactate concentrations (FBLC) from 3200-m running performance in women. Int J Sports Med 1990 October;11(5):373–8.

[111] Spurway NC. Aerobic exercise, anaerobic exercise and the lactate threshold. Br Med Bull 1992 July;48(3):569–91.

[112] Altenburg TM, Degens H, van MW, Sargeant AJ, de HA. Recruitment of single muscle fibers during submaximal cycling exercise. J Appl Physiol 2007 November;103(5):1752–6.

[113] Rosler K, Hoppeler H, Conley KE, Claassen H, Gehr P, Howald H. Transfer effects in endurance exercise. Adaptations in trained and untrained muscles. Eur J Appl Physiol Occup Physiol 1985;54(4):355–62.

[114] Donovan CM, Brooks GA. Endurance training affects lactate clearance, not lactate production. Am J Physiol 1983 January;244(1):E83–92.

[115] Bergman BC, Wolfel EE, Butterfield GE, Lopaschuk GD, Casazza GA, Horning MA, et al. Active muscle and whole body lactate kinetics after endurance training in men. J Appl Physiol 1999 November;87(5):1684–96.

[116] Radda GK. Control of energy metabolism during muscle contraction. Diabetes 1996 January;45(Suppl. 1):S88–92.

[117] Stainsby WN, Brooks GA. Control of lactic acid metabolism in contracting muscles and during exercise. Exerc Sport Sci Rev 1990;18:29–63.

[118] Stainsby WN, Brechue WF, O'Drobinak DM. Regulation of muscle lactate production. Med Sci Sports Exerc 1991 August;23(8):907–11.

[119] Holloszy JO. Adaptation of skeletal muscle to endurance exercise. Med Sci Sports 1975;7(3):155–64.

[120] Maddaiah VT. Exercise and energy metabolism. Pediatr Ann 1984 July;13(7):565–72.

[121] Sahlin K, Katz A, Broberg S. Tricarboxylic acid cycle intermediates in human muscle during prolonged exercise. Am J Physiol 1990 November;259(5 Pt 1):C834–41.

[122] Koike A, Wasserman K, Taniguchi K, Hiroe M, Marumo F. Critical capillary oxygen partial pressure and lactate threshold in patients with cardiovascular disease. J Am Coll Cardiol 1994 June;23(7):1644–50.

[123] Svedahl K, MacIntosh BR. Anaerobic threshold: the concept and methods of measurement. Can J Appl Physiol 2003 April;28(2):299–323.

[124] Spencer MR, Gastin PB. Energy system contribution during 200- to 1500-m running in highly trained athletes. Med Sci Sports Exerc 2001 January;33(1):157–62.

[125] Brooks GA. Anaerobic threshold: review of the concept and directions for future research. Med Sci Sports Exerc 1985 February;17(1):22–34.

[126] Seip RL, Snead D, Pierce EF, Stein P, Weltman A. Perceptual responses and blood lactate concentration: effect of training state. Med Sci Sports Exerc 1991 January;23(1):80–7.

[127] Weltman A, Snead D, Stein P, Seip R, Schurrer R, Rutt R, et al. Reliability and validity of a continuous incremental treadmill protocol for the determination of lactate threshold, fixed blood lactate concentrations, and VO$_2$max. Int J Sports Med 1990 February;11(1):26–32.

[128] Nichols JF, Phares SL, Buono MJ. Relationship between blood lactate response to exercise and endurance performance in competitive female master cyclists. Int J Sports Med 1997 August;18(6):458–63.

[129] Wiswell RA, Jaque SV, Marcell TJ, Hawkins SA, Tarpenning KM, Constantino N, et al. Maximal aerobic power, lactate threshold, and running performance in master athletes. Med Sci Sports Exerc 2000 June;32(6):1165–70.

[130] MacRae HS, Dennis SC, Bosch AN, Noakes TD. Effects of training on lactate production and removal during progressive exercise in humans. J Appl Physiol 1992 May;72(5):1649–56.

[131] McAllister RM. Adaptations in control of blood flow with training: splanchnic and renal blood flows. Med Sci Sports Exerc 1998 March;30(3):375–81.

[132] McArdle WD, Katch FI, Katch VL. Training for anaerobic and aerobic power. In: McArdle WD, Katch FI, Katch VL, editors. Exercise physiology: energy, nutrition and human performance. Baltimore/Philadelphia: Lippincott Williams & Wilkins; 2007. p. 469–507.

[133] Yoshida T, Chida M, Ichioka M, Suda Y. Blood lactate parameters related to aerobic capacity and endurance performance. Eur J Appl Physiol Occup Physiol 1987;56(1):7–11.

[134] Tanaka K, Matsuura Y. Marathon performance, anaerobic threshold, and onset of blood lactate accumulation. J Appl Physiol 1984 September;57(3):640–3.

[135] Astrand PO, Hultman E, Juhlin-Dannfelt A, Reynolds G. Disposal of lactate during and after strenuous exercise in humans. J Appl Physiol 1986 July;61(1):338–43.

[136] Stuewe SR, Gwirtz PA, Agarwal N, Mallet RT. Exercise training enhances glycolytic and oxidative enzymes in canine ventricular myocardium. J Mol Cell Cardiol 2000 June;32(6):903–13.

[137] Brooks GA. Lactate: link between glycolytic and oxidative metabolism. Sports Med 2007;37(4–5):341–3.

[138] Bolser DC, Remmers JE. Synaptic effects of intercostal tendon organs on membrane potentials of medullary respiratory neurons. J Neurophysiol 1989 May;61(5):918–26.

[139] Bajic J, Zuperku EJ, Tonkovic-Capin M, Hopp FA. Interaction between chemoreceptor and stretch receptor inputs at medullary respiratory neurons. Am J Physiol 1994 June;266 (6 Pt 2):R1951–61.

[140] Li Z, Morris KF, Baekey DM, Shannon R, Lindsey BG. Responses of simultaneously recorded respiratory-related medullary neurons to stimulation of multiple sensory modalities. J Neurophysiol 1999 July;82(1):176–87.

[141] Mitchell JH, Sproule BJ, Chapman CB. The physiological meaning of the maximal oxygen intake test. J Clin Invest 1958 April;37(4):538–47.

[142] Wagner PD. Why doesn't exercise grow the lungs when other factors do? Exerc Sport Sci Rev 2005 January;33(1):3–8.

[143] Dempsey JA, Wagner PD. Exercise-induced arterial hypoxemia. J Appl Physiol 1999 December;87(6):1997–2006.

[144] McKenzie DC, Lama IL, Potts JE, Sheel AW, Coutts KD. The effect of repeat exercise on pulmonary diffusing capacity and EIH in trained athletes. Med Sci Sports Exerc 1999 January;31(1):99–104.

[145] O'Toole ML, Douglas PS. Applied physiology of triathlon. Sports Med 1995 April;19(4):251–67.

[146] McArdle WD, Katch FI, Katch VL. Functional capacity of the cardiovascular system. In: McArdle WD, Katch FI, Katch VL, editors. Exercise physiology: energy, nutrition and human performance. Baltimore/Philadelphia: Lippincott Williams & Wilkins; 2007. p. 351–63.

[147] Rowell LB. Integration of cardiovascular control systems in dynamic exercises. In: Rowell LB, Shepard J, editors. Handbook of physiology. New York: Oxford University Press; 1996.

[148] van HG. Lactate kinetics in human tissues at rest and during exercise. Acta Physiol (Oxf) 2010 August;199(4):499–508.

[149] Bevegard BS, Shepherd JT. Regulation of the circulation during exercise in man. Physiol Rev 1967 April;47(2):178–213.

[150] Clement DL, Shepherd JT. Regulation of peripheral circulation during muscular exercise. Prog Cardiovasc Dis 1976 July;19(1):23–31.

[151] Shepherd JT. Circulatory response to exercise in health. Circulation 1987 December;76 (6 Pt 2):VI3–10.

[152] Kayser B. Exercise starts and ends in the brain. Eur J Appl Physiol 2003 October; 90(3–4):411–9.

[153] Lepers R, Theurel J, Hausswirth C, Bernard T. Neuromuscular fatigue following constant versus variable-intensity endurance cycling in triathletes. J Sci Med Sport 2008 July; 11(4):381–9.

[154] Di Prampero PE, Boutellier U, Pietsch P. Oxygen deficit and stores at onset of muscular exercise in humans. J Appl Physiol 1983 July;55(1 Pt 1):146–53.

[155] Sahlin K, Ren JM, Broberg S. Oxygen deficit at the onset of submaximal exercise is not due to a delayed oxygen transport. Acta Physiol Scand 1988 October;134(2):175–80.

[156] McArdle WD, Katch FI, Katch VL. Energy transfer in exercise. In: McArdle WD, Katch FI, Katch VL, editors. Exercise physiology: energy, nutrition and human performance. Baltimore/Philadelphia: Lippincott Williams & Wilkins; 2007. p. 165–81.

[157] Drummond MJ, Vehrs PR, Schaalje GB, Parcell AC. Aerobic and resistance exercise sequence affects excess postexercise oxygen consumption. J Strength Cond Res 2005 May;19(2):332–7.

[158] Sedlock DA, Lee MG, Flynn MG, Park KS, Kamimori GH. Excess postexercise oxygen consumption after aerobic exercise training. Int J Sport Nutr Exerc Metab 2010 August;20(4):336–49.

[159] Gaesser GA, Brooks GA. Metabolic bases of excess post-exercise oxygen consumption: a review. Med Sci Sports Exerc 1984;16(1):29–43.

[160] Quinn TJ, Vroman NB, Kertzer R. Postexercise oxygen consumption in trained females: effect of exercise duration. Med Sci Sports Exerc 1994 July;26(7):908–13.

[161] LaForgia J, Withers RT, Gore CJ. Effects of exercise intensity and duration on the excess post-exercise oxygen consumption. J Sports Sci 2006 December;24(12):1247–64.

[162] Stone MH, Sanborn K, O'Bryant HS, Hartman M, Stone ME, Proulx C, et al. Maximum strength-power-performance relationships in collegiate throwers. J Strength Cond Res 2003 November;17(4):739–45.

[163] Stone MH, Sands WA, Pierce KC, Carlock J, Cardinale M, Newton RU. Relationship of maximum strength to weightlifting performance. Med Sci Sports Exerc 2005 June;37(6): 1037–43.

[164] de Souza EO, Tricoli V, Franchini E, Paulo AC, Regazzini M, Ugrinowitsch C. Acute effect of two aerobic exercise modes on maximum strength and strength endurance. J Strength Cond Res 2007 November;21(4):1286–90.

[165] Spineti J, de Salles BF, Rhea MR, Lavigne D, Matta T, Miranda F, et al. Influence of exercise order on maximum strength and muscle volume in nonlinear periodized resistance training. J Strength Cond Res 2010 November;24(11):2962–9.

[166] de Souza EO, Tricoli V, Paulo AC, Silva-Batista C, Cardoso RK, Brum PC, et al. Multivariate analysis in the maximum strength performance. Int J Sports Med 2012 December;33(12):970–4.

[167] Sharkey B, Gaskill S. Fitness and health. Champaign: Human Kinetics; 1997.

[168] Carlson K, Magnusen M, Walters P. Effect of various training modalities on vertical jump. Res Sports Med 2009;17(2):84–94.

[169] Kannas TM, Kellis E, Amiridis IG. Incline plyometrics-induced improvement of jumping performance. Eur J Appl Physiol 2012 June;112(6):2353–61.

[170] Stark T, Walker B, Phillips JK, Fejer R, Beck R. Hand-held dynamometry correlation with the gold standard isokinetic dynamometry: a systematic review. PM R 2011 May;3(5):472–9.

[171] Murphy AJ, Wilson GJ. The assessment of human dynamic muscular function: a comparison of isoinertial and isokinetic tests. J Sports Med Phys Fitness 1996 September; 36(3):169–77.

[172] Kannus P. Isokinetic evaluation of muscular performance: implications for muscle testing and rehabilitation. Int J Sports Med 1994 January;15(Suppl. 1):S11–8.

[173] Verdijk LB, van LL, Meijer K, Savelberg HH. One-repetition maximum strength test represents a valid means to assess leg strength in vivo in humans. J Sports Sci 2009 January 1;27(1):59–68.

[174] Franchini E, Nunes AV, Moraes JM, Del Vecchio FB. Physical fitness and anthropometrical profile of the Brazilian male judo team. J Physiol Anthropol 2007 March;26(2):59–67.

[175] Shaw CE, McCully KK, Posner JD. Injuries during the one repetition maximum assessment in the elderly. J Cardiopulm Rehabil 1995 July;15(4):283–7.

[176] Ikai M, Fukunaga T. Calculation of muscle strength per unit cross-sectional area of human muscle by means of ultrasonic measurement. Int Z Angew Physiol 1968;26(1):26–32.

[177] Hunter SK, Critchlow A, Shin IS, Enoka RM. Fatigability of the elbow flexor muscles for a sustained submaximal contraction is similar in men and women matched for strength. J Appl Physiol 2004 January;96(1):195–202.

[178] Andersen LL, Andersen JL, Zebis MK, Aagaard P. Early and late rate of force development: differential adaptive responses to resistance training? Scand J Med Sci Sports 2010 February;20(1):e162–9.

[179] Holtermann A, Roeleveld K, Vereijken B, Ettema G. The effect of rate of force development on maximal force production: acute and training-related aspects. Eur J Appl Physiol 2007 April;99(6):605–13.

[180] Aagaard P, Simonsen EB, Andersen JL, Magnusson P, Dyhre-Poulsen P. Increased rate of force development and neural drive of human skeletal muscle following resistance training. J Appl Physiol 2002 October;93(4):1318–26.

[181] Dorossiev DL. Methodology of physical training, principles of training and exercise prescription. Adv Cardiol 1978;24:67–83.

[182] Zaryski C, Smith DJ. Training principles and issues for ultra-endurance athletes. Curr Sports Med Rep 2005 June;4(3):165–70.

[183] Cross DL. The influence of physical fitness training as a rehabilitation tool. Int J Rehabil Res 1980;3(2):163–75.

[184] Kang J, Robertson RJ, Denys BG, DaSilva SG, Visich P, Suminski RR, et al. Effect of carbohydrate ingestion subsequent to carbohydrate supercompensation on endurance performance. Int J Sport Nutr 1995 December;5(4):329–43.

[185] James AP, Lorraine M, Cullen D, Goodman C, Dawson B, Palmer TN, et al. Muscle glycogen supercompensation: absence of a gender-related difference. Eur J Appl Physiol 2001 October;85(6):533–8.

[186] Angeli A, Minetto M, Dovio A, Paccotti P. The overtraining syndrome in athletes: a stress-related disorder. J Endocrinol Invest 2004 June;27(6):603–12.

[187] Winsley R, Matos N. Overtraining and elite young athletes. Med Sport Sci 2011;56:97–105.

[188] Kreher JB, Schwartz JB. Overtraining syndrome: a practical guide. Sports Health 2012 March;4(2):128–38.

[189] Pelliccia A, Culasso F, Di Paolo FM, Maron BJ. Physiologic left ventricular cavity dilatation in elite athletes. Ann Intern Med 1999 January 5;130(1):23–31.

[190] Caselli S, Di PR, Di Paolo FM, Pisicchio C, di GB, Guerra E, et al. Left ventricular systolic performance is improved in elite athletes. Eur J Echocardiogr 2011 July;12(7):514–9.

[191] Saltin B, Henriksson J, Nygaard E, Andersen P, Jansson E. Fiber types and metabolic potentials of skeletal muscles in sedentary man and endurance runners. Ann N Y Acad Sci 1977;301:3–29.

[192] Beneke R, Hutler M. The effect of training on running economy and performance in recreational athletes. Med Sci Sports Exerc 2005 October;37(10):1794–9.

[193] Piacentini MF, De Ioannon G, Comotto S, Spedicato A, Vernillo G, La Torre A. Concurrent strength and endurance training effects on running economy in master endurance runners. J Strength Cond Res 2013 August;27(8):2295–303.

[194] Marriott BM. Food components to enhance performance. Washington: National Academy Press; 1994.

[195] Markiewicz K, Lutz W, Cholewa M. Changes in concentrations of basic energy-yielding substrates in plasma during maximal effort and postexercise restitution in relation to training level. Acta Physiol Pol 1980 September;31(5):453–62.

[196] Narita K, Sakamoto S, Mizushige K, Senda S, Matsuo H. Development and evaluation of a new target heart rate formula for the adequate exercise training level in healthy subjects. J Cardiol 1999 May;33(5):265–72.

[197] Knuttgen HG. Strength training and aerobic exercise: comparison and contrast. J Strength Cond Res 2007 August;21(3):973–8.

[198] Wenger HA, Bell GJ. The interactions of intensity, frequency and duration of exercise training in altering cardiorespiratory fitness. Sports Med 1986 September;3(5):346–56.

[199] McArdle WD, Katch FI, Katch VL. Muscular strength: training muscles to become stronger. In: McArdle WD, Katch FI, Katch VL, editors. Exercise physiology: energy, nutrition and human performance. Baltimore/Philadelphia: Lippincott Williams & Wilkins; 2007. p. 509–53.

[200] Sale DG. Neural adaptation to resistance training. Med Sci Sports Exerc 1988 October;20(Suppl. 5):S135–45.

[201] Liu Y, Heinichen M, Wirth K, Schmidtbleicher D, Steinacker JM. Response of growth and myogenic factors in human skeletal muscle to strength training. Br J Sports Med 2008 December;42(12):989–93.

[202] Heinemeier KM, Olesen JL, Schjerling P, Haddad F, Langberg H, Baldwin KM, et al. Short-term strength training and the expression of myostatin and IGF-I isoforms in rat muscle and tendon: differential effects of specific contraction types. J Appl Physiol 2007 February;102(2):573–81.

[203] Yan Z. Skeletal muscle adaptation and cell cycle regulation. Exerc Sport Sci Rev 2000 January;28(1):24–6.

[204] MacDougall JD, Tuxen D, Sale DG, Moroz JR, Sutton JR. Arterial blood pressure response to heavy resistance exercise. J Appl Physiol 1985 March;58(3):785–90.

[205] Morin KH. Infant nutrition and global poverty. MCN Am J Matern Child Nurs 2007 November;32(6):382.

[206] Redinger RN. The physiology of adiposity. J Ky Med Assoc 2008 February;106(2):53–62.

[207] Bjorntorp P. Visceral obesity: a "civilization syndrome". Obes Res 1993 May;1(3):206–22.

[208] Centers for Disease Control and Prevention. Estimated county-level prevalence of diabetes and obesity – United States, 2007. MMWR Morb Mortal Wkly Rep 2009 November 20;58(45):1259–63.

[209] Berghofer A, Pischon T, Reinhold T, Apovian CM, Sharma AM, Willich SN. Obesity prevalence from a European perspective: a systematic review. BMC Public Health 2008;8:200.

[210] Muller-Wieland D, Kotzka J, Knebel B, Krone W. Metabolic syndrome and hypertension: pathophysiology and molecular basis of insulin resistance. Basic Res Cardiol 1998;93(Suppl. 2):131–4.

[211] Rosolova H. Hypertension and diabetes mellitus–pathophysiology and risk. Vnitr Lek 1999 November;45(11):655–60.

[212] Despres JP. Is visceral obesity the cause of the metabolic syndrome? Ann Med 2006;38(1): 52–63.

[213] Belcaro G. Once we were hunters. Imperial College Press; 2001.

[214] Spurway N. Origins. In: Spurway N, Wackerhage H, editors. Genetics and molecular biology of muscle adaptation. London: Churchill Livingstone Elsevier; 2006. p. 1–23.

[215] Manson JE, Greenland P, Lacroix AZ, Stefanick ML, Mouton CP, Oberman A, et al. Walking compared with vigorous exercise for the prevention of cardiovascular events in women. N Engl J Med 2002 September 5;347(10):716–25.

[216] Blair SN, Kohl III HW, Paffenbarger Jr. RS, Clark DG, Cooper KH, Gibbons LW. Physical fitness and all-cause mortality. A prospective study of healthy men and women. JAMA 1989 November 3;262(17):2395–401.

[217] Ganten D, Spahl T, Deichmann T. Die Steinzeit steckt uns in den Knochen: Gesundheit als Erbe der Evolution. 2nd ed. München: Piper; 2009.

[218] Paul S, Nyncke H. Paläopower. München: Beck C. H.; 2012.

Chapter 4

Pressure Environment

Oliver Opatz* and Hanns-Christian Gunga†

*Postdoctoral Research Associate, Center for Space Medicine and Extreme Environments, Institute of Physiology, CharitéCrossOver (CCO), Charité University Medicine Berlin, Berlin, Germany

†Professor, Center for Space Medicine and Extreme Environments, Institute of Physiology, CharitéCrossOver (CCO), Charité University Medicine Berlin, Berlin, Germany

4.1 HYPOBARIC ENVIRONMENT

To start with, we decided to restrict ourselves here to the most general forms of physiological and pathophysiological adaptations of humans to high altitude, because in the last two decades comprehensive monographs have covered this topic in depth [1–3]. Instead, we took the opportunity to focus on rapidly flourishing recent topics such as intermittent hypoxic training and intermittent high altitude shift working, scientific issues that are important not only for basic science but also for occupational health as well.

4.1.1 Introduction

4.1.1.1 Historical Aspects

The historical aspects will only be briefly covered here. The more interested reader is referred to the detailed descriptions given by other authors such as Bert [4], Heller et al. [5], Zuntz et al. [6], Loewy [7], and more recently by West [8] and Gunga [9]. However, it is at least interesting in the frame of this chapter to note that, according to these sources, four decisive phases emerge in the development of high-altitude physiological and medical research. The first phase stretches back to the sixteenth century. Well-known nature researchers such as Scheuchzer, de Saussure, and v. Humboldt, who at the same time had an important influence on the history of other sciences [10], provided the first precise observations of high-altitude physiological phenomena in their broader scientific studies. These were descriptions of altered physical sensations (nausea, headache, etc.) that the authors observed in themselves and their companions during their expeditions in the high mountain ranges of the Alps, the Andes, and the Himalayas [5,7]. The second phase of the history of high-altitude physiology began in the 1870s in France with the numerous laboratory

Human Physiology in Extreme Environments. http://dx.doi.org/10.1016/B978-0-12-386947-0.00004-6

physiological studies performed by Bert (1833-1886) in the pneumatic chamber. This laboratory research was complemented by increasingly systematic high-altitude field studies. These were carried out in newly built research stations such as Capanna Regina Margherita at the top of Monte Rosa. Here a few recognized researchers, such as Mosso (1846-1910), Zuntz (1847-1920), Durig (1872-1961), and v. Schroetter (1870-1929), performed their experiments. Later other research facilities were added such as Pico de Teide (Tenerife) or Pikes Peak (USA) [8,9]. Figure 4.1 shows some of these remarkable scientists on their way to the Pico de Teide high-altitude research station, a photograph taken on a ship during the International Tenerife Expedition (1910) guided by the German physician Pannwitz (1861-1926) [9].

Some remarkable contributions of their research to our basic understanding of physiology should be mentioned. For example, Zuntz, Durig, and v. Schroetter should receive credit for their comprehensive metabolic studies with ingenious laboratory and field methodologies at high altitude, hypobaric chambers, and balloons. Through these experiments, they highlighted the importance of the drop in oxygen partial pressure and its impact on human physiology at high altitude [8,9]. Douglas, on the other hand, developed the famous "Douglas bag method" for the measurement of pulmonary ventilation and respiratory gas exchange, which allows the estimation of energy expenditure (indirect calorimetry). And finally, Barcroft's contributions include describing different reasons why the body lacks oxygen, that is, hypobaric hypoxia via impaired gas exchange by the lungs, anemic hypoxia, decreased red cell/haemoglobin mass, and ischemic hypoxia, impaired perfusion of the tissue. Only later was a

FIGURE 4.1   Participants of the Pannwitz expedition. Back row from left to right: Douglas, Carl Neuberg, Mascart, Gotthold Pannwitz, Arnold Durig, Piasse, Hermann von Schroetter; front row from left to right: Johannes Orth, Neuberg, Bachmann, Joseph Barcroft, Nathan Zuntz, front: Pannwitz, Carriere. *Source: Mascart: Impressions et observations dans un voyage à Tenerife, Flammarion, Paris, 1912.*

fourth category added, cytotoxic hypoxia, impaired utilization of oxygen by the cell at a mitochondrial level.

In summary, the first and second phases of high-altitude research can be characterized by methodological and fundamental scientific breakthroughs in the understanding of the impact of high altitude exposure on the human body.

The third phase started when the first attempts were made to climb Mount Everest without supplementary oxygen. In 1924, Edward Felix Norton (1884-1954) climbed up to 8500 m on the north side of Mount Everest without supplementary oxygen. Some physiologists could hardly believe that humans could reach such altitude due to the low P_{iO_2} at the summit of Mount Everest. Theories were developed that the lungs would actively secrete oxygen into the blood, which, however, turned out to be wrong [8]. Nevertheless, as illustrated in Figure 4.2, for the final 300 m up to the summit, it took 54 years more until Messner and Habeler reached the top of Mount Everest without supplementary oxygen in 1978.

This successful attempt by Messner and Habeler in 1978 stimulated the fourth, still-running phase. This phase is characterized by highly sophisticated research expeditions, such as the American Medical Research Expedition to Everest (AMREE) in 1981 [11] or more recently the Caudwell Xtreme Everest (CXE) in 2007 [12]. The primary objective of this kind of highly equipped research expedition was to obtain information on human physiology and medical issues at the highest possible altitudes, including the summit of Mount Everest. In addition to the camp-type research, extremely high-altitude research stations >5000 m, such as the Pyramid station in Nepal or in extreme isolated areas of

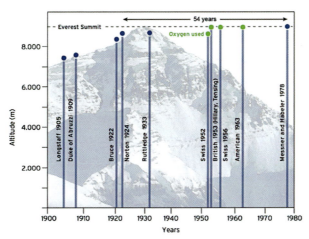

FIGURE 4.2 The major attempts to reach the summit of Mount Everest in the last century with and without artificial oxygen supply. Please note that it took 54 years from the Norton's attempt to the summit in 1924 until 1978, when Messner and Habeler reached the summit of Mount Everest "by fair means". *Adapted from [8].*

the world such as the Concordia Station in Antarctica, were built. In the latter environment, humans can be exposed over months to a high-altitude environment combined with the most extreme coldness, dryness, isolation, and confinement known on Earth (see Chapter 5 for further details).

4.1.1.2 Geographical Aspects

According to Hackett and Roach [13], altitude regions above sea level (a.s.l.) are defined as high altitude (1500-3500 m a.s.l.), very high altitude (3500-5500 m), and extreme altitude (>5500 m). The major distributions of these high altitude mountains ranges are shown in Figure 4.3.

As outlined in Chapter 1, there is strong evidence that our ancestors, the Australopithecins, originated from the East African Rift Valley System, and it can be fairly assumed that some traits or physiological adaptations, for example, physical endurance performance, have their genetic origin in that time in evolution (for further details, see Chapter 1) [14]. Today, the number of permanent high-altitude residents and annual high-altitude visitors worldwide at altitudes >2400 m a.s.l. is estimated to be 80 million people, that is, nearly 1% of the human population on Earth [15]. Meanwhile, the number of tourists at high altitudes has obviously increased sharply. So, Burtscher et al. [16] estimated that in 2001, visitors alone to >2500 m a.s.l. totalled more than 100 million. Actually, today some regions such as the Alps strongly depend economically

FIGURE 4.3 The major distribution of high altitude mountains ranges around the world.

on such altitude tourism as could be seen recently at the International Tourists Exhibition 2014 in Berlin (Germany). Since altitude climate plays a decisive role in the health and well-being of those who expose themselves to it, be it for private or professional/occupational reasons, it deserves a closer look now.

4.1.1.3 Gases in Our Atmosphere

As shown in Figure 4.4 the composition of Earth's atmosphere changed distinctly over time; namely, with the occurrence of life, the atmosphere shifted from a reducing to an oxidative environment.

Today the atmosphere consists of mainly nitrogen (78.084%) and oxygen (~20.946%). All additional gas fractions are shown in Figure 4.5. Number three in the ranking of atmospheric gases is not, as often guessed, carbon dioxide, but argon (0.934%). Carbon dioxide only represents 0.035% of the molecules in our inspired air. Depending on altitude, temperature, and geography, our atmosphere also contains a variable amount of water vapor, on average around 1%.

The structure and the atmospheric processes are physically shaped and influenced by gravity. Gravity causes the density and pressure of air to decrease exponentially as one moves away from the surface of the Earth toward space. The different physical laws related to this will be described in detail in the following section on the hyperbaric environment (Figure 4.6).

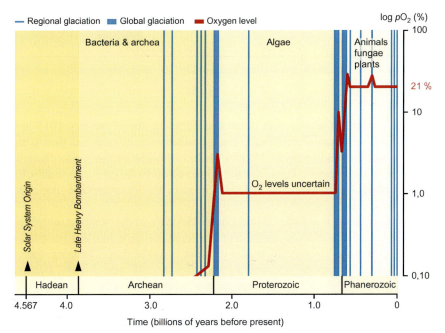

FIGURE 4.4 The development of oxygen concentration in the atmosphere of the Earth over time. Please note (i) the sharp change 2.5 billion years ago that reflects the starting point of photosynthesis by cyanobacteria and (ii) the quite stable oxygen concentrations thereafter.

Main gas fractions	Trace gases
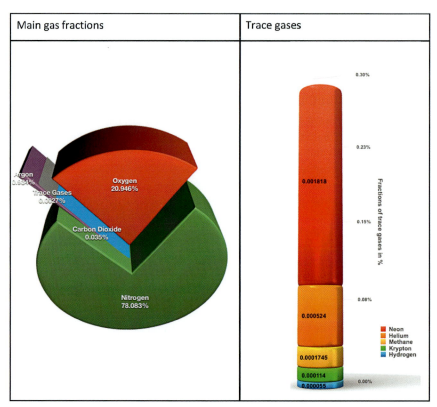	

FIGURE 4.5 Terrestrial atmosphere in gas fractions. Left: the main components, right: most important trace gases.

Air pressure at sea-level is illustrated as being approximately 1013 millibars (mb) or $1\,kg/cm^2$ of surface area. Furthermore, Figure 4.6 shows that the troposphere reaches up to 12 km altitude, followed by the stratosphere up to 50 km with its ozone layer, the mesosphere (up to 80 km), and finally the thermosphere (up to 100 km). The outer edge of the thermosphere marks the border to space.

The International Standard Atmosphere (ISA) is a model of how atmospheric pressure varies with altitude. As an approximation, it can be said that per 1000 m altitude difference, the oxygen partial pressure is lowered by 17 mmHg, the air temperature decreases by 6 °C and air humidity by 25%, respectively. Because with increasing altitude less atmosphere is available to absorb ultraviolet (UV) radiation, UV levels increase by approximately 7% per 1000 m in the troposphere [17]. However, it was once again the German physiologist Nathan Zuntz (1847-1920) (see below), who first recognized that for the different altitude regions around the world and for every subdivision of the specific local high altitude, climate correction factors have to be applied to determine the actual barometric

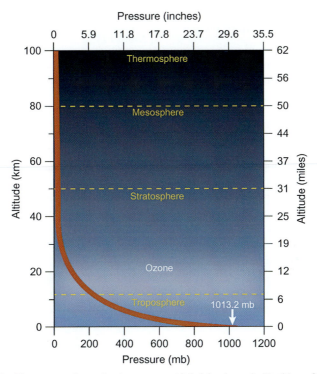

FIGURE 4.6   The average change in air pressure with height above the Earth's surface (ISA). In this graph, air pressure at the surface is illustrated as being approximately 1013 millibars (mb) or 1 kg/cm² of surface area.

conditions due to general physical changes such as temperature [6].[1] Zuntz included these environmental parameters in his famous barometric formula. When it comes to the issue of whether it is possible for humans to reach the summit of Mount Everest (8848 m a.s.l.) without artificial oxygen, this slight correction of the ISA by the Zuntz's formula becomes a crucial issue in addition to the extremely low alveolar CO_2 partial pressures found in humans at the summit [8,11]. Zuntz et al. incorporated the following parameters into this formula:

---

1. Specifically, Zuntz et al. said: "However, the climate does not shift with the altitude at the same pace in every zone. The closer we are to the poles, the lower the elevation at which the characteristics of the high altitude climate are already present, while the closer we are to the equator, the higher we must climb from the upper surface of the ocean in order to reach a climate which corresponds to our conceptions of high altitude. In central Europe, an elevation of approximately 1000 m suffices for this, in Norway this is as low as 500-600 m, in the Bolivian and Peruvian Andes and in the Himalayan mountains, an elevation of over 1500-2000 m is required and even at 4000 m elevation close to the equator, the climate will not evidence all the characteristics of the climate which has already been reached in our Alps at 1800-2000 m" [6, p. 36].

$$\log b = \log B - \frac{h}{72(256.4 + t)} \tag{4.1}$$

h: altitude difference in meters; t: mean temperature of an air column with the height h; B: barometric pressure at the lower level; b: barometric pressure at the upper level.

4.1.2 Physiology

Depending on the latitude and special geographical situations from a physiological and pathophysiological point of view, the altitude can be divided into roughly four zones and three thresholds, which are given in Figure 4.7.

The following physiological changes are characteristic, if a human organism is acutely exposed to a severe hypobaric-hypoxia equivalent to 5000 m. Shortly, in minutes after such an acute exposure, a significant drop in oxygen saturation (SaO_2), the concentration of oxygen in the blood, which is at sea level between 95% and 100%, will be observed, approximately to 75%. If the subjects stay at high altitude, in the following days the (SaO_2) will increase back to 85% as a result of a successful adaptation. Usually, blood oxygen levels should remain >80%, otherwise organ function of the brain and heart might be altered. Concomitantly with the drop in SaO_2, as one piece of an adaptive process, ventilation will be nearly doubled at altitude as compared to sea level. The hyperventilation will lead to lower alveolar CO_2 concentrations, increasing thereby the level of oxygen in the lungs. The reduced CO_2 concentrations in the lungs will attenuate the acid-base balance of the organism, leading to an alkalosis. This alkalosis triggers in the following hours and days an increased bicarbonate excretion via the kidneys to normalize the pH of the blood. This normalization of the acid-base status is a hallmark for living organisms and therefore, as *conditio qua non*, has to be kept

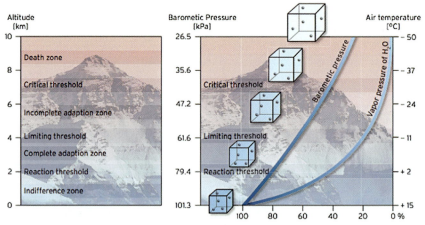

FIGURE 4.7 The four major zones and the three thresholds of high altitude adaptation in humans with respect to the changes of altitude (left panel), barometric pressure, water vapor pressure, and air temperatures changes with altitude (right panel).

by the organism within very narrow limits. Actually, in the human blood the pH only varies from around 7.35-7.45; the ideal pH for blood is 7.4. If blood pH moves <6.8 or >7.8, cells stop functioning and a person will die. Furthermore, besides increased ventilation, the autonomic nervous system will stimulate cardiovascular responses, that is, an increase in heart rate and blood pressure as well as erythropoiesis. This will be triggered via an increased erythropoietin (EPO) production and release, which will increase the number of red blood cells and the hemoglobin (Hb) concentration, leading to a higher oxygen transport capacity of the blood. It has been found that an increase of hemoglobin by 3 g/l leads at sea level to an increase of about 1% in $\dot{V}O_2max$. The carotid bodies play an important role in this regulatory loop because they are designed to distinctively sense any change in arterial oxygen partial pressure. The changes are mediated to the brainstem via signal processing in which the hypoxia-inducible-factor (HIF) is involved. It increases (i) a hypoxic ventilatory response (HVR) of the lungs and (ii) cardiac output by increasing heart rate. Briefly, it can be said that to avoid any rapid and dangerous disturbances in brain functions, the inspiratory P_{O_2} should always be kept above a critical threshold of >35 mmHg. This can be assured (i) with a normal ventilation up to 4000 m a.s.l., (ii) with hyperventilation up to 7000 m, (iii) with pure O_2-inspiration up to 12,000 m, and with pure O_2-inspiration and hyperventilation up to 14,000 m. However, the latter can only be sustained for a limited time span due to the toxic effects of oxygen. Above this altitude, pressured cabins have to be used. By exceeding the hypoxic threshold (death zone, Figure 4.7), continuous reductions of body functions occur. These acute responses to hypoxic exposure differ largely between individuals and can have a profound effect on the $\dot{V}O_2max$. As shown in a field study (Figure 4.8), the reduction in $\dot{V}O_2max$ after an acute altitude exposure to 4325 m a.s.l. can vary from 8% up to 56% between individuals [18].

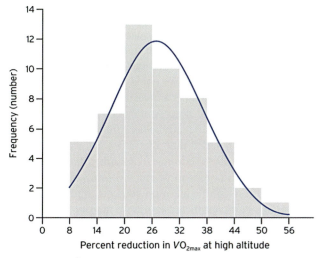

FIGURE 4.8 Reduction of $\dot{V}O_2max$ in 54 lowlanders after acute altitude exposure to 4325 m a.s.l. *Adapted from [18].*

If the adaptation processes are slow and long enough and sleeping altitudes during the night are kept decisively lower than maximum altitudes reached during the day, then a complete adaptation can be reached in the indifferent and complete adaptation zone. However, above 5500 m, regardless of how long and smooth the adaptation process might be, no acclimatization will be possible. Furthermore, at lower altitudes some of the acute responses of the cardiovascular and ventilatory systems described above will be attenuated, and even disappear the longer the subjects are exposed to the hypobaric-hypoxic environment. As typical examples of that reaction, we might have a closer look at some studies we performed on red blood cell formation, specifically the EPO production and release, and metabolic changes observed at moderate and high altitudes. In the adult human, the kidneys are the main organ for the production and release of EPO. Stimulated by tissue hypoxia and again involving the hypoxia-inducible-factor (HIF-1-alpha), specialized cells located between the tubuli in the interstitium of the renal cortex, probably in the endothelium of peritubular capillaries, or interstitial cells such as fibroblasts, start to produce EPO [19]. EPO in itself then stimulates erythropoiesis by increasing the proliferation, differentiation, and maturation of the erythroid precursors in the bone marrow. Only a few studies had been performed in humans at moderate and high altitude when we started our own research on this topic [20]. So, it was a little bit of a surprise when our early field studies in young males revealed that even at an altitude of 2300 m, that is, an altitude at which the haemoglobin is still fully saturated, EPO showed a remarkable increase 48 h after the ascent and a gradual decrease during the following days at altitude afterwards (Figure 4.9).

Later on, similar observations on EPO production and release were made in older, untrained subjects with metabolic syndrome. In that specific study, we focused the research not only on EPO regulation but on cardiovascular parameters and glucose metabolism at moderate altitude as well [22–24]. Such studies are really quite rare to find in literature; nevertheless, they are urgently needed because, as mentioned above, the number of altitude sojourners of all ages is still increasing. Specifically, we investigated the changes in resting blood pressure and heart rate, glycemic parameters, and lipid metabolism during a 3-week sojourn at 1700 m in the Austrian Alps [22]. We could show that the Homeostasis Model Assessment (HOMA) index, which is a measure of insulin resistance, decreased significantly in the subjects, and glucose concentrations obtained after an oral glucose tolerance test were significantly lower after the stay at altitude compared to the basal values. We concluded that after a 3-week exposure to moderate altitude, patients with metabolic syndrome (i) tolerated their sojourn without any physical problems, (ii) exhibited short-term favorable effects on the cardiovascular system, and (iii) had significant improvements in glycemic parameters that were paralleled by a significant increase in high-density-lipoprotein cholesterol. Athletes and their physicians have for years been extolling the beneficial aspects of a intermittent high-altitude training. However, the facts about the efficiency to increase performance are far from clear, as will be shown in the following section.

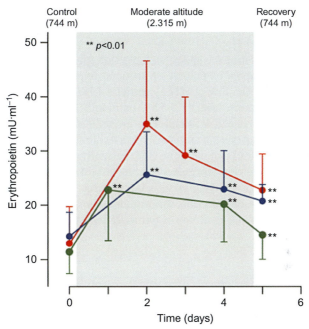

FIGURE 4.9 Time course of erythropoietin (EPO) in three different studies performed at moderate altitude in the Alps in young, healthy male subjects (red $N=20$-21, blue $N=29$, black $N=18$). Please note that the EPO concentrations after an initial increase started to decrease during the stay at altitude. *Adapted from [21].*

4.1.2.1 Intermittent Hypoxia and High-altitude Training

Altitude training, that is, living and/or training in hypoxic environments with different training schemes and at different altitudes, has been used worldwide for many decades mainly as a stimulus to increase endurance performance, especially since the 1968 Olympic Games in Mexico City. As hypoxia triggers EPO, which increases in red cell mass and Hb concentrations, concomitantly, the oxygen transport capacity of the blood will increase [19]. Furthermore, as has been recently found, hypoxia alters muscle metabolism in a positive manner [25]. Not surprisingly, altitude training gained interest from sport scientists and endurance athletes. Especially in the last several years, new training protocols were tested to improve physical performance of laymen, sub-elite and elite sportsmen, for example, intermittent hypoxia (IH) high-altitude training. IH is defined as repeated episodes of hypoxia interspersed with normoxic periods [26]. The hypoxic exposures happen in (i) a natural (high altitude) or (ii) an artificial environment by sojourns in hypobaric chambers or by breathing hypoxic gas mixtures in normobaric conditions. It is assumed that IH initiates, via a progressive increase in ventilation, adaptations of the hematopoietic and cardio-circulatory systems, oxygen delivery to the tissues, optimization of the utilization of oxygen in the tissues, and/or improvement of the immune system.

Furthermore, it might facilitate acclimatization processes in those who intend to go to altitude and might be useful in the prevention and treatment of various illnesses. Bonetti and Hopkins [27] made the first large meta-analytic review of the effects on performance and related physiological measures following adaptation to six protocols of natural or artificial hypoxia: live-high train-high (LHTH), live-high train-low (LHTL), artificial LHTL with daily exposure to long (8-18 h) continuous, brief (1.5-5 h) continuous or brief (<1.5 h) intermittent periods of hypoxia and artificial live-low train-high (LLTH). The findings of this large meta-analysis are summarized in Table 4.1.

Mainly, it was found that natural LHTL currently provides the best protocol for enhancing endurance performance in elite and sub-elite athletes, while some artificial protocols are effective in sub-elite athletes. Likely mediators include $\dot{V}O_2max$ and the placebo, nocebo, and training-camp effects. Modification of the protocols presents the possibility of further enhancements, which should be the

TABLE 4.1 The enhancements of maximal oxygen uptake ($\dot{V}O_2max$) and maximal aerobic power output in different natural and artificial protocols performed by sub-elite and elite athletes ("likely" changes are marked in bold letters) [27]

	Sub-elite athletes	Elite athletes
Enhancement of maximal oxygen uptake ($\dot{V}O_2max$)		
Natural hypoxia protocols		
1. Live-high, train-high (LHTH)	1. **"Likely"** (4.3%)	1. "Reduction" (−1.5%)
2. Live-high, train-low (LHTL)	2. **"Very likely"** (6.4%)	2. **"Very likely"** (6.4%)
Enhancement of maximal aerobic power output		
Natural hypoxia protocols		
1. Live-high, train-high (LHTH)	1. "Unclear" (0.9%)	1. "Unclear" (1.6%)
2. Live-high, train-low (LHTL)	2. **"Likely"** (4.2%)	2. **"Likely"** (4.0%)
Artificial altitude protocols:		
3. LH (long: 8-18 h/day—cont), TL	3. "Possible" (1.4%)	3. "Unclear" (0.6%)
4. LH (brief: 1.5-5 h/day—cont), TL	4. "Unclear" (0.7%)	4. –
5. LH (brief: <1.5 h/day—intermit), TL	5. "Possible" (2.6%)	5. "Unclear" (0.2%)
6. Live-low, train-high—IHT (0.5-2 h/day)	6. "Unclear" (0.9%).	6. –

focus of future research. Specifically, researchers observed that the 51 qualifying studies provided 11-33 estimates for effects on power output with each protocol and up to 20 estimates for effects on maximal oxygen uptake ($\dot{V}O_2max$) and other potential mediators. Substantial enhancement of maximal endurance power output in controlled studies of sub-elite athletes was very likely with artificial brief intermittent LHTL, likely with LHTL, possible with artificial long continuous LHTL, but unclear with LHTH, artificial brief continuous LHTL, and LLTH. In elite athletes, enhancement was possible with natural LHTL, but unclear with other protocols. They concluded that there was evidence that these effects were mediated (i) at least partly by substantial placebo, nocebo, and training-camp effects with some protocols; (ii) by enhancing protocols by appropriate manipulation of study characteristics that produced clear effects with all protocols in sub-elite athletes, but only with LHTH and LHTL in elite athletes. For $\dot{V}O_2max$, increases were very likely with LHTH in sub-elite athletes, whereas in elite athletes a "reduction" was possible with LHTH. The changes with other protocols were unclear, and, remarkably, effects on erythropoietic and other physiological mediators provided little additional insight into mechanisms [27]. The aerobic performance decreases with increasing altitude, that is, approximately 1% for every 100 m altitude gain from 1500 m [28]. That is why the training intensities at high altitude have to be reduced if the training involves large muscle groups such as in the case of long-distance running, cross-country skiing, and cycling (compare Chapter 3, exercise physiology) [28]. Furthermore, this is also the reason why the speed (in % sea-level speed) decreases rapidly in all mid- and long distance runs as shown in Figure 4.10, but increases in all short distances up to 400 m because of the lower air density at high altitude.

Based on the reported IH-effects, at least five IH protocols can be distinguished [26]: (i) short cycles of 30-90 s hypoxia for 7-8 h/day over weeks to years; (ii) IH at rest with a cycle length of 2-10 min for 1-2 h/day over 2-4 weeks; (iii) 1-2 h hypoxia at rest per day for 5 days; (iv) IH at rest for more than 90 min (to h) per day over 2-6 weeks, and (v) IH with exercise for 30 min to 2 h/day over 2-6 weeks.

In conclusion, according to the above mentioned studies, it seems that the various types of IH protocols evoke different adaptation effects also depending on health and training state. However, as recently pointed out in a review by Lundby et al., "The general practice of altitude training is widely accepted as a means to enhance sport performance despite a lack of rigorous scientific studies. For example, the scientific gold-standard design of a double-blind, placebo-controlled, cross-over trial has never been conducted on altitude training. Given that few studies have utilised appropriate controls, there should be more scepticism concerning the effects of altitude training methodologies" [30].

4.1.2.2 *Intermittent Hypoxia and Occupational Health*

Over decades high-altitude research focused on the acute and long-term adaptations of humans exposed to altitude. However, besides the already ongoing research activities in sports medicine mentioned above, occupational health

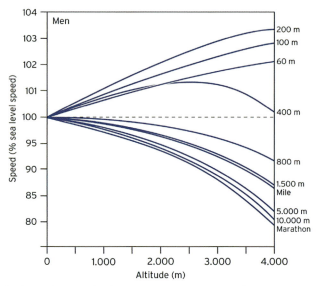

FIGURE 4.10 Speed (in % sea-level speed) decreases rapidly in all mid- and long distance runs but increases in short distances up to 400 m; in the latter because of the lower air density at high altitude [29].

risks related to shift working at high-altitude astronomical observatories such as ALMA (5000 m a.s.l.) or mines received growing interest [2,31–34]. In the latter, the shortness of natural resources has led to activities to exploit such resources even at very high altitudes above 4000 m (Figure 4.11). For a better understanding of how economically important mining activities are for countries such as Chile, a few facts should be given here apart from the physiological aspects of that topic. Multinational private companies, especially from Canada and Australia, are currently investing billions of U.S. dollars to keep and develop lithium, gold, and copper mines, for example, in the South American Andes (The Economist, April 27th-May 3rd 2013).

Beside Brasilia and Peru, Chile is the most important natural resource producer in South America, which actually has about 30% of the world's copper resources and 33% of the world's copper industry. In 2009 the value of the entire Chilean mining industry was estimated to be US $31.5 billion, of which only about US $140 million comes from oil, gas, and coal mining. In Chile, metal mining can be differentiated into three main groups: (i) mining in a large scale by state or multinational companies such as BhP Billiton ltd., Anglo American Plc., Antofagasta Plc., Xstrata (copper); Freeport-McMoran Copper & Gold corporation (copper and gold); Barrick Gold corporation, and Kinross Gold corporation (gold); (ii) national companies such as Codelco; (iii) mining by middle-sized private and (iv) small private companies. (http://www.deutsche-rohstoffagentur.de/DERA/DE/Downloads/Laenderstudie_Chile_Dez2011.pdf?__blob=publicationFile&v=6).

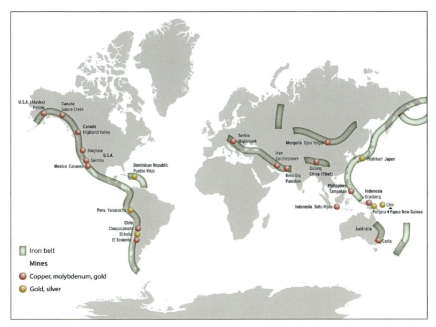

FIGURE 4.11 Iron belts and major mines around the world.

The big international and national mining companies are high-tech facilities, running 24h, usually surface mines, with several thousand employees in a wide age range between 20 and 55 years, usually coming from different South American countries. About 60,000 people are currently working in this industrial sector in Chile. Sometimes special team members are flown in from countries such as Canada, Australia, and China. Only a small number of the employees are indigenous people from South America such as Aymara and Quechua; the overwhelming number of workers come from coastal areas of South America and are therefore non-high altitude adapted [14]. So far, in Chile, the mines are not obliged to follow any special rules or laws to ensure a healthy high-altitude working place or shift cycles apart from the usual occupational guidelines. The mining companies, in cooperation with the miners' professional insurances/associations, decided what to do. However, most recently, *per decretum* No. 28 Modif. Decreto 594, Diario Oficial from November 8, 2012, stipulated that continuous health surveillance has to be ensured by the mining companies for their employees at high altitude.

Our own early research at high-altitude mines began with studying the impact of a 5-year lasting intermittent hypobaric-hypoxic exposure on miners' erythropoietic and cardiovascular functions. We studied miners who had a seven-day shift cycle, which means that they were seven days at sea-level and afterwards seven days at altitudes (>3600 m). We compared their data with the acute exposure of our own group (non-adapted Caucasians) (Figure 4.12).

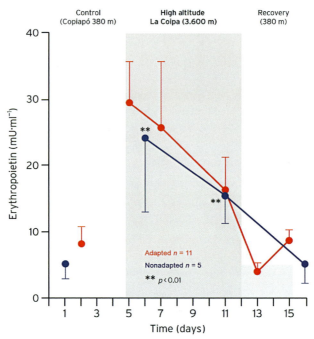

FIGURE 4.12   High-altitude shift working in the Chilean Andes (3600 m) and its influence on erythropoietin [32].

As can be seen, the 5-year lasting intermittent high-altitude exposure did not change the acute response of erythropoietin production and release from the kidneys as compared to a group of non-adapted Caucasians [32]. The absolute increase of EPO production and release was similar to that already observed at moderate altitude (compare Figure 4.9). Furthermore, the miners also showed a decrease in EPO concentrations while staying still at altitude. In respect to the cardiovascular adaptations, we observed that the changes in central venous pressure (CVP) were different from those in Caucasian lowlanders insofar as the CVPs were generally higher in the shift workers than in the Caucasians [32]. This might explain to some extent observations published by Jalil et al [31] on the isotonic work performance in 34 workers of similar age. These miners had working shifts of 4 days at 4500m a.s.l. and resting periods of 4 days at sea level over a period of at least 2 years. Interestingly, Jalil et al. found no differences in the cardiovascular response to exercise between the first and fourth day of the stay at high altitude in workers chronically exposed to intermittent hypobaric hypoxia. That they found no changes might be due to a large central blood volume, i.e., an increased CVP as observed in our study. Most frequently in Chilean mines 7- to 10-day cycles are performed with 12-h shifts, followed by rest cycles at sea level lasting 3-5 days. The latter is dependent mainly on the mining company and the absolute altitude reached by the mines [31–33]. Usually the miners live at high altitude in special camps that are equipped with dormatories, canteens, and recreation areas/sports facilities and located between 2800 and 3200 m, so that their sleeping altitude is normally at a lower level than their work place in the surface mines. This means that the miners have weekly changing day-and-night schedules and hypobaric-hypoxic shifts during their stay at altitude. This type of shift-work puts additional stress on the human body [35]. Accidents are therefore forseeable, and the long-term effect on the human body has not been sufficiently studied so far. Several studies on shift workers at sea level have already shown that such work is associated with a large number of deficiencies and diseases in humans [36–41]. Furthermore, recent studies point to the fact that genes related to the control and function of the cellular "clock" are attenuated by hypoxia alone [42,43].

A broader analysis of the currently available literature reveals that the following organs and respiratory organ systems might be attenuated by hypobaric-hypoxic exposure: circadian rhythms [34,44–48], energy metabolism and supply [35,49–52], cardiovascular parameters [53], neuro-endocrinological functions [54,55], and even basic functions of the cell such as cell divison [56,57]. In conclusion, it seems obvious that the entire spectrum of the impact of intermittent work at high altitude is not known. Definitely, there is a need for further research and technological developments such as an oxygen-enriched atmosphere at the working places at high altitude. Currently, such plans are ongoing; for example, at ALMA, now the highest year-round-operated astronomical site in the world (Figure 4.13).

The inspired P_{O_2} at San Pedro de Atacama is only 53% of sea level, and occupational health problems and an increased number of accidents can be foreseen. Already, some workers on site are taking artificial oxygen via a cannula from a portable backpack reservoir. Currently, in cooperation with the management of

FIGURE 4.13 Atacama Large Millimeter/sub-millimeter Array (ALMA) site near San Pedro de Atacama in Chile at 5000 m a.s.l. *Photo courtesy O. Opatz*

the European Southern Observatories at ALMA, we are planning pilot studies to analyze more thoroughly the acclimatization process, sleep and exercise performance, as well as mental abilities in these very high-altitude shift workers during their stay at ALMA and at home in the coastal areas of Chile.

4.1.2.3 Pathophysiology

As outlined above, the adaptation process, as far as the limiting threshold, depends on (i) velocity of ascent, (ii) absolute altitude reached, (iii) the relative accomplished altitude difference, and (iv) the state of health but not on the state of endurance training. Above 3000 m a.s.l. acclimatization should always be stepwise. It has become apparent that very important for acclimatization process is the respective sleeping altitude during an ascent of several days. In general, the night rest should always be as low as possible, and for every 500 meters altitude one should sleep two nights at the same altitude. This means that during ascent, the sleeping altitude should not be increased more than 1000 m/week; for an altitude of 5000 m a.s.l., at least 2-3 weeks' acclimatization are necessary. If the ascent is too high too fast, then signs of high altitude illnesses may occur, that is, high altitude headache (HAH), acute mountain sickness (AMS), high-altitude pulmonary edema (HAPE), or even high-altitude cerebral edema (HACE). AMS, HAPE, and HACE are sometimes also subsumed as severe high-altitude illness (SHAI). The pathophysiological nucleus of the illnesses is the lowering barometric pressure and concomitantly reduced partial pressure of oxygen, which leads to a reduced arterial P_{O_2} and concomitantly reduced tissue oxygenation [58,59]. The severity of these high-altitude illnesses might be altered significantly by other factors such as a reduced HVR; decreased oxygen desaturation, especially during sleep; an impaired gas exchange; uneven pulmonary vasoconstriction due to the Euler-Liljestrand reflex; fluid retention; increased sympathetic drive; increased intracranial pressure; and oxidative stress and/or inflammatory processes [60,61]. The prevalence of SHAI, that is, AMS, HAPE, and HACE, was reported recently in a prospective cohort study in which a total of 1326 subjects went through a hypoxic exercise test before a sojourn above 4000 m [62]. Richalet et al. found that severe AMS occurred in 314 subjects (23.7%), HAPE in 22 subjects (1.7%), and HACE in 13 subjects (0.98%). Slightly lower results for HAPE (0.49%) and HACE (0.28%), respectively, were observed in railway workers in Qinghai-Tibet [63,64]. However, a remarkable higher prevalence of HAPE was observed (15%) in soldiers who were flown directly from sea level to 3500 m a.s.l. in the Himalayas [1].

4.1.3 Altitude Diseases

4.1.3.1 High-Altitude Headache (HAH)

HAH is the most frequent symptom, afflicting up to 80% of high-altitude sojourners [3]. Besides hypoxia, risk factors such as hypo-hydration, overexertion, and insufficient energy intake can trigger the development of HAH in

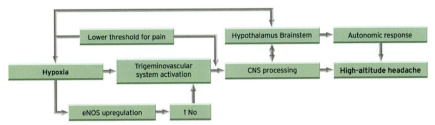

FIGURE 4.14 Proposed pathophysiological pathway of the mechanisms leading to HAH. CNS, Central nervous system; eNOS, endothelial nitric oxide synthase; NO, nitric oxide [3]. *Adapted from [114].*

susceptible subjects. According to Sanchez del Rio and Moskowitz [65], the presumable pathway for the development of HAH is as follows: Hypoxic exposure activates the trigeminovascular system via mechanical and chemical stimuli, that is, eNos (endothelial nitric oxide synthase) up-regulation and an increase in NO (nitric oxide) concentration. The subsequent vasodilation and compression of intracranial tissues in combination with a change of the threshold for pain under hypoxia attenuates CNS (central nervous system) processing in specific regions in the brainstem and hypothalamus, leading to autonomic responses that culminate in the HAH (see Figure 4.14). If HAH develops, stop the ascent, rest, and acclimatize at the same altitude. Furthermore, symptomatic treatment with analgesics, antiemetics, or azetolamide (125-250 mg two times a day) can speed acclimatization. Descending 500 m or more may also help [3].

4.1.3.2 Acute Mountain Sickness (AMS)

Acute Mountain Sickness (AMS) is the most common discomfort that can be frequently observed in high-altitude sojourners going higher than 2500 m a.s.l. It is usually benign but can rapidly progress to the more severe and potentially fatal forms, such as HACE and HAPE, which will be described later. The cerebral effects of ascent to high altitudes seem to be triggered by the subjects' relative hypoventilation, impaired gas exchange by interstitial edema, fluid retention, and redistribution, as well as an increased sympathetic drive [3,66]. Similar to HAH, mechanical, chemical, and P_{O_2}/P_{CO_2} mediators are involved in the presumed pathophysiology and cerebral effects of AMS (Figure 4.15). The major symptoms, which can vary individually to a large extent, are headache, fatigue or weakness, anorexia, nausea or vomiting, dyspnea, hyperventilation, difficulty sleeping, gastrointestinal symptoms (anorexia, nausea, or vomiting), and decreased thirst. Symptoms are most intense on the second and third days after arrival. Most frequently, AMS accompanied by high-altitude sleep disturbances that include (i) prolonged periods of apnea, (ii) sleep fragmentation with several arousals and alterations of sleep stages, leading to (iii) decreased arterial oxygen content and (iv) eventually cardiac rhythm disturbances. The AMS symptoms usually disappear 5-8 days after arrival, if one does not ascend to higher altitudes [1,67]. Cited in Netzer et al. [60], AMS has a prevalence of about 10%

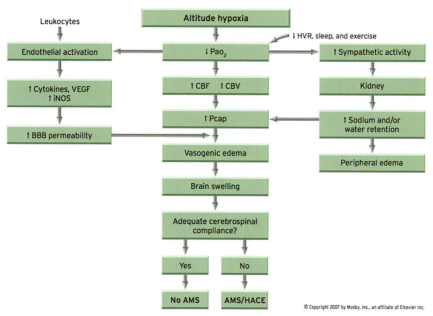

FIGURE 4.15 Proposed pathophysiological pathway of the mechanisms leading to AMS [3]. BBB, blood-brain barrier; CBF, cerebral blood flow; CBV, cerebral blood volume; HACE, high-altitude cerebral edema; HVR, hypoxic ventilatory response; iNOS, inducible nitric oxide synthese; Pcap, capillary pressure, VEGF, endothelial growth factor. *Adapted from [114].*

for those going directly from sea level to 2500 m a.s.l.; however, this number will increase to 30-40% if those people ascend to 3500 m in the Alps or Tibet. As there is a wide range of AMS symptoms that are also typical features in other high-altitude diseases such as HAH, HAPE, and HACE, the Lake Louise Score was developed and adopted at the 1991 International Hypoxia Symposium held at Lake Louise in Alberta (Canada). This score was found to be a practical tool to differentiate between the most severe high altitude illnesses—AMS, HAPE, and HACE [68]. The scale is grouped into specific symptoms of AMS (headache, gastrointestinal, fatigue, sleep, overall activity), and each of these categories is subjectively rated from 0 (showing no symptoms) to 3 (having severe symptoms). In this system mild AMS will have a total score of 3 and severe AMS 6, for example [69]. Moderate-to-severe AMS could be treated in the field by low-flow oxygen, that is, 0.5-1 l/min by mask or cannula (if available), acetazolamide, 125-250 mg two times a day, with or without dexamethasone (4 mg orally, intramusculary, or intravenous every 6 h), hyperbaric therapy ("Gamowbag"), or again immediate descent to a lower altitude [3].

4.1.3.3   High-Altitude Pulmonary Edema (HAPE)

High-altitude pulmonary edema (HAPE) is a life-threatening form of non-cardiogenic pulmonary edema that was misdiagnosed for centuries as a "pneumonia" [2]. HAPE is characterized by fluid accumulation in the lungs that occurs in otherwise healthy, sometimes even well-acclimatized mountaineers at altitudes typically above 2500 m (Hackett & Roach; Ward Milledge West). In some rare cases, HAPE has been recorded at lower altitudes (1500-2500 m) in highly susceptible subjects. Besides individual susceptibility, rate of ascent, altitude reached, degree of cold, physical exertion, and certain underlying medical conditions are all important factors in the development of HAPE. In the literature, the incidence of HAPE varies from 1 in 10,000 to 15 in 100 in lowlanders going to high altitude, mainly depending on the rapidness of ascent and absolute altitude reached. Women seem to be less susceptible [3]. According to the Lake Louise consensus, the presence of at least two of the following symptoms are characteristic of HAPE: dyspnea at rest, cough, weakness or decreased exercise performance, chest tightness or congestion, or at least two of the following signs: crackles or wheezing in at least one lung field, central cyanosis, tachypnea, or tachycardia [68].

In general, susceptible persons are those who, at sea level, already show an excessive increase in pulmonary artery pressure and pulmonary vascular resistance during hypoxic exposure at rest and during exercise, and sometimes even during exercise in normoxia. It has been suggested that this abnormal increase in pulmonary artery pressure is induced by an overreactivity of the pulmonary circulation to both hypoxia and exercise [3]. HAPE remains the major cause of death related to high-altitude exposure, with a high mortality rate in the absence of adequate emergency treatment. In Figure 4.16, assuming a normal function of the left side of the heart, the other two major players in the development of HAPE are excessive pulmonary hypertension and high protein permeability leakage in the lung, and the interactions are shown. The drop in P_{aO_2} increases sympathetic drive and leads to an uneven hypoxic pulmonary vasoconstriction, which in itself causes an overperfusion locally, as well as an increase in capillary pressures, capillary stress failures, and concomitantly a capillary leak (Figure 4.16). According to researchers, currently the following treatments should be applied: minimize exertion, keep warm, descend immediately or begin hyperbaric therapy, 4-6 l/min until improving, then 2-4 l/min. If these treatments are unavailable, one of the following should be taken: Nifedipine, 30 mg extended release every 12 h; Sildenafil, 50 mg every 8 h; or Tadalafil, 10 mg every 12 h [3].

4.1.3.4   High Altitude Cerebral Edema (HACE)

HACE is a severe medical condition that can be found in 0.5-1% of high-altitude sojourners, usually those suffering from AMS and HAPE. As can be seen in Figure 4.15 AMS and HACE show a similar pathophysiology. But whereas HACE is an encephalopathy, a disease that affects the function or structure of the brain, AMS is not. Typical clinical signs of HACE are confusion, changes in behavior, fatigue, ataxia, headache, difficulty speaking, vomiting, hallucinations, blindness, paralysis of a limb, seizure, unconsciousness, total paralysis, and in the final stage, coma [68]. The treatment includes immediate descent or evacuation, oxygen 2-4 l/min, dexamethasone, 8 mg orally, intramuscularly, or intravenously 4 mg every 6 h, or hyperbaric therapy [3].

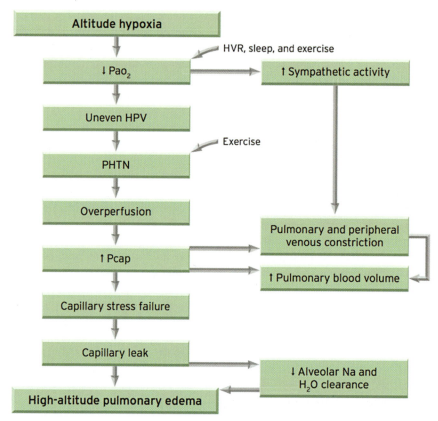

FIGURE 4.16 Proposed pathophysiology of the development of HAPE. HPV, hypoxic pulmonary vasoconstriction; HVR, hypoxic ventilatory vasoconstriction; HVR, hypoxic ventilatory response; Pa_{O_2}, arterial partial pressure of oxygen; P_{cap}, capillary pressure; PHTN, pulmonary hypertension [3]. *Adapted from [114].*

4.2 HYPERBARIC ENVIRONMENT

Compared to other extreme environments such as microgravity, heat and cold, or hypobaric environments, the hyperbaric environment (i.e., underwater) is more accessible and tolerable for most people. Today, one can easily complete a SCUBA (self contained underwater breathing apparatus) certification course, which enables hundreds of certified divers to explore the vast underwater wilderness. Nevertheless, the underwater environment is a hazardous environment for the human body. The quickly changing pressures in a medium of different density than air equates to most of the problems, which primarily occur during submerging. In this chapter we will not describe all of the possible hazards related to diving; rather we will concentrate on the main changes due to elevated gas pressure.

4.2.1   History

The detailed history of diving has been described often in the common text-books of diving medicine. A short, but very comprehensive history of diving medicine/physiology was written and published by Acott [70]. Therefore, we will focus on one man, who was a pioneer in the field of modern diving physiology and may be rightfully called the *father of diving medicine* (Figure 4.17).

John Scott Haldane was a Scottish physiologist (1860-1936) who pioneered today's understanding of the effects of different gases on the mammalian/human physiology. Most of these findings derived from experiments performed on himself and his family.

Moreover, due to his unique knowledge about pressure and gases, he was consulted by the British government, whenever there was need for a specific prevention method against mining accidents, technical innovations regarding the combination of physical gas theories and human physiology, or testing a new deep diving technology [71].

Haldane described exactly the phenomena of afterdamp, blackdamp, and whited damp and discussed the toxicological phenomena of carbon dioxide, carbon monoxide, and hydrogen sulfide. Furthermore, he provided useful countermeasures for miners, such as the safety lamp and the use of sentinel animals. The famous canary bird, which was used in British pits until the late 1980s, was the prime candidate for this line of work. Those birds were known

FIGURE 4.17   British physiologist Dr. John Scott Haldane (1860-1936) using an apparatus for penetrating a noxious atmosphere. *Photo by Hulton Archive/Getty images; picture taken on January 1, 1910.*

to be about 20 times as sensitive to toxic gas than men. Thus, they were taken to the mines in a specialized cage, which closed the front door automatically as soon as the bird fell from his perch, having breathed toxic gas.

Aside from his activities with coal miners, Haldane was asked to assist in the development of the first gas masks. With the advent of World War I, a vital countermeasure was needed to prepare the British soldiers for the appearance of the first (chemical) weapons of mass destruction. Zuntz was also involved at the same time on such safety issues for soldiers and miners in Germany. Actually, Haldane visited the Zuntz laboratory in Berlin at the Landwirtschaftliche Hochschule before World War I and the two men maintained a scientific exchange [9]. Haldane's contributions to modern scuba diving were manifold: He discovered that carbon dioxide, not oxygen, as it was previously believed, was the strongest stimulus for breathing. He was also a trailblazer in the analysis of blood gases. Through his research, he was able to provide a much clearer picture of the physiology of decompression disease, which he researched by order of the British government. Because he did this in the employ of the British Admiralty, rather costly and complex experiments could be carried out by using a steel pressure chamber and sending divers to great depths in Scottish deep water lochs. However, it was these experiments that brought us the first decompression table. The concept was published in 1905 in his report on the prevention of air sickness [72]. As a consequence, the first decompression table was presented in 1907. This table was used until the 1950s.

The so-called Haldane effect, which was published in 1914, describes the affinity of haemoglobin to carbon dioxide to be dependent on the concentration of oxygen in the respective tissue. It can be viewed as the counter mechanism to the Bohr effect, which describes the dependency of oxygen binding to the concentration of carbon dioxide. The Haldane effect also has great relevancy for apnea diving, where it leads to a lower partial pressure of CO_2 due to an elevated solubility of it in deoxygenated blood.

4.2.2   Gas Laws

4.2.2.1   The Ideal Gas Law and Its Counterparts

Ideal Gas Law equation

$$p \cdot V = n \cdot R \cdot T \tag{4.2}$$

Almost every baro-physiological question can be answered using this *Ideal Gas Law*, thus it is presented above as the ideal starting point for this chapter. It was described first in 1834 by Emile Clapeyron, who combined Boyle's Law, which describes the correlation of gas volume and pressure, and the *Laws of Gay-Lussac*, which correlate temperature, volume, and pressure [73].

The variables in this equation behave inverse proportionally to each other. The higher the pressure p, the smaller the volume V.

This is what we call the Law of Boyle-Marriotte.

Law of Boyle-Marriotte

$$\frac{p_2}{p_1} = \frac{V_1}{V_2}$$ (4.3)

The higher the number of particles n, the lower the temperature T, when the volume remains constant. Therefore, when the particle number n remains constant and either volume or pressure go up, the temperature also increases. That is what we call the Laws of Gay-Lussac.

Laws of Gay-Lussac; law I correlates temperature and volume, whereas law II describes the dependencies of temperature and pressure

$$\frac{T_1}{T_2} = \frac{V_1}{V_2} \qquad \frac{T_1}{T_2} = \frac{p_1}{p_2}$$ (4.4)

Gay-Lussac I   Gay-Lussac II

When these principles are transferred to hyperbaric physiology, this means the following:

i) Because temperature is kept in a very small range in a normothermic organism, the temperature does not play a significant role in pressure or volume, because, for instance, a change in body temperature of 1 °C would only have an impact of $1/(273+37)=0.003$. The Boyle-Marriotte law plays a role in the following situations:
   - Shrinking of the lungs during apnea diving. That is, a doubling of the aquatic pressure, for example, from 0 to 10 m depth at sea level, means a halving of the lung volume (or a doubling of the concentration of gases in the lung parenchyma).
   - In cases of decompression, micro gas bubbles develop in the bloodstream. According to the Boyle-Marriotte law, these micro-bubbles double their size with every 10 m, or 1 ATM, ascent to the surface in cases of a rapid ascent, this can be potentially hazardous to the diver.

ii) In the scuba diver, the number of particles *n* plays a decisive role; as with descent, volume decreases. Therefore, the higher the pressure of the inspired gas (that is adapted to the water pressure by stage II of the lung automatically to prevent lung edema), the higher the number of particles will be stored in the tissues in the body (Figure 4.18).

Because temperature remains rather constant in the diver, the particle number doubles if the pressure doubles and the volume (of the lung) stays constant. Thus, with a constant pulmonary volume one can imagine that an increase in pressure would lead to a higher number of gas molecules diffusing through the alveolar wall, which then leads to higher concentration of the inspired gases in the pulmonary circulation.

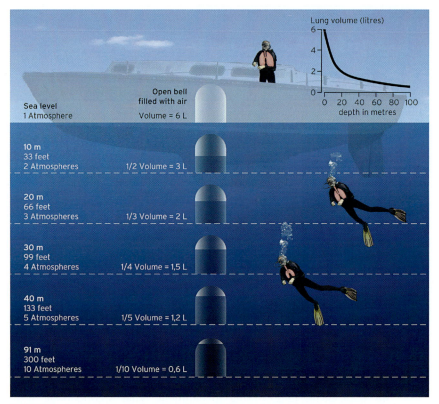

FIGURE 4.18   Boyle's Law describes the dependency of the volume on pressure. In the apnea diver, the lung volume shrinks, the deeper he/she dives.

4.2.2.2 The Law of Henry

William Henry's law (1803) states that: "…at constant temperature, the amount of a given gas that dissolves in a given type and volume of liquid is directly proportional to the partial pressure of that gas in equilibrium with that liquid." [74].

Law of Henry

$$p = k_H \cdot c \tag{4.5}$$

where p is the partial pressure of the gaseous solute above the solution, k_H is a constant with the dimensions of pressure divided by concentration and c is the concentration of the dissolved gas.

Henry's law can be observed when one opens a bottle or can containing a carbonated beverage. As soon as the pressure drops inside the bottle, the partial pressure of carbon dioxide decreases. The dissolved gas begins to escape from its liquid phase. The drink starts to bubble.

The same effect can be observed in human physiology. After having breathed pressurized air during scuba diving, the inert gases dissolved in the tissue can behave in the same manner as our carbonated beverage example. During and after a rapid ascent, the dissolved gases leave the liquid phase in arteries, muscles, joints, and/or the brain. This can lead to decompression injuries in the scuba diver.

4.2.3   Comparative Physiology

Aquatic mammals and human beings share a common ancestor [75]. Human beings do retain the ability to express the "mammalian dive reflex," which is very strongly expressed in animals such as diving birds, seals, and whales [76,77]. Therefore, these mammals provide the best way to begin our discussion concerning comparative physiology.

4.2.4   Mammalian Diving Reflex

The mammalian diving reflex allows animals as well as humans to extend their apnea time underwater through the conservation of inspired and dissolved oxygen [76,78,79]. Primarily aquatic mammalians and birds use this effect on a daily basis. This phylogenetic relic, when expressed in humans during submersion in water, can lead to different physiological responses [80]. The trigger of this mechanism is currently hypothesized to be the stimulation of the trigeminal nerve through contact with cold water on the face [81]. Bradycardia is also one of the effects of immersion. As so eloquently described by the German physiologist, Otto Gauer (1909-1979), this reflex is characterized by a redistribution of volume to the central circulation from the peripheral circulation of the body. The ventricular muscles of the heart distend, allowing for a greater stroke volume to accommodate the increase in filling volume of the atria [82]. The diving response also has therapeutic uses; it has been previously described as an effective means to terminate paroxysmal tachycardia [83]. The four main characteristics of the mammalian dive reflex are the following:

i) Upon submersion, apnea, or breath holding, prevents the aspiration of water during submersion. This is an effect that is present very early in life [81]. It has been shown to diminish later in life. Therefore, newborns are protected against aspiration during or shortly after their birth.

ii) Bradycardia improves the economy of energy consumption by the cardiovascular system [79].

iii) Peripheral vasoconstriction supports the redistribution and centralization of oxygen carriers to the central nervous system and the heart. The main trigger is the cooling of the skin, which elicits peripheral vasoconstriction, by which the body isolates its core against cold exposure [84].

iv) Splenic blood redistribution has been known to be practiced by aquatic mammalians, such as seals and whales [85]. It has also been recently observed

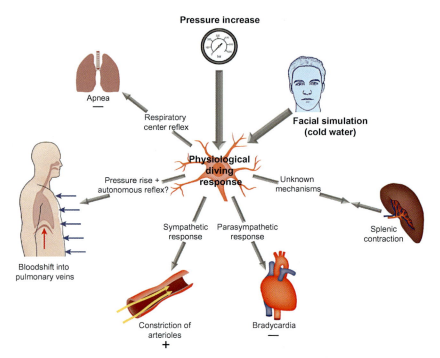

FIGURE 4.19   The basic function of the mammalian diving response. *Adapted and modified from Gooden [78].*

by Hurford et al. and Schagatay et al. during their human immersion experiments [86,87]. The effect was recently demonstrated in an MRI study with short respiratory manipulations of 30 s duration [88], although the effect itself was described a long time ago (Figure 4.19).

4.2.5 Humans in the Hyperbaric Environment

The scuba diving industry has exploded in popularity in the recent years and so has the market for diving gear. Due to the affordability and rather easy accessibility to water for recreational diving, the frontiers of the underwater world, a long forbidden environment for human beings, have now become immensely popular. Nonetheless, diving remains an activity in an extremely dangerous environment. In this extreme environment, the aforementioned gas laws cannot be underestimated. There are several specific hazards that make the hyperbaric environment different from an activity on dry land.

4.2.5.1 Sensory Systems

Vision: When submerging, our human optical sensor system, namely the eyes, is not equipped to adapt to the underwater environment. In the water, light is scattered by floating particles. Additionally, water has a 33% higher refraction coefficient than air. Therefore, proper sharp vision can only be obtained by wearing a diving mask. As a drawback, everything seen through this refraction aid appears about 33% larger and about 25% closer than reality. This can lead to wrong estimations of object sizes and distances. The third optical effect is wavelength-dependent light absorption, which causes a loss of color vision the deeper one dives [89].

Hearing: The aquatic environment, due to its higher density, transmits sound differently. Sound waves, which are transmitted about four times faster, can be perceived over miles. Man's stereo-acoustic sensory system, adapted to life on land, does not work properly under these circumstances. When submerged, it is almost impossible to determine which direction a sound or noise comes from [90,91].

Proprioception and labyrinth: The human proprioceptive sensory system is adapted to be surrounded by air. Dietz et al. described that the sensing of the position of limbs and body underwater is hardly possible without vision or feeling the buoyancy [92]. This handicap gets even worse in the case of losing the labyrinth function, located in the inner ear, due to barotrauma.

For more information about the physics of diving, please read *The Physics of Scuba Diving* by Anderson [93].

4.2.5.2 Apnea Diving in Humans

Apnea diving, or free diving, has evolved rapidly over the past 20 years. New materials and new breathing techniques led to records that were beyond anybody's imagination a few years ago. This is such a fast-developing area that all very recent information regarding this issue can be found on the website of the Association Internationale pour le Développement d'Apnée (AIDA, http://www.aidainternational.org). Table 4.2, which is taken from the AIDA website, records the different disciplines of free diving.

Of particular interest is the recent record of Herbert Nitsch, who reached a diving depth of 214 m. This feat leads to asking the question: Is there a physical frontier for human free divers? Some very optimistic authors postulate that diving depths up to 1000 m can be reached in the future [94]. The hypothesis [95] that diving depth might be dependent on the functional residual capacity is, at least for the trained apnea diver, non-applicable. The calculations as described by Klingmann andTetzlaff [95] are good for the understanding of the physiological and anatomical consequences of apnea diving. Calculating the maximal depth using this equation (Eq. 4.6), one will get a depth limit between 30 and 40 m. It is based on the simple argument that as soon as the total lung capacity has shrunk to the size of the residual volume due to water pressure (law of Boyle-Marriotte), the limit of reduction in lung size might have been reached. Eq. 4.6 DD, diving depth; ATA-1, atmosphere at sea level; TLC, total lung capacity; RC, residual capacity

$$DD(ATA\text{-}1) = \frac{TLC}{RC} \qquad (4.6)$$

TABLE 4.2 Records in apnea diving (2013). Modified according to Association Internationale pour le Développement d'Apnée (AIDA).

Type of record	♂	♀
Constant weight apnea without fins (CNF)	**101 m** Name: William TRUBRIDGE (NZL) Date: 2010-12-16 Place: Long Island, Bahamas	**68 m** Name: Natalia MOLCHANOVA (RUS) Date: 2013-04-25 Place: Blue Hole, Dahab
Constant weight apnea (CWT)	**126 m** Name: Alexey MOLCHANOV (RUS) Date: 2012-11-20 Place: Long Island, Bahamas	**101 m** Name: Natalia MOLCHANOVA (RUS) Date: 2011-09-23 Place: Kalamata, Greece
Dynamic apnea without fins (DNF)	**218 m** Name: Dave MULLINS (NZL) Date: 2010-09-27 Place: Naenae & Porirua, New Zealand	**182 m** Name: Natalia MOLCHANOVA (RUS) Date: 2013-06-27 Place: Belgrade, Serbia
Dynamic apnea (DYN)	**281 m** Name: Goran COLAK (CRO) Date: 2013-06-28 Place: Belgrade, Serbia	**234 m** Name: Natalia MOLCHANOVA (RUS) Date: 2013-06-28 Place: Belgrade, Serbia
Static apnea (STA)	**11 min 35 s** Name: Stéphane MIFSUD (FRA) Date: 2009-06-08 Place: Hyères, France	**9 min 02 s** Name: Natalia MOLCHANOVA (RUS) Date: 2013-06-29 Place: Belgrade, Serbia
Free immersion apnea (FIM)	**121 m** Name: William TRUBRIDGE (NZL) Date: 2011-04-10 Place: Long Island, Bahamas	**88 m** Name: Natalia MOLCHANOVA (RUS) Date: 2011-09-24 Place: Kalamata, Greece
Variable weight apnea (VWT)	**142 m** Name: Herbert NITSCH (AUT) Date: 2009-12-07 Place: Long Island, Bahamas	**127 m** Name: Natalia MOLCHANOVA (RUS) Date: 2012-06-06 Place: Sharm el-Sheikh, Egypt
No limits apnea (NLT)	**214 m** Name: Herbert NITSCH (AUT) Date: 2007-06-14 Place: Spetses, Greece	**160 m** Name: Tanya STREETER (USA) Date: 2002-08-17 Place: Turks & Caicos

Nevertheless, as soon as the rise of the surrounding pressure leads to compression of the lung parenchyma, different mechanisms effectively support the structural integrity of the lung tissue. These are the previously described blood shift, which replaces the volume of the compressed air in the lungs [96] as well as chest and diaphragmatic elasticity.

4.2.5.3   Scuba Diving

That we cannot breathe underwater is caused by the higher density of the medium and lower content of oxygen. Thus, we are forced to take our air atmosphere with us when we want to attempt to stay underwater for an extended time.

4.2.5.3.1   Equipment

Scuba diving really began as an industry in the 1960s and 1970s, after the development of the so-called "Aqua-lunge" in the 1940s, which was the first commercially available underwater breathing system. The renowned Jean-Jaques Cousteau (1910-1997) and his engineer Emile Gagnan (1900-1979) participated in the development of this regulator. The Aqua-lunge made it comfortable to breathe underwater, but it was still a very exhaustive and hazardous experience for the diver because it was not an effective buoyancy control device. The development of the Buoyancy Control Device (BCD) in the 1970s marked the second milestone in the rise of the diving industry. With a BCD, it was possible to hover free at a certain depth without uncontrolled sinking or ascending. It gave the diver greater control over the third dimension underwater without wasting too much energy.

Today scuba divers use a redundant two-stage regulator and different types of BCDs. Moreover, diving computers have been established and mostly replace diving tables. With a diving computer, it is possible to calculate the blood gas saturation of nitrogen and other used gases in real time and thereby determine the allowed duration of stay at a certain depth. In other words, they calculate the speed with which decompression leads to the formation of gas bubbles in the blood. To do this accurately, the different tissue compartments in the human body have to be taken into account in this calculation. One of the most popular models is the reduced gradient bubble model (RGBM) from Wienke [97], which was based on the algorithms of Haldane [72], Workmann [113], and Bühlmann [98,99].

4.2.5.3.2   Technical Diving

Technical diving means to max out the limits of normal scuba diving in matters of water depth, remote locations, exceeding decompression times, or water temperatures. All technical diving activities share a common characteristic: A direct ascent is not granted to the diver, because (i) diving depth requires a multiple-stage decompression (deep diving), and (ii) the diver enters an overhead environment, where a horizontal distance has to be covered to get back to the surface (e.g., wreck diving, ice diving, cave diving).

For these activities, it is necessary to practice the techniques and use specific technologies and devices to minimize the undoubtable higher risk.

4.2.5.3.3   Saturation Diving

Because the diver's body is only threatened by decompression sickness (DCS) when decompression actually happens, the technology of saturation diving has been developed. Initiated in the 1930s and 1940s [100], the Genesis project in 1958 led to the final establishment of this technology in order to carry out professional activities (research stations, drill rigs—offshore diving) in greater diving depths up to 500 m.

After their working day underwater, the divers retreat to a so-called *habitat*—a pressurized room where they stay, sleep, and eat until the next shift. This habitat contains the same pressure as the divers' aquatic working environment. Thus, during their stay in this habitat, their bodies' compartments (e.g., blood, muscle, fat, bone, brain) become completely saturated with the inspired inert gases, such as nitrogen, helium, and hydrogen. Therefore, it is absolutely impossible for a saturation diver to get back to the normobaric environment immediately. Several days of decompression have to be allowed so that they can return to the surface alive, although with the established decompression algorithms, acute impairments are rather seldom. However, middle and long-term damages concerning bone marrow have been observed [101,102].

4.2.5.3.4   Gases Used for Diving

To extend dive times or to prevent the effects of nitrogen narcosis, different breathing gases may be used.

4.2.5.3.4.1   Air

The most common gas mixture used for diving. It contains about 20% oxygen and 80% nitrogen. Because of the high nitrogen content, divers are prone to suffer nitrogen narcosis at a depth of approximately 30 m and deeper. There is also a risk of higher saturation of nitrogen in all tissues during one or more dives and thus a higher risk of DCS when used for long or repetitive dives. The maximal diving depth with normal air is 30-40 m taking into consideration the nitrogen effects, and 60 m concerning oxygen toxicity. For seizure prevention, a partial pressure of oxygen should not be higher than 1.4-1.6 bar [103].

4.2.5.3.4.2   Nitrox (also Enriched Air Nitrogen—EAN)

The second most used type of diving gas. It is also a gas mixture composed from nitrogen and oxygen and represents a variable mixture out of both. The most frequently used combinations are EAN32 and EAN36, with 32% and 36% oxygen respectively. The main advantage of using Nitrox is the reduced concentration of nitrogen in relation to the higher oxygen content. Therefore, longer non decompression diving times (NDC) are granted. However, due to the higher partial pressure of O_2, there is a higher risk of oxygen toxicity [103], so the maximal diving depth with EAN is reduced compared to normal air.

4.2.5.3.4.3   Trimix

Trimix is a combination of oxygen, nitrogen, and helium. The abbreviated form is Tx[oxygen fraction]/[helium fraction], for instance, Tx 21/50 for 21% oxygen, 50% helium, and 29% nitrogen.

Normally, Trimix is used at an oxygen concentration of 21% and below. If the concentration is higher than 21%, it is called Triox. Helium has a much lower narcotic potential than nitrogen, so it is suited for much deeper dives. By lowering the oxygen fraction, it is possible to prevent its toxicity effects. For example, a diver who dives at 100 m depth with Trimix (oxygen fraction 10%) has an oxygen partial pressure of only 1.1 bar (1.4 bar is maximally allowed). The main disadvantage of Trimix is its high cost.

Another risk, which does not derive from the mixture itself, but from the fact that it is often breathed at very high pressures, is the so-called high-pressure nervous syndrome (HPNS). HPNS is caused by breathing in compressed gas at depths exceeding 100 m. This massive concentration of inspired gas leads to synaptic disruptions in the central nervous system. Symptoms of HPNS can include headache, myoclonus, tremors, and in experimental animals, convulsions. These symptoms appear to disappear with ascension [104] (see below).

4.2.5.3.4.4   Heliox

Heliox is a mixture of helium and oxygen. It can be used for even deeper dives and is even more expensive than Trimix.

4.2.5.3.4.5   Hydrox

Another rarely used diving gas is hydrogen. It can be mixed with helium and certainly oxygen. Its narcotic potency is as low as helium, but it is even less dense than helium, which can be an advantage when reaching depths below 300 m. Mixed with helium, it is called **Hydreliox**. Using hydrogen for diving was first described by Zetterström in 1948 [105], who died while testing it. It was then reinvented by the U.S. Navy, who used it to reach depths between 500 and 700 m. The major disadvantage is that this gas mixture is highly explosive.

4.2.5.3.4.6   Argon

Argon, due to its high narcotic potency, is not used as a diving gas, but due to its density and thermal properties. it has been used to fill the dry suits [106], especially when used in very cold water.

4.2.6   Diseases Related to Exposure to the Hyperbaric Environment

4.2.6.1   Barotrauma

As described in the previous chapters, exposure of the human organ system to high pressure leads to physiological adaptations. The most apparent of these are the rising pressures in air-filled spaces in the human body, for example, the inner ear. As a consequence of rising pressures, these air-filled cavities can become compressed, which can lead to what is called *barotrauma*. A minor version of this trauma most frequently experienced by divers occurs when diving to the floor of a swimming pool. As the air within the middle ear is compressed, we feel a stinging pain on the tympanic membrane in the middle ear. In cases of a longer stay in an environment with elevated pressure, this may lead to ruptured eardrums and pain and inflammation for days. Thus, pressure equalization, that is, ear clearing, is the most important skill for any kind of diver, regardless of apnea or scuba.

4.2.6.2   Nitrogen Narcosis — Rapture of the Deep

In the past, the relationship of the lipophilic properties of gases and central nervous neuron membranes was held in doubt [107,108]. The unitarity principle, that is, that there is a unified mode of action for all narcotic gases, has been regarded as antiquated and has been discounted. However, there is still no comprehensive theory about the mechanism of action of anesthetic gases. Thus the theory of Meyer and Overton might not be fully abandoned [109].

Because of a rising concentration of nitrogen related to increasing depth, the narcotic effect rises as well because of its lipophilic properties in concordance with the Meyer-Overton law. This is the main reason why scuba diving with air should not be performed deeper than 30-40 m, because the effect of nitrogen narcosis normally begins around 30 m. This is also dependent on individual susceptibility and constitution.

4.2.6.3   Decompression Injury

The word "bend" derives from a fashion trend in the nineteenth century called "The Grecian Bend." Women, as pictured in Figure 4.20 bent their upper bodies forward. To support this look, additional fabric was stuffed around the lower back. During the construction of the Brooklyn Bridge, workers used to work inside so-called *caissons*, the French word for box or case. These were large watertight vessels that were pressurized to allow the workers to prepare and build the bridge foundation. After several hours of work under pressure, due to the *law of Henry*, the inert gases were dissolved in a liquid in proportion to the partial pressure of that gas. However, in modern diving medicine we differentiate stages of a decompression disease based on their clinical severity.

4.2.6.3.1   Decompression Injury (DCI)

As mentioned before, DCI develops by the formation of gas bubbles in the human tissue as a result of decompression after saturation of different types of tissue with inert gases. The severity of this disease depends on many factors, such as diving depth, length of stay, ascent speed, water temperature, and body composition [110]. The term *DCI* describes all types of decompression-related

FIGURE 4.20   The Grecian Bend. *Source: Wiki Commons.*

impairments, as well as DCS type I and II and arterial gas embolism (AGE). The differences in pathophysiology derive from the origin of the formed bubble and the anatomical location of damage. From a clinical point of view, we must first decide between DCS I and DCS II. DCS I can be described as the less severe of the two, characterized by involvement of more peripheral symptoms, such as itching of the skin, erythema, and joint aches (e.g., BENDS). DCS II comprises more severe impairments of the central nervous system, inner ear or labyrinth, heart, lungs, or other organs. However, this classification suggests that in cases of DCS, bubble formation would be locally confined, which might lead to an underestimation of the severity of symptoms.

To decide on possible therapy options, for example, immediate recompression or only 100% oxygen, a clinician can use the diagnostic criteria as shown in *Modern Diving Medicine* by Klingmann and Tetzlaff [111] to make more precise descriptions about the circumstances of the symptoms (Table 4.3). In this kind of DCI classification, the course and dimension of an injury can be better assessed. However, it does not give one a reliable clinical prediction.

TABLE 4.3 The main five diagnostic criteria for initiation of recompression therapy modified after Klingmann and Tetzlaff (Quellle). (Adapted from [111])

Criterion	Symptom(s) or Context
Latency after dive	Fulminant or delayed
Dynamics	Remission, stabile, aggravation
Compromised organs	Central of peripheral nervous system, cardiopulmonary, dermatologic, musculoskeletal system, unspecific
Diving profile	Ascent speed, dive time, maximum diving depth
Additional impairment	E.g., barotrauma

4.2.6.3.2   Open Foramen Ovale

The open foramen ovale (Botalli) persists in about 30% of the general population. This phenomenon is the most frequent reason for paradox embolism. In cases of DCI, gas bubbles formed in the venous system and travelled across the oval foramen from the right to the left side of the heart. This allows the air finally to reach the cerebral arteries and cause a cerebral insult.

4.2.6.3.3   Cerebral Arterial Gas Embolism (CAGE)

Although symptoms of a DCI can also mimic cerebral gas embolism (CAGE), there are fundamental differences regarding the etiology. AGE happens after a barotrauma of the lungs, when alveolar air can get into the pulmonary veins. This embolism travels to the left side of the heart and onward to the carotid arteries, which perfuse the brain. As a consequence, an occlusion of the medial cerebral artery occurs. The etiology of CAGE and a cerebral embolism via the foramen Botalli are completely different. Nevertheless, the symptoms and consequences are similar and so is their therapy by hyperbaric oxygen therapy (HBO).

4.2.6.4   Oxygen Toxicity

Oxygen toxicity plays a role in when the partial pressure of the inspired gas mixture exceeds 1.4/1.6 bar. It is hypothesized that because of its polar properties, oxygen reduces the central nervous seizure threshold. Therefore, at partial pressures beyond this level a seizure due to hyper-polarization of nervous membranes becomes more probable. Under normal circumstances, where scuba diving is practiced at depths of up to 40 m, there is no danger of oxygen toxicity, as long as the diver is using normal air (fraction of oxygen = 0.21).

4.2.6.5 High-Pressure Nervous Syndrome (HPNS)

HPNS describes a pathophysiological effect caused by rapidly increasing barometric pressure. Symptoms comprise dizziness, shivering/tremor, vertigo, nausea, and vomiting. The probability of experiencing HPNS occurs when diving deeper than 120 m. It was first thought to be caused by helium, which is added quite often in those depths to replace oxygen and nitrogen respectively [112]. However, HPNS was found to be caused by the high pressure itself, and the speed of compression also plays a role. To modify and prevent HPNS there are two strategies: (i) reduction of the descending speed and (ii) adding a specific amount of an anesthetic gas to the breathing mixture. Adding a small amount of nitrogen (sometimes also argon) to the breathed mixture can diminish the symptoms or even prevent them. It is believed that this narcotic gas effect impairs the high-pressure effect on the central nervous system.

4.2.7 Treatment

HBO is the therapy of choice regarding DCS and CAGE respectively. A typical setup is shown Figure 4.21. As mentioned above, the recompression scheme by Haldane was one of the first methods employed. However, it had already been described in the nineteenth century in workers during tunnel constructions under pressure. In the 1930s and 1940s the U.S. Navy developed a standardized set of tables with a maximal diving depth of 50 m (6 ATA = 600 kPa). This was modified by Goodman and Workman, who reduced the diving depth to 18 m and replaced the breathed air by 100% oxygen. By doing this, it was possible

FIGURE 4.21 Hyperbaric chamber at the HBO-Center on Sansibar; suited for one patient.

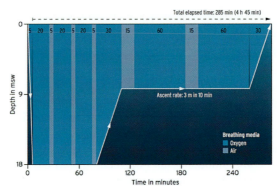

FIGURE 4.22 Hyperbaric therapy using a modified scheme of Navy Table 6 for the treatment of severe symptoms after diving. Patients breathe oxygen and in between normal air to lower the risk of toxic side effects. Table 6 is used for DCS II symptoms.

to remove the inert gas faster and thereby reduce the bubble size in the tissue compartments [113]. A typical decompression scheme for severe symptoms of decompression, DCS2, is given in Figure 4.22.

REFERENCES

[1] Hultgren HN. High altitude medicine. Hultgren Publications Stanford; 1997.
[2] Ward M, Milledge JS, West JB. High altitude medicine and physiology. London: Chapman and Hall Medical; 1989.
[3] Auerbach PS. Wilderness medicine [Internet]. Wilderness medicine, http://www.clinicalkey. com/dura/browse/bookChapter/3-s2.0-C2009039724X; 2012.
[4] Bert P. La pression barométrique: recherches de physiologie expérimentale. Masson; 1878.
[5] Heller R, von Schrötter H, Mager W. Luftdruck-Erkrankungen mit besonderer Berücksichtigung der sogenannten Caissonkrankheit. 1900.
[6] Zuntz N, Loewy A, Müller F, Caspari W. Höhenklima und Bergwanderungen in ihrer Wirkung auf den Menschen. Berlin, Leipzig, Wien, Stuttgart. Deutsches Verlagshaus Bong.; 1906.
[7] Loewy A, Mörikofer W. Physiologie des Höhenklimas. J. Springer; 1932.
[8] West JB. High life: a history of high-altitude physiology and medicine. New York: Oxford University Press; 1998.
[9] Gunga H-C. Nathan Zuntz: his life and work in the fields of high altitude physiology and aviation medicine. Burlington, London. Elsevier; 2009.
[10] Hölder H. Geologie und Paläontologie in Texten und ihrer Geschichte. Freiburg, München. Verlag Karl Alber; 1960.
[11] West JB. American medical research expedition to Everest. High Alt Med Biol 2010;11(2): 103–10.
[12] Grocott MP, Martin DS, Wilson MH, Mitchell K, Dhillon S, Mythen MG, et al. Caudwell xtreme Everest expedition. High Alt Med Biol 2010;11(2):133–7.
[13] Hackett PH, Roach RC. High-altitude medicine. In: Wilderness medicine. St. Louis, MO: Mosby; 1995. p. 1–37.

[14] Hochachka PW, Gunga HC, Kirsch K. Our ancestral physiological phenotype: an adaptation for hypoxia tolerance and for endurance performance? Proc Natl Acad Sci 1998;95(4):1915–20.

[15] Moore LG. Altitude-aggravated illness: examples from pregnancy and prenatal life. Ann Emergency Med 1987;16(9):965–73.

[16] Burtscher M, Bachmann O, Hatzl T, Hotter B, Likar R, Philadelphy M, et al. Cardiopulmonary and metabolic responses in healthy elderly humans during a 1-week hiking programme at high altitude. Eur J Appl Physiol 2001;84(5):379–86.

[17] Alexandris D, Varotsos C, Ya Kondratyev K, Chronopoulos G. On the altitude dependence of solar effective UV. Phys Chem Earth Part C: Sol Terr Planet Sci 1999;24(5):515–7.

[18] Pandolf KB, Sawka MN, Gonzalez RR. Human performance physiology and environmental medicine at terrestrial extremes. Brown & Benchmark; 1988.

[19] Jelkmann W. Regulation of erythropoietin production. J Physiol 2011, March 15;589(Pt 6): 1251–8.

[20] Eckardt KU, Boutellier U, Kurtz A, Schopen M, Koller EA, Bauer C. Rate of erythropoietin formation in humans in response to acute hypobaric hypoxia. J Appl Physiol (1985) 1989, April;66(4):1785–8.

[21] Gunga HC, Kirsch K, Röcker L, Schobersberger W. Time course of erythropoietin, triiodothyronine, thyroxine, and thyroid-stimulating hormone at 2,315 m. J Appl Physiol 1994, March;76(3):1068–72.

[22] Schobersberger W, Schmid P, Lechleitner M, von Duvillard SP, Hörtnagl H, Gunga HC, et al. Austrian Moderate Altitude Study 2000 (AMAS 2000). The effects of moderate altitude (1,700 m) on cardiovascular and metabolic variables in patients with metabolic syndrome. Eur J Appl Physiol 2003, February;88(6):506–14.

[23] Greie S, Humpeler E, Gunga HC, Koralewski E, Klingler A, Mittermayr M, et al. Improvement of metabolic syndrome markers through altitude specific hiking vacations. J Endocrinol Invest 2006, June;29(6):497–504.

[24] Schobersberger W, Greie S, Humpeler E, Mittermayr M, Fries D, Schobersberger B, et al. Austrian Moderate Altitude Study (AMAS 2000): erythropoietic activity and Hb-O(2) affinity during a 3-week hiking holiday at moderate altitude in persons with metabolic syndrome. High Alt Med Biol 2005;6(2):167–77.

[25] Gore CJ, Clark SA, Saunders PU. Nonhematological mechanisms of improved sea-level performance after hypoxic exposure. Med Sci Sports Exerc 2007, September;39(9):1600–9.

[26] Burtscher M. Intermittierende hypoxie: hohenvorbereitung, training, therapie. Schweizerische Zeitschrift für Sportmedizin und Sporttraumatologie 2005;53(2):61.

[27] Bonetti DL, Hopkins WG. Sea-level exercise performance following adaptation to hypoxia. Sports Med 2009;39(2):107–27.

[28] Buskirk ER, Kollias J, Akers RF, Prokop EK, Reategui EP. Maximal performance at altitude and on return from altitude in conditioned runners. J Appl Physiol 1967;23(2):259–66.

[29] Péronnet F, Thibault G, Cousineau DL. A theoretical analysis of the effect of altitude on running performance. J Appl Physiol (1985) 1991, January;70(1):399–404.

[30] Lundby C, Millet GP, Calbet JA, Bärtsch P, Subudhi AW. Does 'altitude training' increase exercise performance in elite athletes? Br J Sports Med 2012;46(11):792–5.

[31] Jalil J, Braun S, Chamorro G, Casanegra P, Saldías F, Beroíza T, et al. Cardiovascular response to exercise at high altitude in workers chronically exposed to intermittent hypobaric hypoxia. Rev Med Chil 1994, October;122(10):1120–5.

[32] Gunga HC, Röcker L, Behn C, Hildebrandt W, Koralewski E, Rich I, et al. Shift working in the Chilean Andes (>3,600 m) and its influence on erythropoietin and the low-pressure system. J Appl Physiol (1985) 1996, August;81(2):846–52.

[33] Richalet JP, Donoso MV, Jiménez D, Antezana AM, Hudson C, Cortès G, et al. Chilean miners commuting from sea level to 4500 m: a prospective study. High Alt Med Biol 2002;3(2):159–66.

[34] Vargas M, Jiménez D, León-Velarde F, Osorio J, Mortola JP. Circadian patterns in men acclimatized to intermittent hypoxia. Respir Physiol 2001, July;126(3):233–43.

[35] Coste O, Beaumont M, Batéjat D, Beers PV, Touitou Y. Prolonged mild hypoxia modifies human circadian core body temperature and may be associated with sleep disturbances. Chronobiol Int 2004, May;21(3):419–33.

[36] Barger LK, Lockley SW, Rajaratnam SM, Landrigan CP. Neurobehavioral, health, and safety consequences associated with shift work in safety-sensitive professions. Curr Neurol Neurosci Rep 2009, March;9(2):155–64.

[37] Brown RE, Basheer R, McKenna JT, Strecker RE, McCarley RW. Control of sleep and wakefulness. Physiol Rev 2012, July;92(3):1087–187.

[38] Czeisler CA, Moore-Ede MC, Coleman RH. Rotating shift work schedules that disrupt sleep are improved by applying circadian principles. Science 1982, July 30;217(4558):460–3.

[39] Folkard S, Tucker P. Shift work, safety and productivity. Occup Med (Lond) 2003, March;53(2):95–101.

[40] Hoshikawa M, Uchida S, Sugo T, Kumai Y, Hanai Y, Kawahara T. Changes in sleep quality of athletes under normobaric hypoxia equivalent to 2,000-m altitude: a polysomnographic study. J Appl Physiol (1985) 2007, December;103(6):2005–11.

[41] Zulley J, Knab B. Unsere innere Uhr. Herder; 2000.

[42] Chilov D, Hofer T, Bauer C, Wenger RH, Gassmann M. Hypoxia affects expression of circadian genes PER1 and CLOCK in mouse brain. FASEB J 2001, December;15(14):2613–22.

[43] Viola AU, James LM, Archer SN, Dijk DJ. PER3 polymorphism and cardiac autonomic control: effects of sleep debt and circadian phase. Am J Physiol Heart Circ Physiol 2008, November;295(5):H2156–63.

[44] Ashkenazi IE, Ribak J, Avgar DM, Klepfish A. Altitude and hypoxia as phase shift inducers. Aviat Space Environ Med 1982, April;53(4):342–6.

[45] Copinschi G, Spiegel K, Leproult R, Van Cauter E. Pathophysiology of human circadian rhythms. In Mechanisms and biological significance of pulsatile hormone secretion: Novartis Foundation symposium 227; 2000. p.143–62.

[46] Bosco G, Ionadi A, Panico S, Faralli F, Gagliardi R, Data P, et al. Effects of hypoxia on the circadian patterns in men. High Alt Med Biol 2003;4(3):305–18.

[47] Mortola JP. Hypoxia and circadian patterns. Respir Physiol Neurobiol 2007, September 30;158(2–3):274–9.

[48] Hastings MH, Reddy AB, Garabette M, King VM, Chahad-Ehlers S, O'Brien J, et al. Expression of clock gene products in the suprachiasmatic nucleus in relation to circadian behaviour. In Novartis Foundation symposium 253; 2003. p. 203–17. Discussion 102–9, 218–22, 281–4.

[49] Vazir A, Hastings PC, Dayer M, McIntyre HF, Henein MY, Poole-Wilson PA, et al. A high prevalence of sleep disordered breathing in men with mild symptomatic chronic heart failure due to left ventricular systolic dysfunction. Eur J Heart Fail 2007, March;9(3):243–50.

[50] Behn C, Araneda OF, Llanos AJ, Celedón G, González G. Hypoxia-related lipid peroxidation: evidences, implications and approaches. Respir Physiol Neurobiol 2007, September 30;158(2–3):143–50.

[51] De Bacquer D, Van Risseghem M, Clays E, Kittel F, De Backer G, Braeckman L. Rotating shift work and the metabolic syndrome: a prospective study. Int J Epidemiol 2009, June;38(3):848–54.

[52] Eckel-Mahan KL, Patel VR, de Mateo S, Orozco-Solis R, Ceglia NJ, Sahar S, et al. Reprogramming of the circadian clock by nutritional challenge. Cell 2013, December 19;155(7):1464–78.

[53] Prommer N, Heinicke K, Viola T, Cajigal J, Behn C, Schmidt WF. Long-term intermittent hypoxia increases O_2-transport capacity but not VO_{2max}. High Alt Med Biol 2007;8(3):225–35.

[54] Coste O, Beers PV, Bogdan A, Charbuy H, Touitou Y. Hypoxic alterations of cortisol circadian rhythm in man after simulation of a long duration flight. Steroids 2005, November;70(12):803–10.

[55] Dadoun F, Darmon P, Achard V, Boullu-Ciocca S, Philip-Joet F, Alessi MC, et al. Effect of sleep apnea syndrome on the circadian profile of cortisol in obese men. Am J Physiol Endocrinol Metab 2007, August;293(2):E466–74.

[56] Kwarecki K, Krawczyk J. Comparison of the circadian rhythm in cell proliferation in corneal epithelium of male rats studied under normal and hypobaric (hypoxic) conditions. Chronobiol Int 1989;6(3):217–22.

[57] Filipski E, Delaunay F, King VM, Wu MW, Claustrat B, Gréchez-Cassiau A, et al. Effects of chronic jet lag on tumor progression in mice. Cancer Res 2004, November 1;64(21):7879–85.

[58] León-Velarde F, Maggiorini M, Reeves JT, Aldashev A, Asmus I, Bernardi L, et al. Consensus statement on chronic and subacute high altitude diseases. High Altitude Med Biol 2005;6(2):147–57.

[59] Bärtsch P, Saltin B. General introduction to altitude adaptation and mountain sickness. Scand J Med Sci Sports 2008;18(s1):1–10.

[60] Netzer N, Strohl K, Faulhaber M, Gatterer H, Burtscher M. Hypoxia-related altitude illnesses. J Travel Med 2013;20(4):247–55.

[61] Burgess KR, Johnson P, Edwards N, Cooper J. Acute mountain sickness is associated with sleep desaturation at high altitude. Respirology 2004, November;9(4):485–92.

[62] Richalet JP, Larmignat P, Poitrine E, Letournel M, Canoüi-Poitrine F. Physiological risk factors for severe high-altitude illness: a prospective cohort study. Am J Respir Crit Care Med 2012, January 15;185(2):192–8.

[63] Wu TY, Ding SQ, Liu JL, Yu MT, Jia JH, Chai ZC, et al. Who should not go high: chronic disease and work at altitude during construction of the Qinghai-Tibet railroad. High Alt Med Biol 2007;8(2):88–107.

[64] Hackett PH, Rennie D, Levine HD. The incidence, importance, and prophylaxis of acute mountain sickness. Lancet 1976, November 27;2(7996):1149–55.

[65] Sanchez del Rio M, Moskowitz MA. High altitude headache. Lessons from headaches at sea level. Adv Exp Med Biol 1999;474:145–53.

[66] Wilson MH, Newman S, Imray CH. The cerebral effects of ascent to high altitudes. Lancet Neurol 2009, February;8(2):175–91.

[67] Coote JH. Medicine and mechanisms in altitude sickness. Recommend Sports Med 1995, September;20(3):148–59.

[68] Sutton JR, Coates G, Houston CS. Hypoxia and mountain medicine: proceedings of the 7th international hypoxia symposium held at Lake Louise, Canada, February 1991. Pergamon; 1992.

[69] Cheung SS. Advanced environmental exercise physiology. Human Kinetics; 2010.

[70] Acott C. A brief history of diving and decompression illness. SPUMS J 1999;29:98–109.

[71] Brubakk AO, Lang MA. JS Haldane, the first environmental physiologist. In Proceedings and Other Publications. 2009.

[72] Haldane JS, Boycott A, Damant G. The prevention of compressed air illness. J Hyg 1908;8:342–443.

[73] Clapeyron E. Mémoire sur la puissance motrice de la chaleur. J. Gabay; 1834.

[74] Henry W. Experiments on the quantity of gases absorbed by water, at different temperatures, and under different pressures. Philos Trans R Soc London 1803;93:29–276.

[75] Uhen MD. Evolution of marine mammals: back to the sea after 300 million years. Anat Rec (Hoboken) 2007, June;290(6):514–22.

[76] Blix AS, Gautvik EL, Refsum H. Aspects of the relative roles of peripheral vasoconstriction and vagal bradycardia in the establishment of the "diving reflex" in ducks. Acta Physiol Scand 1974;90(2):289–96.

[77] Noren SR, Kendall T, Cuccurullo V, Williams TM. The dive response redefined: underwater behavior influences cardiac variability in freely diving dolphins. J Exp Biol 2012, August 15;215(Pt 16):2735–41.

[78] Gooden BA. The diving response in clinical medicine. Aviat Space Environ Med 1982, March;53(3):273–6.

[79] Wein J, Andersson JP, Erdéus J. Cardiac and ventilatory responses to apneic exercise. Eur J Appl Physiol 2007, August;100(6):637–44.

[80] Schagatay E, Haughey H, Reimers J. Speed of spleen volume changes evoked by serial apneas. Eur J Appl Physiol 2005, January;93(4):447–52.

[81] Pedroso FS, Riesgo RS, Gatiboni T, Rotta NT. The diving reflex in healthy infants in the first year of life. J Child Neurol 2012, February;27(2):168–71.

[82] Risch WD, Koubenec H-J, Beckmann U, Lange S, Gauer OH. The effect of graded immersion on heart volume, central venous pressure, pulmonary blood distribution, and heart rate in man. Pflugers Arch 1978;374(2):115–8.

[83] Smith G, Morgans A, Taylor DM, Cameron P. Use of the human dive reflex for the management of supraventricular tachycardia: a review of the literature. Emerg Med J 2012, August;29(8):611–6.

[84] Gooden BA. Mechanism of the human diving response. Integr Physiol Behav Sci 1994;29(1):6–16.

[85] Cabanac A, Folkow LP, Blix AS. Volume capacity and contraction control of the seal spleen. J Appl Physiol 1997, June;82(6):1989–94.

[86] Hurford WE, Hong SK, Park YS, Ahn DW, Shiraki K, Mohri M, et al. Splenic contraction during breath-hold diving in the Korean ama. J Appl Physiol 1990, September;69(3):932–6.

[87] Schagatay E, Andersson JP, Nielsen B. Hematological response and diving response during apnea and apnea with face immersion. Eur J Appl Physiol 2007, September;101(1):125–32.

[88] Inoue Y, Nakajima A, Mizukami S, Hata H. Effect of breath holding on spleen volume measured by magnetic resonance imaging. PLoS One 2013;8(6):e68670.

[89] Jerlov NG. Marine optics. In: Elsevier oceanography series, Elsevier; 1976.

[90] Etter PC. Underwater acoustic modeling and simulation. CRC Press; 2013.

[91] Tolstoy I, Clay CS. Ocean acoustics: theory and experiment in underwater sound. Acoustical Society of America; 1987.

[92] Dietz D, Horstmann H, Trippel T, Gollhofer G. Human postural reflexes and gravity—an under water simulation [Internet]. Neurosci Lett 1989;106(3):350–5, http://www.sciencedirect.com/science/article/pii/0304394089901894.

[93] Anderson M. The physics of scuba diving. Nottingham University Press; 2011.

[94] Ostrowski A, Strzała M, Stanula A, Juszkiewicz M, Pilch W, Maszczyk A. The role of training in the development of adaptive mechanisms in freedivers. J Hum Kinet 2012, May;32:197–210.

[95] Klingmann C, Tetzlaff K. Moderne Tauchmedizin. Gentner Verlag; 2007.

[96] Rowell LB, Murray JA, Brengelmann GL, Kraning KK. Human cardiovascular adjustments to rapid changes in skin temperature during exercise. Circulat Res 1969;24(5):711–24.

[97] Wienke W. Reduced gradient bubble model [Internet]. Int J Bio-Med Comput 1990;26(4): 237–56, http://www.sciencedirect.com/science/article/pii/002071019090048Y.

[98] Bühlmann AA. Experimental principles of risk-free decompression following hyperbaric exposure. 20 years of applied decompression research in Zurich. Schweiz Med Wochenschr 1982;112(2):48.

[99] Bühlmann AA. Decompression: Decompression Sickness. Berlin: Springer-Verlag; 1984.

[100] Van Der Aue OE, White Jr. WA, Hayter R, Brinton ES, Kellar RJ. Physiologic factors underlying the prevention and treatment of decompression sickness. 1945.

[101] Lehner CE, Adams WM, Dubielzig RR, Palta M, Lanphier EH. Dysbaric osteonecrosis in divers and caisson workers. An animal model. Clin Orthop Relat Res 1997, November; 344:320–32.

[102] Davidson JK. Dysbaric disorders: aseptic bone necrosis in tunnel workers and divers. Baillieres Clin Rheumatol 1989, April;3(1):1–23.

[103] Farmery S, Sykes O. Neurological oxygen toxicity. Emerg Med J 2012, October;29(10):851–2.

[104] Jain KK. High-pressure neurological syndrome (HPNS). Acta Neurol Scand 1994, July;90(1):45–50.

[105] Zetterstrom A. Deep-sea diving with synthetic gas mixtures. Military Surgeon 1948;103(2):104.

[106] Nuckols ML, Giblo J, Wood-Putnam JL. Thermal characteristics of diving garments when using argon as a suit inflation gas. IEEE; 2008.

[107] Meyer H. On the theory of alcohol narcosis: first communication. Which property of anesthetics determines its narcotic effect? Arch Exp Pathol Pharmakol 1899;425:109–18.

[108] Overton E. Studien über die Narkose. G. Fischer; 1901.

[109] Missner A, Pohl P. 110 years of the Meyer-Overton rule: predicting membrane permeability of gases and other small compounds. Chemphyschem 2009, July 13;10(9–10):1405–14.

[110] Wienke BR. Basic decompression: theory and application. Best Publishing Company; 2008.

[111] Klingmann C, Tetzlaff K. Moderne Tauchmedizin. Dtsch Z Sportmed 2012;63(3).

[112] Hunger Jr WL, Bennett PB. The causes, mechanisms and prevention of the high pressure nervous syndrome; 1974.

[113] Workman RD. Calculation of decompression schedules for nitrogen-oxygen and helium-oxygen dives; 1965.

[114] Hackett P, Roach R. High-altitude medicine. In: Auerbach P, editor. Wilderness medicine. Philadelphia: Mosby-Elsevier; 2007. p. 2–36.

Chapter 5

Desert and Tropical Environment

Hanns-Christian Gunga

Professor, Center for Space Medicine and Extreme Environments, Institute of Physiology, CharitéCrossOver (CCO), Charité University Medicine Berlin, Berlin, Germany

5.1   INTRODUCTION

When thinking of a hot environment, one usually starts with the impressions of the vast, sandy desert areas around the world. One thinks of the imminent danger of dehydration and death due to lack of potable water.

However, in case of the high altitude deserts in Chile (Atacama) or central Antarctica, extremely cold temperatures can prevail, especially during winter and night-time. From a meteorological-geographical perspective, deserts can be differentiated as extremely arid, semi-arid, and arid areas. Thus, not only the ambient temperature but also the amount of precipitation must be taken into account (Figure 5.1).

From a physiological point of view and in respect to the availability of freshwater, the largest deserts for humans are the oceans. A shipwrecked human, although immersed and surrounded by a gigantic mass of water, cannot drink a drop of it because it would vastly increase dehydration due to its high salt content (3.5%) [1]. Another general misconception—at least for those people who grew up in the moderate latitude of this planet such as the Caucasians—is that living in the tropics is often taken as a synonym for an easy life style and a place for vacations. Due to the permanent (12 months) mean high ambient temperatures (18 °C and higher) combined with an average precipitation of at least 60 mm of rain (high relative humidity), and less air movement [2], the tropical rainforest environment is a very stressful climate for nonadapted humans [3–5]. From a physiological point of view, the cardiovascular system is permanently stressed, day and night. Therefore, it is not surprising that the tropical climate is also called the "white man's grave."

Because thermoregulation in humans is quite complex, a basic physical understanding of heat transfer, that is, convection, conduction, radiation, and evaporative heat loss, is needed, as well as knowledge about the multiple autonomic pathways involved in maintaining core body temperature.

Human Physiology in Extreme Environments. http://dx.doi.org/10.1016/B978-0-12-386947-0.00005-8

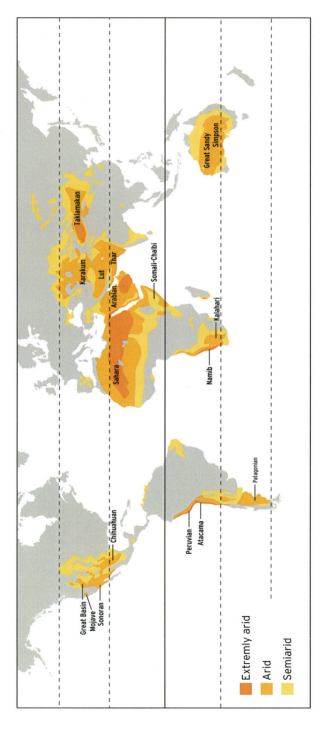

FIGURE 5.1 The main great deserts around the world, with exception of the polar deserts.

5.2 THERMAL BALANCE

5.2.1 Ectothermic and Endothermic Metabolism

Humans are endothermic organisms. This means that in contrast to the ectothermic (poikilothermic) animals such as fishes and reptiles, humans are less dependent on the external environmental temperature [6,7]. Endothermic organisms have much higher basal energy consumption, which is mainly necessary to keep their body temperature constant within a wide range of different environmental temperatures. In the body core (cranial, thoracic, abdominal cavities), the human body temperature is around 37 °C. In the periphery (extremities), it is lower and exhibits regional differences (28-36 °C) (Figure 5.2). Under ambient environmental conditions, the core body temperature of 37 °C is maintained by

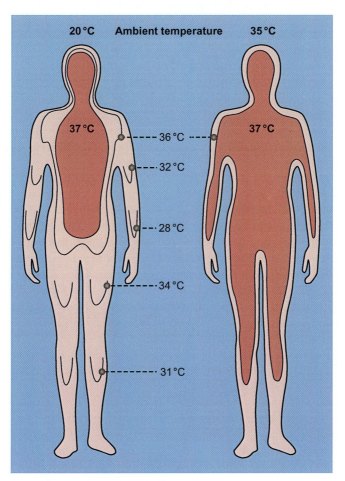

FIGURE 5.2 Body core and body shell temperatures under different ambient temperatures in a cold (20 °C, left panel) and a hot environment (35 °C, right panel). *Adapted from [94].*

the permanent metabolic active internal organs such as brain, heart, liver, and gastrointestinal tract through a fine-tuned thermoregulatory system that mainly adjusts peripheral perfusion of the skin and evaporation by the sweat glands to the thermal needs of the body [6,8]. Heat loss, under resting conditions, occurs mainly via radiation. In a warm/hot environment and under strenuous exercise, the organism depends on the evaporative pathway (Figure 5.2).

However, the core body temperature is not consistently regular. In the course of the day, the core body temperature shows cyclic changes (circadian rhythm), and in women a monthly pattern occurs due to the menstrual cycle. In ecto-thermic organisms (amphibia, reptiles, fishes), the temperature gradient compared to the environment is low (<5 °C). Their body temperature, and thus their activity, depends largely on the prevailing external environmental conditions (thermo-conform). Thus, these organisms remain viable (eurytherm) over a wide temperature range and, due to a low metabolism (bradymetabolism), they can overcome longer phases of food shortage. The temperature field of humans can be subdivided into a body shell and a body core (Figure 5.2) [6,8,9]. A core body temperature of more than 37.5 °C is defined as hyperthermia; below 35.5 °C is known as hypothermia. By definition, the parts of the body where the tissue temperatures exhibit predominantly 37 °C are considered the body core (homeothermic). Particularly in the metabolically active organs such as the heart, brain, and liver, heat is produced permanently. In the limbs and skin, mean tissue temperature can vary extensively (poikilothermic body shell). The tissue temperature of the body shell decreases with increasing distance from the body core, and the shell crowds around it like onion layers (isotherms). The body shell functions on the one hand as heat insulation of the body core. On the other hand, the heat exchange with the environment takes place at the surface. The relation of body core to the shell is not kept constant. If heat is to be emitted, the body core expands to the skin level, therefore facilitating heat transfer via the enlarged subcutaneous vascular bed. In a cold environment, heat loss has to be reduced to prevent hypothermia. Thus, the body shell has to be enlarged via vasoconstriction, which leads to a better insulation of the core (Figure 5.2). Humans require a constantly high core body temperature (homiothermic = equally warm) between 36.4 and 37.4 °C. Variations of the core body temperature are tolerated only in a very small range (stenothermic). In order to assure this constant core body temperature in a cold environment, insulating layers are necessary to reduce heat losses [8]. Furthermore, a correspondingly high heat production via endothermia and tachymetabolism is required. This causes the metabolic rate of endothermic organisms to be three to four times higher than that of ectothermic organisms. Fluctuations in air temperature, relative air humidity, and airflow, as well as internal heat production require an effective system of thermoregulation. Through these mechanisms, endothermic organisms maintain a high temperature gradient compared to environmental temperatures. This enables them to be more active than ectothermic organisms at the cost of a higher energy consumption [8,10].

5.2.2   Neutral Temperature Zone

The neutral or indifferent temperature zone is defined as a range of environmental temperatures in which a naked, resting person can keep an even heat balance through changes in skin blood flow. It is standardized to basal metabolic conditions, that is, a healthy, unclothed, resting adult at a relative air humidity of 50% minimal air movement and a temperature of approximately 27-31 °C (Figure 5.3) [10].

Above an ambient temperature of 31 °C, heat dissipation is mainly regulated by evaporation, whereas the contributions of radiation, conductance, and convection decrease. At an air temperature >37 °C, the body actually gains heat from the environment (reversal heat flux) [8]. As shown in Figure 5.3, the indifference temperature for an unclothed, resting human under atmospheric conditions lies around 27-31 °C. In water, however, the indifferent temperature is defined as an even smaller range (34.5-35.5 °C). This is caused by the special physical properties of water: Water has a thermal conductivity 25 times higher and a specific heat capacity 4000 times higher than air [6,8]. In contrast to air, heat transfer for convection and conductance is increased in water. Thus, submerging in water temperatures below 25 °C, heat losses cannot be compensated for by means of internal heat production [6]. Once hypothermia begins, the viscosity of blood and musculature activity increase. In addition, the heat

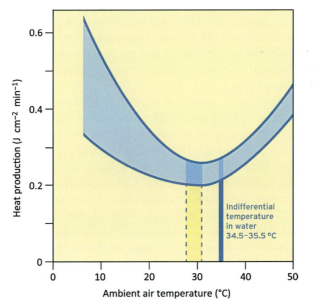

FIGURE 5.3   The thermal neutral or indifferent zone in a resting human under atmospheric conditions (scattered range) and the changes in metabolic heat production. As indicated, in water (solid line) the thermal neutral zone is much higher and very narrow. For further information, see text. *Adapted from Ref. [10].*

loss coefficient is approximately tripled with active swimming movements. The velocity in which nerve signals are conducted are reduced. A falling blood temperature shifts the oxygen binding curve to the left, thus improving the uptake of oxygen in the lungs, but the delivery of oxygen to the tissue, however, is thereby strongly diminished. On the other hand, hot water immersion leads to a rapid increase in core body temperature.

5.3   HEAT PRODUCTION

5.3.1   Metabolic Heat Production and Metabolic Rate

Living cells absorb high-energy nutrients, metabolize them, and finally excrete substances of lower energy [10]. The process of transformation of the energy from nutrients into internal energy forms and their utilization is described as basal energy rate. Two terms have to be distinguished in this process: (i) metabolic heat production=transformation of chemical energy into heat, and (ii) metabolic rate=rate of transformation of chemical energy into heat and mechanical work. Under resting conditions, the metabolic rate corresponds to the metabolic heat production. The chemical energy of the ingested food products can thus be utilized for heat production, muscle activity, and for the synthesis of adenosintriphosphate (for further details see Chapter 3 and Ref. [10]). Because the efficiency factor of mechanical muscle work lies only around 20-30%, approximately 70-80% of the chemical binding energy of the nutrients burned for external work accumulates as additional heat. This strongly influences the thermal comfort of the subject [8]. In order to remain within thermal balance, heat loss also has to be increased. The higher the basal energy rate and insulation value of the clothing, the lower is the ambient temperature of thermal comfort. For the basal energy rate during different activities, the unit MET (metabolism) is used: 1 MET designates the energy rate during sitting activity, approximately 300 mL of O_2 consumption, equivalent to an energy need of 400 kJ/h. At the same time, the influence of air movement and the kind of clothing have to be taken into account, because the boundary air layer lying around the body (microclimate) contributes to the heat insulation. An insulation value of 1 clo (deduced from the English word "clothing") corresponds to $0.155 \, m^2 \, K/W$. One might use as a rough approximation: thermal insulation (clo)=0.08 × total number of layers +0.51 [5]. This insulation value quickly decreases with higher wind velocities. This is mainly because the boundary layer is thinner and the laminar heat convection along the body axis passes into a turbulent current [6].

5.3.2   Thermal Balance under Different Physical and Environmental Settings

Approximately 80% of the heat within the internal organs is produced at physical rest, whereas the other parts of the body contribute only 20%. Under resting conditions, the organs with high heat production (60-140 J $100 \, g^{-1} min^{-1}$) are situated exclusively in the cranial, thoracic, and abdominal cavities, thus within the body core. The body shell, on the other hand, contributes only slightly to heat production. With physical work, heat dissipation changes fundamentally. Up to 90% of the entire heat production can then be ascribed to the working musculature, and thus the tissue temperature in muscles can be distinctly above core body temperature. Because subcutaneous fat in limb muscles is rather sparse, and the tissue is well supplied with blood, heat can be dissipated rather easily. Thus, perfusion of the working muscles not only serves for delivery and removal of metabolic products, but also for thermoregulation. Because the thermal conductivity of nonperfused tissue has the same dimension as fat tissue, the dissipation of heat depends mainly on the variation in this perfusion, specifically by opening arteriovenous anastomoses in the skin serving as as "thermal windows" of the body [8].

5.3.3 Influencing Factors

A constant core body temperature presupposes that heat production and heat losses are in balance. For other mammals and birds, this core body temperature is independent of their weight and volume, lying between 36 and 40 °C. The necessary basal metabolism increases with the body mass. This metabolism of an endothermic organism is determined both by the heat-producing body mass and the heat-emitting body surface. The specific metabolic rate of an organism is obtained by dividing the metabolic rate by the body mass, which means, for example, that the specific metabolic rate of the Etruscan shrew (0.002 kg) per time unit is approximately 175 times larger than that of the elephant (10,000 kg; law of metabolic reduction). This is due to the surface-volume ratio. As the body mass, i.e. body volume, increases with the third power (m³), the body surface, however, only with the second power (m²), accordingly small organisms (i.e., newborn infants) have an unfavorable surface-volume relation [10].

5.4 METHODOLOGIES OF CORE BODY TEMPERATURE MEASUREMENT

The temperature measurements practiced most frequently, sublingual and axillary, exhibit the largest source of errors. As a rule, by using axillary temperature measurement, one determines only the peripheral temperature and not the core body temperature. For a correct recording of the core body temperature, the sensor would have to be tightly pressed into the axilla for 30-40 min, which is the time it takes the body core to extend into the axilla (Figure 5.2). When measuring sublingual temperature, the position of the thermometer at the root of tongue varies by approximately 0.6-0.8 °C, according to [6,8]. Under experimental conditions, the core temperature is measured by inserting a thermosensor in the esophagus, nasopharynx, rectum, or tympanum/auditory meatus. However, none of these methods are really applicable outside the laboratory, especially during field research in extreme environments. This is due to the fact that the requirements for a method to accurately measure core body temperature are demanding. Basically, a thermosensor has to be (i) noninvasive, (ii) easy to handle, (iii) meeting basic hygiene standards, and (iv) not biased toward various environmental conditions, while (v) changes should quantitatively reflect small changes in arterial blood temperature and (vi) last but not least, the response time of the thermosensor to temperature changes should be as short as possible [11,12]. These requirements are essential. Previous studies in humans have shown that when high environmental temperatures and humidity prevail, the heat load will cause a rapid rise in core body temperature. This may result in heat stress-related injuries such as orthostatic collapse or even heat stroke [8]. If there are thermoregulatory impairments due to fever or drugs, the deleterious effects may occur even faster [13,14].

The tissue temperature of each organ depends on local metabolism and perfusion of the respective organ. Both factors are able to evoke measurable temperature differences within one organ. Thus, it becomes evident that the temperature of the organs in cranial, thoracic, and abdominal cavities deviates approximately 0.4 °C from the arterial blood temperature as reference value. Currently, there is no accurate and easy method to measure core body temperature in a field setting. The relative advantages and disadvantages of core temperature measurements and measurement sites, including the time response of the different kind of sensors, have been intensively discussed ever since the first benchmark investigations on this topic by Claude Bernard in 1876 [15–20]. Most thermal physiologists agree that the esophageal temperature is close to the best noninvasive index of core temperature for humans. It correlates well with changes in central venous blood temperature [12,16,21]. The esophageal temperature is obtained by inserting a catheter containing a thermocouple or thermistor through the nasal passage into the throat and then swallowing it. Although appropriate for research settings and during surgery, it is highly inappropriate in ambulatory or field assessments. This holds true for other sites such as rectal or tympanic probes as well, because they are all impractical for use in the field. Rectal temperature is obtained by inserting a temperature sensor at a minimal depth of 8 cm past the anal sphincter [14,22]. During exercise it takes approximately 25-40 min to achieve a steady-state rectal temperature value [15,22,23]. This is 0.2-0.3 °C higher than simultaneously measured nasopharyngeal and esophageal temperatures under resting and thermoneutral (27 °C) environmental conditions [17,19,23,24]. Rectal as well as esophageal temperatures are largely independent of the environmental temperatures [19,25]. As a result, the steady-state rectal temperature provides a good index to assess body heat storage [24,26]. The main problem with the rectal temperature is that it shows a slow response in comparison to the other measurement sites, a fact that recently was proven again in 60 patients who underwent a post-operative rewarming [26]. The reason for the slow response is probably (i) a low rate of blood flow to the rectum compared to other measurement sites [21,27] and (ii) the mass of organs located in the body cavity. This greater mass of tissue in the lower abdominal cavity requires a far greater amount of energy to cause a rapid temperature change. Tympanic temperature is obtained by inserting a small temperature sensor into the ear canal in direct contact with the tympanic membrane. Some subjects perceive this to be uncomfortable [8]. In addition, there are reports of the temperature sensor's perforating the tympanic membrane [28–30]. Because of the potential discomfort and trauma, as well as placement problems associated with tympanic measurements, some investigators chose to measure the temperature at the external auditory meatus. For this measurement, a temperature sensor is placed in an ear plug. Proper placement of the probe is vital because there is a substantial (~0.5 °C) temperature gradient along the wall of the meatus. In addition, several studies have shown that the auditory meatus temperature measurements do not provide a reliable index of the level of core body temperature during either rest or exercise [11,22,23,31] In regard to environmental conditions, these temperature values might be lower or higher than simultaneously measured steady-state rectal [22] and esophageal temperature values [31–33]. In addition, local heating of the head and/or increased air flow to the face will bias the tympanic temperature [22,23,31]. The best and most reliable method of assessing thermal state in operational environments is direct measurement of core body temperature using a network-enabled ingestible core temperature sensor [34]. However, the use of the latter is impractical for routine application. Thus, these devices are reserved for use during high thermal stress missions, while encapsulated in nuclear, biological, and chemical protective suits, and/or if use is indicated by medics during combat casualty care such as cooling interventions in case of heat injuries [20]. Skin surface can be accessed much easier than other core body temperature measurement sites mentioned above, although this site has many flaws because skin temperature is largely affected by cutaneous blood flow and/or sweat evaporation. Furthermore, environmental changes such as air temperature, humidity, wind speed, and radiation will alter skin temperature as well. Therefore, thermal physiologists prefer to determine mean skin temperature as a sum of weighted individual skin temperatures taken at different skin surface areas up to 16 sites, and for certain questions even more [8]. It is obvious that for most cases such a complex, heavy wired temperature measurement setup is currently highly impractical. Therefore, most recently Gunga et al. have introduced a combined skin temperature and heat flux sensor (Double Sensor) [35,36] (Figure 5.4).

In contrast to similar methodological attempts in the past [37,38], this new miniaturized heat flux sensor can be used without extra heating and comes in a special capsule [35]. It is placed at the vertex or at the front of the head and has been tested under various physical and environmental conditions, such as changing workloads and ambient temperatures of 10, 25, and 40 °C. A measurement comparison of the new sensor with the rectal temperature revealed that the Double Sensor (i) differed up to 0.06 °C from the average of the rectal temperature, (ii) showed with increasing ambient temperatures increasing concordance

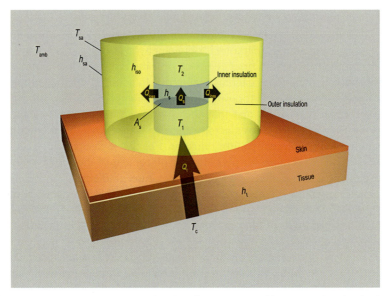

FIGURE 5.4   The technical scheme of the heat flux sensor (Double Sensor) placed at the front of the head to monitor continuously noninvasive core body temperature (T_c) under different environmental and physical conditions. T_c is calculated by measuring the heat flux dQ/dt from the body core to the outer sensor T_2. Heat loss (Q_{loss}) is taken into account (see Refs. [35,36]).

correlation coefficients, and (iii) exhibited a more rapid response to the core body temperature changes for all resting periods and at all ambient conditions as compared to rectal temperature. However, we observed limitations of the heat flux sensor in cold environments, which have to be investigated further.

5.5   BASICS OF HEAT TRANSFER

5.5.1   Internal Heat Transfer

The heat exchange between two objects is proportional to the difference of their temperatures. For the process of heat loss, internal heat has to be transported from the various tissues and organs to the cooler body surface. Heat transfer from the body core to the shell is called internal heat transport. With regard to thermoregulation, heat loss is under sympathetic control. Local mechanisms are also involved in addition to the systemic regulation of blood circulation. Under high physical strain, the skin vessels, for example, in the thorax, are dilated to excess [8]. This increased vasodilation is induced by bradykinin and other mediators excreted with sweat. An increase in internal heat production and/or external heat stress therefore also leads to an increase in skin perfusion. Due to their geometry, these parts of the body possess a large surface in relation to their volume, which allows an easier dissipation of heat. Under heat stress the deep arteriovenous anastomoses in the acrae are largely closed, and the arterial blood from the body core is redirected into the opening venous vascular bed of the skin [6,9]. Thus, this anatomical particularity promotes heat loss to the environment.

On the other hand, closure of deep arteriovenous anastomoses can inhibit the return of heat to the body core and prevent overheating. The high efficiency of the acrae to regulate the internal heat current lies, above all, in the large variability of their blood circulation (finger 1:600, hand 1:30, trunk 1:7) [10].

5.5.2 External Heat Transfer

Environmental conditions such as ambient air temperatures, radiation, relative air humidity, or wind velocity, as well as the skin temperature and the effective body surface are parametric for heat transfer from the body surface to the environment. In this external heat transfer, four different pathways play a role: convection, conduction, radiation, and evaporation (Figure 5.5) [8,10].

5.5.2.1 Convection

The convective heat exchange between body surface and environment takes place mainly in an air layer (boundary layer) that is merely a few millimeters thick and lies above the skin. In convective heat transfer, two forms have to be distinguished: natural and forced convection.

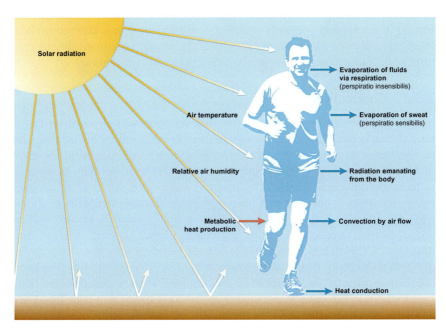

FIGURE 5.5 Parameters affecting heat balance during exercise and different heat transfer mechanisms.

5.5.2.1.1 Natural Convection

If a warm body is situated in a cooler medium, for example, air or water, a mass movement along the body axis of cooler parts toward warmer parts occurs in an upward direction in the medium. In doing so, the medium transports heat convectively. Under atmospheric conditions in an unclothed person, this convective mass transport amounts to approximately 600 L air per minute. In space, that is, weightlessness, this mass transport does not take place in astronauts, which contributes to a thermal discomfort (for further details, see Chapter 7). The amount of heat emitted in humans can be estimated by the assumption that 3 W of heat are dissipated in $1 m^2$ effective exchange surface area per centigrade temperature difference between skin and ambient temperature. Thus, at 25 °C ambient temperature and a mean skin temperature of 33 °C a total heat loss of $24 W/m^2$ will be achieved. With a surface-related basal metabolic rate of an adult of $0.1-0.2 J/(cm^2 min)$, this amounts to 4000-4350 kJ/day ($45-50 W/m^2$) [6].

5.5.2.1.2 Forced Convection

Forced convection exists when a body is transferred into a moving medium (wind, water current) or it is moved through this medium. Thus, the size and shape of the object play an important role in heat loss, meaning that the heat loss per square unit in small organisms (such as a mouse) by far exceeds that of large organisms (such as an elephant). In small organisms, forced convection can thus quickly lead to disturbances of the heat balance, especially when the thickness of the boundary layer is decreased by the forced convection and a laminar current in the boundary layer is transferred to a turbulent one. Convective heat losses also develop via the respiratory tract. However, in humans this mechanism of heat loss is small compared to other mammals (dog, horse).

5.5.3 Conduction

The direct heat transport between two solid substances in physical contact is termed conduction. The heat flows from the substance with a higher temperature to that with a lower. On an atomic level, heat is exchanged in the form of kinetic energy among the neighboring atoms. This means that, in contrast to the convective heat transport, no mass is transported. The rate of conductive heat transport between two objects depends on their temperature difference, the effective exchange area, and the material properties, as well as their special thermal conductivity. Silver, for example, has a very high thermal conductivity (430 W/m K), whereas air possesses merely a very low conductivity (0.024 W/m K) [6,8]. Based on its low heat conductivity, air can function as an excellent insulator. The amount of heat transported by the blood from the body core to the skin surface is conductively absorbed in the resting boundary layer and dissipated convectively by the air movement. In the external heat transport, conduction only takes place when skin surface areas come into direct contact with materials of high heat conductivity (e.g., metals) [6]. In such cases, an unprotected contact can lead to immediate burning or freezing, respectively. The thicker the resting boundary layer around the body, the lower the heat transport between body surface and air. By choosing proper clothing, humans can increase or decrease the thickness of their boundary layer. Here the amount of the air enclosed within the clothing is decisive for the insulation value. The higher the latter, the larger the enclosed amount of air [10].

5.5.4   Radiation

Each substance with a temperature above the absolute zero point ($-273.15\,°C$) emits an electromagnetic radiation of a certain wavelength. The wavelength emitted depends on the surface temperature and is inversely proportional to the latter. Short wavelengths are thus radiated by hot objects, and long wavelengths are radiated by cool objects [8]. Humans and animals are relatively cool objects within the temperature spectrum, thus they radiate in the long-wave infrared range [6]. The surface temperature is not only decisive for the wavelength emitted, but also for the rate at which a body emits radiation energy. Under resting conditions with an air temperature of 20-25 °C, relatively low air humidity and low windspeed, humans can emit approximately 50-60% of their total heat production via infrared radiation to the environment (Figure 5.6) [6].

The remaining portions are distributed more or less equally to convection/conduction and evaporation (Figure 5.6) [6]. For the amount of heat loss or heat

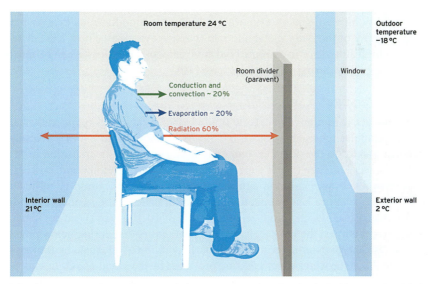

FIGURE 5.6   Heat loss pathways and their percentage in a lightly clothed human (0.5-0.6 clo) under resting conditions. Conduction and convection (green), radiation (red), and evaporation (blue) at a room temperature of 24 °C are shown in the figure. A screen decreases the heat loss by radiation. Further details are given in the text. *Adapted from Ref. [10].*

gain via radiation, the surface temperatures of the objects and walls nearest to the body surface are decisive, because the greater the temperature difference between the two, then the greater the heat loss or heat gain would be, respectively. With cool external or window temperatures, a screen or a curtain placed before it with a surface temperature aligning with the mean room temperature can considerably decrease the heat loss (Figure 5.6) [6]. An unprotected exposure of the body to significantly cooler surfaces leads to a decrease of the local skin temperature, which leads to the activation of cold receptors in the skin and to vasoconstriction. Then the concerned skin areas and the musculature lying below the skin proceed to cool further, producing a "cold sensation." Chilling and muscle tension can occur as a consequence of longer exposure. On the other hand, objects with a surface temperature higher than the skin (mean human skin temperature 32-33 °C) supply the organism with heat by means of radiation (radiant heater, furnace) [6,8].

5.5.5   Evaporation

Passively, the organism loses water by diffusion through the skin and the mucosa of the respiratory system (perspiratio insensibilis, extraglandular release of water) [10].

5.5.5.1   Perspiratio Insensibilis

The amount of fluid lost in an adult human by perspiratio insensibilis adds up to approximately 20-30 mL/h, that is, 400-600 mL/day at 33 °C ambient temperatures. Under thermoneutral environmental conditions and with 50% relative air humidity, this amount of evaporation leads to a simultaneous heat loss of about 20-30% of the daily metabolic rate (passive evaporative heat loss) at sea level. At high altitude or in extreme cold environments these losses can be much higher, especially during physical exercise [10].

5.5.5.2   Perspiratio Sensibilis

Physically active humans can excrete fluid via the sweat glands (perspiratio sensibilis, glandular release of water). By means of the evaporation of sweat, the organism can lose a considerable amount of heat, because in the transition from a liquid to a gaseous state (water vapor) energy is required (endothermic process). With complete evaporation, an amount of sweat of about 2 g/min is sufficient to dissipate the entire heat production of an adult under resting conditions (80-90 W). Because adults can produce a maximum of 10-15 g/min sweat per square meter, evaporation is the key mechanism of heat dissipation during heavy physical exercise and/or external heat stress [10]. In addition to adequate hydration, it is essential that the water vapor pressure produced by the sweat glands lies above that of the environment. The higher the water vapor pressure in the environmental air (high air humidity, tropical climate), the more difficult the heat loss via evaporation gets. That is why tropical climates are regarded as very stressful environments for nonadapted Caucasians, for example. With a low relative humidity in the air (dry desert climate), however, humans can also tolerate extremely high air temperatures and external heat load for a short time, because the gradient of the water vapor pressure from the skin to the environment is very large. According to Folk and Semken [39], this special anatomical-physiological feature (sweat glands) found in primates evolved in humans during the Tertiary Age in concert with bipedalism and a smooth hairless skin. Today, the following kinds of glands can be distinguished: (i) sebaceous (oily), (ii) apocrine, and (iii) eccrine (sweat) glands. The sebaceous glands can be found evenly distributed along the body axis with exception of the palms and soles. In some areas of the body, such as the scalp and face, the number of sebaceous glands can be extremely dense (900/cm^2). As shown in the schematic cross cut of the human skin, sebaceous and apocrine glands are anatomically close together (Figure 5.7) [39]. The latter, however, is usually located deeper in the dermis and invariably linked with a hair follicle, mainly in adults in the axilla (axillary organ) and in the pubic, anal, and auditary meatus (ear) canals as well. This organ, characterized by a nearly equal number of eccrine and apocrine glands in the axilla, is a feature that interestingly can only be found in our closest relatives in nature, the chimpanzees and gorillas.

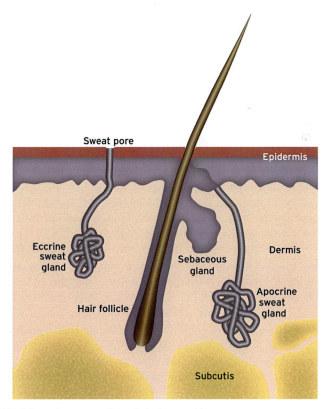

FIGURE 5.7   Schematic cross cut through the human skin.

They are also rudimentary in orangutans and are completely missing in all other mammals. As described by Folk and Semken [39], they have a tubular form and secrete a milky, viscid, gray, or reddish fluid. In the human fetus, apocrine glands can be found everywhere over the body up to the 5-5.5 month, then they disappear except those regions mentioned above. Eccrine glands (sweat glands) (i) have also a tubular form, (ii) are smaller than apocrine glands, and (iii) are located in the outer region of the dermis (Figure 5.7).

The number of sweat glands varies between 2-5 million over the whole body surface area with the highest density at the palms/soles. A smaller number of sweat glands can be found at the head/trunk, and the lowest numbers at the extremities. On the average 150-340/cm^2 can be found in the human skin [39]. However, the absolute number of eccrine glands can vary distinctly between different ethnic groups [35]. In general, populations living in colder environmental conditions have a smaller number of sweat glands than those living permanently in a hot and/or tropical environment. So, for example, the Ainu in the Arctic regions have in total only 1.4 million sweat glands, whereas Caucasians have >3 million [4,40]. Similar observations in different ethnic groups living in the circumpolar regions were made by other authors, as well. So, for example, Schäfer et al. conducted specific research on the activity and number of sweat glands in *Eskimos* and Caucasians [41]. Specifically, they studied the functioning sweat glands in 17 skin sites on the face, body, and limbs of 37 adult male *Eskimos* and 21 Caucasian controls. They found that *Eskimos* showed greater numbers and greater activity of functioning sweat glands on exposed parts of the face such as nose and cheeks, while responding with significantly less sweat gland activity on all body surfaces that are normally heavily clothed in winter. According to Schaefer et al., trunks, arms, hands, legs, and feet showed a progressive reduction of sweat gland response in the order of one-half on the trunk to one-fifth on feet when comparing mean sweat gland counts per square centimeter in *Eskimos* and controls. The comparative reduction of sweat gland response in the *Eskimos* progressed in the same order as the distance of the part from the body core and as the risk of the part to freezing. They concluded from the study that this reduction of sweat gland activity may represent a morphological and/or functional adaption to environmental conditions including climate and clothing [41]. In addition, during the process of heat acclimatization, the sweat glands are targets of specific metabolic/biochemical changes. Before heat acclimatization, the sweat has an osmolality of about one-third (100 mosm/L) of that of plasma (285-295 mosm/L) [35]. Normal sweat composition consists of water, minerals, lactate, and urea. On average, the mineral (including trace minerals) composition of sweat is as follows: sodium (0.9 g/L), potassium (0.2 g/L), calcium (0.015 g/L), magnesium (0.0013 g/L), zinc (0.4 mg/L), copper (0.3-0.8 mg/L), iron (1 mg/L), chromium (0.1 mg/L), nickel (0.05 mg/L), and lead (0.05 mg/L) [42]. After an acclimatization period of about two weeks, (i) the threshold of sweating will be decreased (earlier starting of sweating), (ii) the regional pattern of sweating will change (Figure 5.8), (iii) the absolute amount of sweat produced by the eccrine sweat

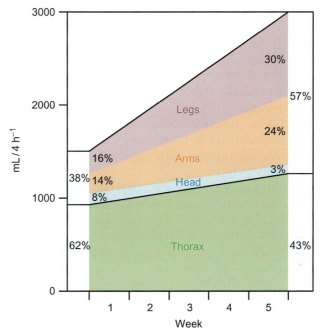

FIGURE 5.8   Changes of regional sweating pattern after a four-week lasting exposure to a tropical climate. Please note that especially legs and arms show a remarkable increase. *Adapted from Ref. [43].*

glands will increase from 1.5 to 4L/h, and (iv) the sweat composition will be attenuated, showing a lower osmolality [20,44]. In fully heat-acclimatized subjects, the latter can be reduced to as low as 10mosm/L, thereby (i) saving essential electrolytes for the human body and (ii) avoiding eventually deleterious hyponatraemia during prolonged exercise in the heat [43,45–47].

However, although the anatomy and physiology of human sweat glands have been studied intensively by different research groups in the past [48–60], such research has focused mainly on young males [51]. Indeed, much less information is available regarding females, children, and the elderly [51]. Therefore, recently, Havenith et al. [61] and Smith and Havenith [62,63] studied the regional variation in sweating over the body under resting conditions and exercise. They used a modified absorbent technique to collect sweat at two exercise intensities [55% (I1) and 75% (I2) $\dot{V}O_2$max in moderately warm conditions (25 °C, 50% relative humidity, 2m/s air velocity). At I1 and I2, highest sweat rates were observed on the central (upper and mid) and lower back, with values as high as 1197, 1148, and 856g/(m h), respectively, at I2. Lowest values were observed on the fingers, thumbs, and palms, with values of 144, 254, and 119g/(m² h), respectively at I2. Sweat mapping of the head demonstrated high sweat rates on the forehead (1710g/(m² h) at I2) compared with low values on the chin (302 g/(m² h) at I2) and cheeks (279 g/(m² h) at I2). Sweat rate increased significantly in all regions from the low to high exercise intensity, with exception of the feet and ankles. They observed no significant correlation between regional sweat rates and regional skin temperature, nor did regional sweat rate correspond to known patterns of regional sweat gland density. Aside from a detailed mapping of regional sweat rates over the whole body, their study demonstrated a large intra- and intersegmental variation and the presence of consistent patterns of regional high versus low sweat rate [63]. In Caucasian females [62] regional sweat rates were determined at two exercise intensities (60% (I1) and 75% (I2) $\dot{V}O_2$max) in moderately warm conditions (25 °C, 45% relative humidity, 2 m/s air velocity). A comparison was made between the females and males. The results are summarized in Figure 5.9 [63].

They found that female I1 regional sweat rate was highest at the central upper back, heels, dorsal foot, and between the breasts (223, 161, 139, and 139 g/(m²h), respectively). Lowest values were over the breasts and the middle and lower outer back (<16 g/(m²h)). At I2, the central upper back, bra triangle, and lower back showed the highest regional sweat rate (723, 470, and 333 g/(m²h), respectively). Regions of the breasts and palms had the lowest regional sweat rate at I2 (<82 g/(m²h)). Significantly greater gross sweat loss and thus regional sweat rates were observed in males versus females at both exercise intensities. For the same metabolic heat production (male I1 vs. female I2), absolute and normalized regional sweat rates showed a significant region-sex interaction,

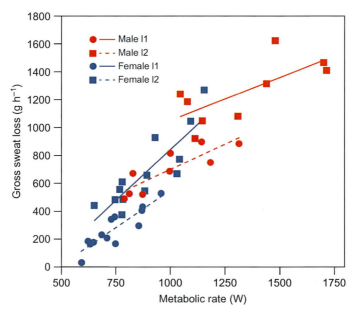

FIGURE 5.9   Absolute mean gross sweat rate (g/h) and absolute mean metabolic rate (*W*) for trained females and males at exercise intensity 1 (I1) and intensity 2 (I2). *Male data have been adapted from Smith and Havenith [63].*

with a greater distribution toward the arms and hands in females versus males. They concluded from their studies that despite some differences in distribution, both sexes showed highest regional sweat rates on the central upper back and the lowest toward the extremities. Furthermore, regional variation in sweat rates neither correlates with regional skin temperature nor does it correspond with regional sweat gland densities reported in the literature [62,63].

5.6 THERMOREGULATION

The task of the thermoregulatory system is to keep the body temperature constant within narrow limits, so that a balance exists between heat production and heat losses. The control variable is the core body temperature, an integrative value resulting from the local temperatures of many parts of the body. The autonomic mechanisms of temperature regulation (skin blood flow, sweating) permit a certain degree of acclimatization under extreme environmental conditions, although to a limited extent. Most important are adequate cognitive behavioral adaptations. Otherwise, under extreme environmental conditions, a failure of the regulatory mechanisms can rapidly occur, leading to hypo- and/or hyperthermia. This pertains especially to children and elderly people. For the registration of external environmental and internal core body temperatures, the skin is supplied with cold and heat receptors, unequally distributed over the human body (Table 5.1). Via afferent sensory nerve fibers, these external (body shell) and internal (body core) thermoreceptors are connected with the spinal cord and the hypothalamus, which is regarded as the regulatory center of temperature regulation. Most cold receptors are located approximately 0.2 mm below the skin and are, with exception of the scrotum, more numerous than the heat receptors [6]. Following fast increases of the skin temperature, the heat receptors in the first instance react with an excessive discharge rate of approximately 10-50 impulses/s [6,35]. Afterwards their impulse rate decreases quickly (to approximately 20 impulses/s). Cold receptors have a considerably lower spontaneous discharge rate (2-10 impulses/s) than heat receptors. During the acute excitation (excitatory) phase of either heat or cold receptors, the other receptors in the skin are inhibited. To date, not much is known about the mechanisms of signal transduction of the temperature signal. There are indications that changes of the sodium pump and the passive Na^+/K^+ conductivity might play an important part [6,35].

Anatomically, the preoptic area of the anterior hypothalamus (POAH) seems to be where the actual temperatures of the body shell and the body core are compared to a so-called setpoint value. In technical systems (e.g., air-conditioning systems) this setpoint value is set by means of a temperature reference signal placed within the control circuit. In the hypothalamus, special neurons are supposed to exist producing this signal independently of the temperature [6,8]. Such neurons in the hypothalamus, however, until recently could not be verified in a sufficient number. When actual temperature and setpoint value deviate from each other, various control elements in the control

TABLE 5.1 Distribution of Warm and Cold Points in Different Body Areas

Body Region	Distribution of Warm and Cold Points	
	Cold Points per cm²	Warm Points per cm²
Forehead	5.5-8	2
Nose	8-13	1
Mouth	16-19	–
Remaining face	8.5-9	1.7
Chest	9-10.2	0.3
Forearm	6-7.5	0.3-0.4
Palm of hand	1-5	0.4
Finger, outer face	7-9	1.7
Finger, inner face	2-4	1.6
Thigh	4.5-5.2	0.4

Data are Taken from Kunsch and Kunsch, 2007 [95]

circuit (motor system, brown adipose tissue, vasomotor activity, sweat secretion, pilomotor activity) are changed by the autonomic nervous system via efferent vegetative nerve fibers in the circuit of positive and/or negative feedback [6,35]. A decrease of the core body temperature below the setpoint value set by the hypothalamus leads to (i) a vasoconstriction of the skin and shell vessels (negative feedback), whereby the heat release via the body shell is reduced; (ii) a piloerection of the hair (goose bumps), which enlarges the insulating boundary layer above the skin and thus decreases the heat loss; and (iii) an increased heat production by shivering. When the actual value, on the other hand, lies above the setpoint value, all those mechanisms that might evoke a further increase of the body temperature (motor system) are extenuated (negative feedback), and the mechanisms of heat loss are strengthened (vasodilatation in the body shell, increase of sweat secretion). These different defense mechanisms for the maintenance of the core body temperature are reflexes and cannot be influenced entirely arbitrarily (autonomic control). The conscious sensations of thermal comfort or discomfort are generated in the sensory cortex, which receives the excitations of the internal and external cold and heat receptors via the tractus spinothalamicus and the unspecific medial thalamic regions. With distinct thermal discomfort, not only a stimulation of the autonomic countermeasures is initiated, but also, mediated via the cortex, changes in behavior, which leads to selection of warmer clothing or taking shelter in a heated room in a cold environment [6,8,35].

5.6.1 Cardiovascular Regulation

Particularly under extreme environmental conditions such as high ambient temperature and/or high relative air humidity, the simultaneous demands of metabolically active musculature and the thermoregulatory system (increased blood circulation of the skin for heat dissipation) put high stress on the cardiovascular system. The skin blood circulation can then increase to several liters per minute and constitute a considerable part of the cardiac output per minute. This is illustrated in Figures 5.10 and 5.11 [20,64].

With a simultaneous increase of the body core and body shell temperature, the compliance in the venous system of the skin is augmented and vasomotor tone decreases. This entails a decrease of the central blood volume, which is

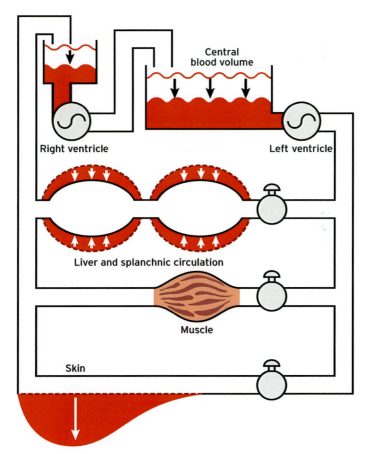

FIGURE 5.10 A schematic view of the different parts of the cardiovascular system. Please note the skin as a highly variable volume compartment of the cardiovascular system. If volume is markedly increased during, for example, heat stress, concomitantly the blood volume in the right ventricle will be decreased. *Adapted from Ref. [64].*

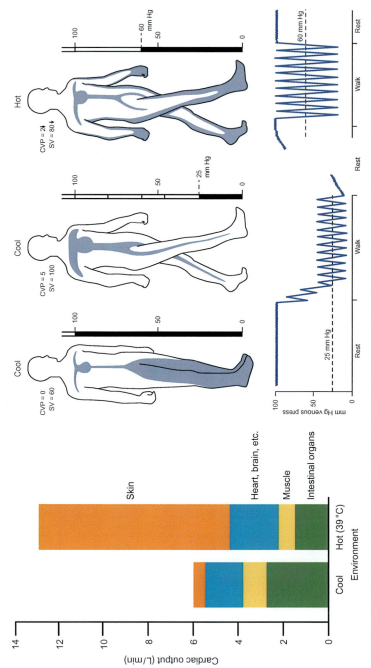

FIGURE 5.11 Circulatory changes in blood flow under different environmental conditions (left panel) and changes in central blood volume and central venous pressure in a cold and a hot environment (right panel). *Adapted from Ref. [64].*

essential for the physical capacity. In a human at rest in an upright position and at an indifferent ambient air temperature, approximately 70% of the blood volume is below the heart and 85% thereof in the low pressure system [65]. Physical work under the extreme environmental conditions thus leads to a disturbance of the volume distribution (heat collapse) with a descent of arterial pressure, filling pressure, and stroke volume [64,65].

An orthostatic intolerance thus has to be considered as an emergency reaction leading to unconsciousness. Placement into a horizontal position leads to a new redistribution of the blood volume from the periphery to the center. In the case of a heat collapse, this redistribution from the periphery to the center can be supported by an elevation of the extremities. Furthermore, as can be seen in Figures 5.10 and 5.11, the additional blood volume to maintain a higher perfusion of the skin is taken from the liver and splanchnic region. Because all food (i) is absorbed by the intestines and (ii) processed in the liver, any reduction in blood supply to these tissues and organs is as well accompanied by a reduction in the efficiency of local immune system. That is why (i) parasites, apart from the fact that the liver is a very nutrient-rich organ, prefer to go and stay there; and (ii) the immune deficiency in the intestines can very rapidly lead to a systemic infection (endotoxemia). The latter effect is illustrated in Figure 5.12, which shows that a decline in consumed plasma antilipopolysaccharide (LPS) concentrations occurred already at rectal temperatures as low as 39-40 °C [66].

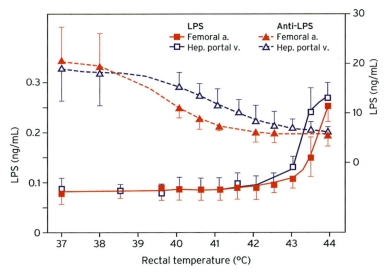

FIGURE 5.12 Endotoxemia following intestinal ischemia in an anesthetized nonhuman primate. Please note that (i) at a rectal temperature of 42-43 °C, plasma lipopolysaccharide (LPS) concentrations increase first in the hepatic portal vein, (ii) about 10-15 min later also in the systemic circulation; however, (iii) the decline in consumed anti-LPS antibodies already occurred at rectal temperatures as low as 39-40 °C. *Adapted from Ref. [66].*

5.6.2   Water/Salt Homeostasis

Evaporation, as outlined before, is a significant mechanism of heat loss for humans and can amount to up to several liters per hour. The organism also loses electrolytes with sweat in addition to body fluid. This can exert a substantial influence on the water/salt balance. If the loss in fluid and electrolytes is not compensated for, there is a risk of dehydration and hyponatremia/hypocalcemia. In the early stages, plasma volume is used for the production of sweat in which strong evaporation reduces the circulating blood volume with the adverse effects on the cardiovascular system (decrease of filling pressure and stroke volume). A hypotonic sweat facilitates the necessary fluid shifts for the maintenance of the circulating blood volume, because the loss entails an increase of the intravasal colloid-osmotic pressure. This is the driving force for the influx of water from the interstitial space and, with more extensive evaporation, also from the intracellular compartment [64]. The body temperature plays a decisive role in the function of fine motor skills as well as gross motor skills. If the body temperature drops, fine motor skills are restricted. With the onset of shivering, gross motor skills are likewise seriously disturbed. Due to this fact, sporting events are held preferably in the afternoon and evening hours because during this period the circadian core body temperatures are highest and the environmental conditions rather moderate—important prerequisites for peak motor performance.

5.7   SPECIAL TEMPERATURE REGULATION

5.7.1   Heat Production During Physical Work

Heat production during physical work maximum oxygen uptake ($\dot{V}O_2max$) depends on the maximum heart rate (HF_{max}), the maximum stroke volume (SV_{max}) and the maximum arteriovenous oxygen difference ($AVo_2\,diffmax$):

$$\dot{V}O_2\,max = HF_{max} \times SV_{max} \times AV_{O_2 diffmax}$$

The environmental conditions (PO_2, air temperature and radiation temperatures, air humidity, air velocity) can have a substantial impact on human performance. If in the course of a physical endurance exercise, for example, blood volume decreases due to strong sweating and insufficient fluid intake, the stroke volume decreases, thus $\dot{V}O_2max$. If blood circulation in the body shell has to be increased for thermoregulatory reasons, this blood volume is lost for the metabolically active musculature, for example, in the course of a physical endurance exercise, the blood volume decreases due to strong sweating and insufficient fluid intake, stroke volume, thus $\dot{V}O_2max$, likewise declines. If for thermoregulatory reasons the blood circulation in the body shell has to be increased, this blood volume is lost for the metabolically active musculature, and the maximum $AVo_2diffmax$ is diminished. Under resting conditions, the oxygen uptake of the skeletal muscle lies around 1.5 mL/(min kg min) and can be increased to tenfold during physical activity. The entire heat production of the adult human under resting conditions corresponds to a performance of approximately 80 W and can increase to up to 1000 W [8]. With sufficient energy reserves and/or continuous fluid supply, this physical exercise can be maintained over several hours, for example, in a marathon or triathlon. If these augmented mechanisms of heat loss, such as evaporation, had not evolved in humans, internal heat production would limit the endurance capacity to approximately 20 min because every 5-8 min the core body temperature would increase by about 1 °C, resulting in a rapidly occurring lethal hyperthermia [20].

5.7.2   Heat Loss

The total heat loss from the body core to the skin surface is comprised of two components: (i) fixed value transmitting the heat by conduction (passively) via the inactive musculature and the subcutaneous skin layers, and (ii) strongly variable heat transport by means of the convective heat transport of blood/circulation. Increased heat dissipation from the body core to the body shell is mainly ensured by means of an augmented blood flow from the muscles to the skin and under physical and/or warm environmental conditions by an increased evaporation of sweat [64]. The skin blood circulation can vary in different parts of the body. Thus, for example, it can be increased in the trunk by a factor of 7, at the hand by a factor of 30, and at the fingers by a factor of 600. Although the hands are merely a small part of the total body surface, they play a decisive role in thermoregulatory system of the human body because they serve as "thermal windows." Furthermore, per gram of evaporated sweat, an ultrafiltrate of the plasma, the organism loses approximately 2.5 kJ, the sweat glands being able to produce about 2-4 L/h, thus an amount of 30 g/min of sweat. The maximum sweat production, as well as the composition of the sweat, is variable. Besides a generally lower normal temperature, trained organisms adapted to heat have a lower sweating threshold. The trained athlete begins to sweat earlier so that his or her body temperature is lower under comparable conditions than that of less well-trained persons. Under the same environmental conditions and load, the athlete is thus able to maintain a performance for a longer period of time. Furthermore, the cooling of the body shell by sweating is important to maintain a lower skin temperature and thus a heat gradient from the body core to the body shell. If the body shell is warmer than the body core, the body core is supplied with heat from the body shell via the vessels (sauna effect). The heat flux is converted, and the core body temperature increased. The amount of heat that can be emitted by respiration to the environment plays only a minor role in the total balance of heat fluxes to prevent hyperthermia. However, considerable amounts of heat can thus be withdrawn from the body at very cold air temperature and heavy exercise, i.e. high respiratory breathing rates.

5.7.3   Age-Dependent Temperature Regulation

With an environmental temperature of 32-34 °C (relative air humidity approximately 60%), the thermal neutral zone of the newborn is distinctly higher than that of adults. The reason for this is that newborns have a very unfavorable surface-volume relation (threefold larger) compared to adults and only a very thin subcutaneous fat tissue. Newborns, however, can activate the shiver-free heat production in the brown adipose tissue via the sympathetic nervous system. Six to eight weeks after birth, distinct circadian variations of the core temperature can already be observed with lowest values between 2 and 4 o'clock in the morning. Sweat production increases distinctly with the beginning of puberty being about $350\,g/(m^2 h)$ prior to puberty (ages between 7 and 11 years old) and between $500\,g/(m^2 h)$ after puberty (ages 13-15 years old) [51]. Falk [67], Falk and Dotan [68], and Sinclair et al. [69] have addressed this topic in reviews and different studies. They conclude that prepubertal children's ability to thermoregulate when exposed to hot and humid environments is deficient compared to adults. However, definitely—although difficult to perform for ethical reasons—more research is urgently needed in this field. Not only is the younger population endangered in hot environments but also elderly people (Larose J 2013[70]; Popkin BM [71]) (Figure 5.13). Much like newborns, the elderly have a larger requirement for warmth. This might be due to their overall decreasing metabolic rate, the decrease in water content of the skin, a thin subcutaneous fat tissue, and/or a reduced vasomotor activity. Elderly people are endangered especially

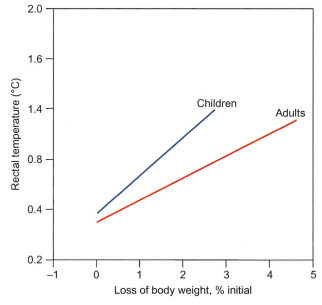

FIGURE 5.13 Comparison of core temperature changes in children and young healthy adults. *Adapted from Ref. [51].*

in the first 1-3 days of heat wave during the summer period (Leon LR 2012 [96]; Larose J [70]; Popkin BM [71]).

5.8 ADAPTATION, ACCLIMATIZATION, ACCLIMATION, AND HABITUATION

5.8.1 Definitions

An organism must be adequately adapted to all kind of environmental stressors in order to survive. Four major distinct terms are used to describe this process in biology and have to be decisively distinguished: adaptation, acclimatization, acclimation, and habituation [9].

An adaptation is a trait with a current functional role in the life history of an organism. Adaptation is an evolutionary process by which an organism becomes adjusted to its environment and becomes fitted through natural selection for some special activity. Adaptations contribute to the overall fitness and survival of the organism. Such changes can encompass genotypic and phenotype adaptations as well as behavioral responses to environmental stressors such as heat, cold, and hypoxia, to mention a few conditions.

The process of acclimatization takes a short period of time (days to weeks). Acclimatization, compared to adaptation, occurs within the organism's lifetime. The individual organism adjusts gradually during this time span to the natural environment, allowing it to maintain performance across a range of environmental conditions.

Acclimation is a term used to describe adaptive physiological responses to experimentally induced changes, in particular, climatic factors such as heat or hypoxia by means of an artificial exposure, for example, in a hypobaric chamber [35]. In general, organisms can adjust their morphological, behavioral, physical, and/or biochemical traits in response to changes in environmental challenges [9]. Among those responses might be changes in the biochemistry of cell membranes, making them more rigid in cold temperatures and more porous in warm temperatures by increasing the number of membrane proteins. Specific proteins, those so-called chaperones, are, for example, the well-known heat shock proteins (HSP 70, HSP 72, HSP 90). The molecular substances (i) bind to normal proteins to prevent them from deforming and (ii) can even unfold damaged proteins back to their original shapes. The number of heat shock proteins increases within about an hour or so after heat stress and expand the range of survival of the cellular structures by approximately 1.5-2.0 °C. Chaperones are therefore essential in maintaining cell functions under periods of extreme stress, especially in organisms living in the hot deserts (Kamler, 2004) [99]. It has been shown that organisms that are acclimated to high or low temperatures display relatively high resting levels of heat shock proteins so that when they are exposed to even more extreme temperatures, the proteins are readily available (Leon, 2012) [96]. Other examples of biochemical adaptations are neurohormonal factors such as dopamine or 5-hydroxytryptamine.

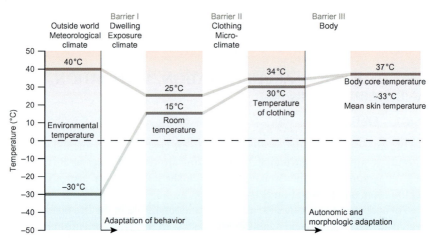

FIGURE 5.14   The range of the autonomic adaptations in humans in comparison to behavioral adaptations. *Adapted from Ref. [10].*

However, under extreme environmental conditions behavioral adaptations are the main avenue by which humans are able to survive (Figure 5.14). For example, the autonomic responses to overcome extreme thermal strains, heat and cold, are very limited, and without other counter measures, will lead rapidly to a fatal hypo- or hyperthermic state of the organism.

As humans evolved in a relatively dry, moderate altitude environment with a limited food and water availability, they were forced to run about 20 km daily (for details, see Chapters 1 and 3) in order to acquire resources. Thus, it is not surprising that humans possess hardly any natural biochemical, physiological, and/or anatomical-morphological protective mechanisms against the cold (hypothermia), such as thick fur or thick subcutaneous adipose tissue. Humans however, are equipped against heat strain (hyperthermia). The major mechanisms to avoid the latter include (i) a decrease in the sweating threshold, (ii) an increased amount of sweat (Figure 5.15), (iii) a decrease of the electrolyte content of the sweat (Figure 5.16), (iv) a decrease in heart rate, and finally, (v) an increase in total blood volume. The Zeitgang (time course) of these major adaptations is shown in Figure 5.16.

In humans, the maximum sweat production in relation to the body surface area is markedly higher than in any other organisms. The lower electrolyte content of the sweat has several effects: (i) electrolytes are preserved in the organisms, that is, a deficiency in supply of the organism is counteracted; and (ii) the sweat evaporates more easily from the skin. The augmented amount of plasma proteins in the long run increases the plasma volume by 10-20%. Hematocrit decreases correspondingly, which lowers blood viscosity, so that the cardiovascular system is thus in a better initial position to operate much more efficiently and effectively. Therefore, comparable loads during physical activity can be accomplished with lower heart rates. All mentioned factors contribute to the

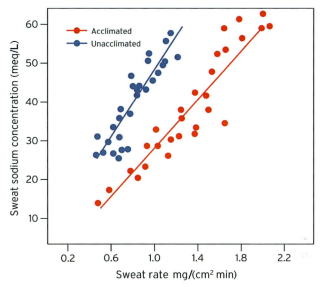

FIGURE 5.15   Relationship between sweat sodium concentration and back sweating rate before and after heat acclimation. *Adapted from Ref. [45].*

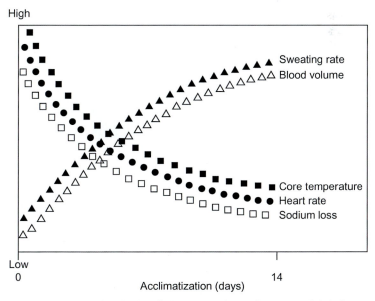

FIGURE 5.16   Thermal acclimatizations in humans to a hot environment and their time course.

fact that increases in body core temperature under physical exercise is slower than in nonadapted persons. The long-term heat acclimatization processes thus bear a strong resemblance to those adaptations observed in endurance athletes.

5.8.2 Special Adaptations

5.8.2.1 Adaptation of Behavior

The capacity of autonomic adaptations is relatively limited in humans as mentioned before. That is why these autonomic adaptations have to be supplemented by corresponding behavioral adaptations. During a stay in the desert, physical activities should be carried out either in the early morning hours, in the late evening hours, or at night. This avoids high load peaks during the day in which the long-wave radiation maxima of the sun is present. Fluid intake to compensate and—with very strong heavy sweat losses and restricted availability of a balanced diet—also an additional salt uptake have to be made consciously and repeatedly. Thirst and appetite for salt are inadequate with strong losses. Nightly cramps in the calves indicate corresponding deficiency symptoms in the electrolyte balance. In Figure 5.17 the different fluid losses associated with different kinds of activities in a hot desert climate are illustrated [9,35].

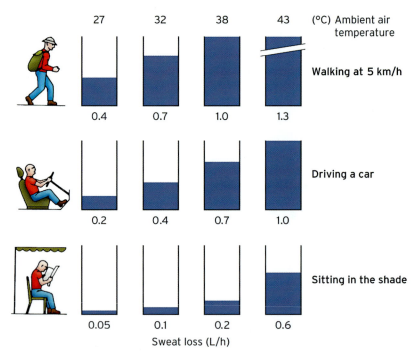

FIGURE 5.17 Fluid losses under different environmental and physical conditions. *Adapted from Ref. [44].*

The clothing should cover as much of the skin surface as possible and be permeable to water vapor and loose-fitting in order to permit air circulation along the body axis (natural convection, pleasant microclimate). Shkolnik et al. [73] investigated whether black robes help the Bedouins to minimize solar heat loads in a hot desert and reported that the amount of heat gained by a Bedouin exposed to the hot desert is the same whether wearing a black or a white robe. The additional heat absorbed by the black robe was lost before it reached the skin. The higher temperature in the loose black clothing as compared to the white clothing increased the natural convection between the different layers of Bedouins coats. The color of the clothing is therefore of less importance than its material and style as nicely shown in Figure 5.18. The important is that in dark, loose-fitting multilayer clothing, the higher absorption of energy in the outer layers increases the convection within the clothing. This effect is so strong that in the end, the surface temperatures on the skin, which are important for a pleasant feeling, that is, the microclimate, is the same as dressed in white clothing, which physically absorbs less energy from radiation.

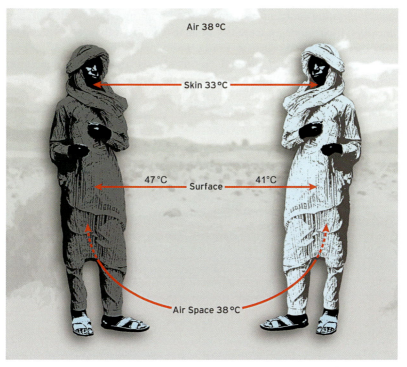

FIGURE 5.18 At an ambient temperature of 38 °C the surface temperature of a black Bedouin coat (left person) is higher compared to a white one. The skin temperature is in both cases 33 °C at the face as indicated by the arrows. *Adapted from Ref. [73].*

A special form of heat adaptation is observed in large animals (giraffes, camels) in hot climate zones, the so-called adaptive heterothermia. These animals increase their sweating threshold and decrease their shivering threshold, thus widening the range in which changes of the core body temperature are tolerated. This procedure reduces the quantity of sweat and subdues an early heat production by shivering during decreasing of the core body temperature. During the day, the body stores large amounts of heat and releases it at night through convection and radiation to the cool environment. In addition, a wide range of change in their core temperatures could be observed during a 24-h period, and this could be attributed to a well and dehydrated state of the organism (Figure 5.19) [74,75].

In addition, large desert animals, such as the camel and donkey, have a very low evaporative water loss in terms of percentage of their body weight in comparison to humans and other animals as summarized in Figure 5.20.

Under arid environmental conditions with low availability of water these are very reasonable strategies to survive. Furthermore, in indigenous humans living in the tropics, an increase in the sweating threshold has been detected. This can also be interpreted in the sense of economization. In general, humans tolerate a dry-hot climate, that is, a desert climate, better than the tropical climate (warm/hot ambient air temperature, high moisture, still air) because physically a high vapor pressure has to be reached to fully evaporate the sweat thereby cooling the body surface. Furthermore, due to the high ambient humidity, not all sweat is fully evaporated, but instead sweat drops fall to the ground. This sweat is lost for cooling because under these circumstances the cooling effect of sweating cannot be achieved and the permanent high skin perfusion, as mentioned above, is a heavy burden for the circulatory system. Older subjects with cardiovascular diseases are especially prone to suffer under these climate conditions.

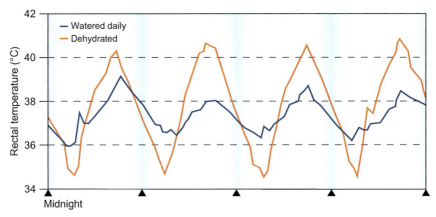

FIGURE 5.19   Circadian core body temperature changes in a hydrated and dehydrated camel. *Adapted from Ref. [74].*

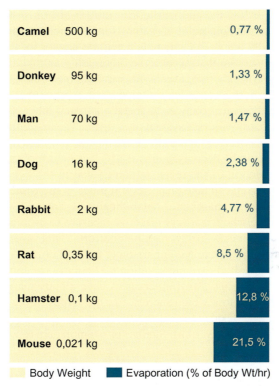

Camel	500 kg	0,77 %
Donkey	95 kg	1,33 %
Man	70 kg	1,47 %
Dog	16 kg	2,38 %
Rabbit	2 kg	4,77 %
Rat	0,35 kg	8,5 %
Hamster	0,1 kg	12,8 %
Mouse	0,021 kg	21,5 %

Body Weight   ■ Evaporation (% of Body Wt/hr)

FIGURE 5.20 Differences of evaporative fluid loss in humans and some animals living in different environmental settings. *Adapted from Ref. [97].*

5.8.2.2 Torpor, Hibernation, and Estivation

We will now focus on adaptations in animals living in extreme environments in order to show the range of extraordinary adaptations enabling them to tolerate and survive under stressful environmental conditions. For example, some mammals and birds fall into dormancy every winter or in very hot summer months. According to its length and the environmental conditions, a distinction is drawn among torpor, hibernation, and estivation. Daily phases of rigor with reduced body temperature and metabolic activity are designated as torpor. This phenomenon can be observed, for example, in bats and colibris. Mammals (hamsters, kangaroo, rats) decrease their metabolic rate and thus body temperature not only for hours, but rather for weeks and months. This state is called hibernation. For that kind of adaptation, it is necessary that the animals built up large energy reserves (body fat) during the summer months. Furthermore, these animals hibernate mostly in deep burrows that provde considerably enlarged thermal insulation to the harsh environment prevailing outside this self-made cave. Nevertheless, the body temperatures of the hibernators fall below the ambient temperature in the cave, which might be barely above the freezing point of water. If the ambient temperature tends to decrease markedly below 0 °C, there is a definite danger for the animal to die by freezing. But before that occurs, the hibernating animal starts to produce heat by nonshivering thermogenesis, for example (see below). The organism then reaches its normal core body temperature (approximately 36 °C) for a short time and afterwards reduces its metabolism and other physiological parameters of the cardiovascular system once again. The regulation of body temperature is thus entirely intact in these organisms, only the threshold of heat production seems to be decreased drastically. It is therefore assumed that this is a defined reduction of the mean setpoint value in the hypothalamus. The process of waking up can take place several times in the winter months. Even though these waking phases are metabolically expensive for the organism, hibernation on the whole is a very effective measure to overcome meager and hostile seasons by means of low metabolic efforts. This saves approximately 80% of the energy required in comparison to normothermia. Small organisms with a large surface-body volume relation particularly use hibernation or torpor as survival strategies. Most hibernators have a body mass of about 85 g. The smallest hibernators weigh merely 5 g, the largest more than 100 kg, if bears are included. Besides torpor and hibernation, some organisms fall into a summer dormancy (estivation). Snails, for example, are thus able to survive longer dry periods [9,35].

5.9 HYPERTHERMIA AND FEVER

5.9.1 Hyperthermia

5.9.1.1 Pathophysiology

Hyperthermia is characterized by a disproportion between heat loss and internal heat production or external heat supply without a shift of the setpoint value as observed with fever [6]. With increased physical exercise or external heat supply (sauna), body temperatures above 41 °C can be observed. After the end of the physical exertion (load) or after the end of the heat exposure, the body temperature returns to its initial value. Recent investigations have shown that especially heat exposure of longer duration and high temperature displays the highest mortality rate. In the past 100 years, the frequency of these extreme climatic conditions has increased significantly in regions with normally moderate climate due to global warming. During the day, these heat waves lead to heating of the densely built inner cities, so that the temperatures at night do not fall below 28-30 °C. The population living in these areas is exposed to a permanent 24-h heat stress, particularly in the attic floors. Children (<1 year), elderly people, diabetics, and persons suffering from cardiovascular diseases are especially affected. For those who are also immobile (bedridden), the convective heat loss is decreased. The circulation is forced to maintain the heat transport from the core to the periphery via a steadily increased cardiac output per minute in order to guarantee heat transmission from the skin to the environment via an increased skin perfusion. As a result of high environmental temperatures, heat cramps, heat collapse, heat exhaustion, and heat stroke may result [72,76].

5.9.1.2   Heat Cramps

With heavy physical work, especially with environmental temperatures above 27 °C and high relative humidity, the fluid loss by evaporation can amount to 5-10 L/day. Electrolytes are also lost in sweating, particularly Na^+, Cl^-, Mg^{2+}, and Ca^{2+}. Drinking low-electrolyte water further augments the decrease of extracellular ion concentrations. The loss of Na^+ and Mg^{2+} may result in heat cramps in the calf musculature. Occasionally, the abdominal muscles are affected by cramps and thus an abdominal emergency situation seems to exist. The loss of Cl^- ions promotes a hypoacidity of the stomach. Especially in a tropical climate, this achylia promotes the incorporation of pathogens into the gastrointestinal tract [72,76]. Daily fluid intake should be sufficient to guarantee a daily urine output of 800-1000 mL/day.

5.9.1.3   Heat Collapse

The most frequently observed heat illness is heat collapse. This is an orthostatic circulatory disorder occurring in particular during the first days of a stay in hot climate. Heat collapse is caused by a failure of the cardiovascular reaction to high environmental temperature, thus creating a disturbance of the volume distribution of circulation. As an expression of the disproportion between heat production, heat absorption, and heat loss, first of all, the skin blood circulation is increased by peripheral vasodilatation. Even at rest, this requires an increase of the cardiac output per minute. Due to peripheral vasodilatation, the intrathoracic volume is diminished, which can lead to a heat collapse following a phase of orthostatic instability. In contrast to the heat stroke (see below), sweat secretion is intact so that the skin is moist. The body temperature is lower (38.5-41 °C) in comparison to a body temperature of above 41 °C in the case of heat stroke. As possible indications of a circulatory collapse, one has to consider blood pressure decrease, bradycardia (weakness, dizziness, tiredness), headache, lack of appetite, nausea, vomiting, and an urge to defecate. Further symptoms, such as piloerection (goose bumps), observed on the chest and at the upper arms, hyperventilation, muscle cramps, and sometimes also neuronal signs such as ataxia and incoherent speech. Laboratory tests reveal, among others, hemoconcentration, hypernatremia, hypophosphatemia, and hypoglycemia [72,76].

5.9.1.4   Heat Stroke

A longer lasting overheating above 40 °C leads to a heat stroke (disorientation, cramps, delirium). Heat stroke is a life-threatening disturbance of temperature regulation occurring particularly often during the first days of a heat wave in elderly people suffering from chronic diseases such as arteriosclerosis, cardiac insufficiency, or diabetes mellitus. Furthermore, the occurrence of heat strokes is well known after application of anticholinergic drugs for the inhibition of sweat production, under diuretic therapy, as well as in persons with skin diseases who have a hindered heat loss (e.g., ectodermal dysplasia, congenital missing of sweat glands, severe sclerodermia). However, young people can also suffer heat strokes after extreme physical exertion and an uncovered head in strong solar radiation. The temperature regulation in the hypothalamus is disturbed so that, in spite of high thermal stress, a vasoconstriction of the peripheral skin vessels occurs and the sweat production is suspended. Patients suffering from heat stroke thus have dry, hot skin [72,76]. For every 1 °C temperature increase, the basal metabolic rate increases by 7%. At a temperature of 41 °C, this means a metabolic increase of almost 40%. Frequently, blood volume is diminished and hematocrit is increased. In the pulmonary vascular bed, a resistance increase takes place. The warning symptoms include the following appearing at short notice: intense headaches, dizziness, feeling of weakness, abdominal pain, confusion, or hyperpnea. Blood pressure is extraordinarily low. The muscles are flaccid, and the tendon reflexes can be diminished. Pulse rate is increased and respiration is fast and weak. Depending on the severity, lethargy, stupor, and coma can exist. Heat stroke can also be fatal. One important criterion for the existence of a heat stroke is the core body temperature above 40.5 °C. Even cases with temperatures of up to 44.4 °C have been reported. The skin is red, dry, and hot. Mortality with heat stroke lies at least at 10%. The patients die within a few hours after being found or in the following days and weeks from the consequences of various complications, for example, renal failure, cardiac infarction, or bronchopneumonia. If no sufficient heat dispersion is taken care of during hyperthermia, temperature will continue to increase. Because the heat-regulating mechanisms fail and perspiration is no longer possible, external support for heat dispersion has to be applied (ice-water immersion) [72].

5.9.1.5 Preventions

Prophylactic fluid supply before exposure, light clothing, frequent cool baths, cool environment, and reduced physical activity (in particular old and very young people) can help to avoid heat diseases, especially heat stroke. In order to avoid heat cramps and heat collapse, physical exertion at high ambient temperatures (>26 °C) and high humidity as well as low air movements should be restricted or avoided. Under such environmental conditions, long-distance adult runners should drink, for example, approximately 250 mL of slightly salted fluids every 3-4 km also during the competition. A small amount of electrolytes (salt) and carbohydrates should be added to the fluids. This is better than a rehydration of the organism by plain water. The reason is that pure water gets rapidly absorbed from the gut and will lead a decrease in sodium in the plasma. This will induce an increase in urine production and thereby increase the body's fluid losses. The small amount of salt is needed to stimulate the thirst drive. The addition of carbohydrates to the fluids can enhance the intestinal absorption of water. So, for example, Gisolfi et al. [77] observed that a 6% carbohydrate-electrolyte (2% glucose, 6% sucrose, 20 meq Na⁺, 2.6 meq K⁺) solution was absorbed sixfold faster than water. However, solutions containing carbohydrate concentrations >10% will cause an osmotic-driven net movement of fluid into the intestinal lumen when such solutions are ingested during exercise. Thus, an effective loss of water from the vascular compartment will occur. This will impair cardiovascular function, leading to a fall in blood flow to muscle, skin, and other tissues [78,79].

5.9.2   Fever

Fever is not hyperthermia because the core temperature setpoint is changed via order of the hypothalamus. Fever is thus the symptom of an acute adjustment of the body to ensure this higher setpoint. The change in the setpoint is initiated by pyrogenic substances [6,80–81]. A distinction is made between exogenous and endogenous pyrogens. Exogenous pyrogens are viruses, bacterial toxins, LPSs, as well as muramyldipeptides of bacterial membranes. These pyrogens stimulate granulocytes and macrophages to release a whole series of large hydrophil polypeptides or proteins (interleukin-1 beta, interleukin-6, TNF-alpha, interferon) into the blood (Figure 5.21).

These cytokines are referred to as endogenous pyrogens. Today it is assumed that not only one single cytokine is responsible for the initiation of fever but that a simultaneous stimulation of various cytokines is necessary (cytokine cocktail). In order to induce fever, these cytokines have to come into contact with the neuronal structures of the preoptic area in the anterior hypothalamus. Due to their size, the mentioned cytokines are actually incapable of passing the blood-brain barrier. The fenestrated capillaries in the organum vasculosum laminae, a structure located in the direct vicinity of the POAH, however, enable a transfer to the endogenous pyrogens. Here the cytokines are supposed to activate monocytes, endothial and glia cells expressed in the tissue that effect an increase in the prostaglandin-E2 production. This prostaglandin-E2 is now able to pass the ependymal blood-brain barrier. It is not entirely clarified whether prostaglandin-E2 will then have a direct effect on the neurons of the POAH or whether also neurons in the organum vasculosum laminae are interposed. The fact that prostaglandin-E2 inhibitors like actylsalicilic acid or indomethacin have a fever-reducing (antipyretic) effect indicates, among other things, that prostaglandin-E2 plays a decisive role in the mediation of an adjustment of the setpoint value. The observation has been made that sometimes after intravenous injection of exogenous pyrogens, an increase of the setpoint value is observed even before a measurable increase of endogenous pyrogens in the blood. This has led to the search for alternative signal transduction pathways. A possible neural pathway could lead from the liver via the nervus vagus (N. vagus) to the POAH. The Kupffer cells possess macrophages that come into contact with the exogenous pyrogens in the circulating blood. Via mediators released by the Kupffer stellate cells a stimulation of

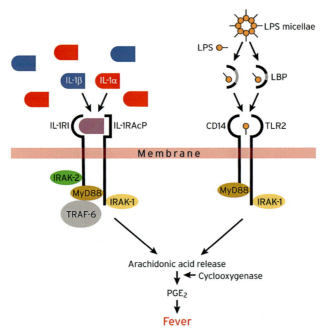

FIGURE 5.21   Fever and presumable pathways of fever induction via IL-1 alpha and IL-1 beta cytokines (pathway A, left axis) as well as lipopolysaccharides (LPSs), CD14, and toll-like-receptor 2 (TLR2) (pathway B, right axis). According to Netea et al. [81]2000 and Dinarello [80]2004, in pathway A IL-1 (IL-1α and β) binds to its cellular receptor type I (IL-1RI) and the IL-1 receptor accessory protein (IL-1RacP). This leads to signal transduction via receptor associated proteins IRAK-1 and -2, MyD88, and TRAF-6 with release of arachidonic acid and prostaglandin-E2 (PGE2) and finally induction of fever. In pathway B the LPS complex acts with LPS-binding protein, which enables binding of LPS to CD14 and toll-like receptor 2 (TLR2). Thereafter, the signal transduction pathway is very similar to that of the IL-1 receptor. Here too, MyD88 and IRAK-1 are activated, arachidonic acid is released, and fever is induced through PGE2. *Adapted from Ref. [81].*

the afferent parts of N. vagus occur. The N. vagus could, via its nuclei in the medulla oblongata and from there via projections to the POAH, change the setpoint directly or via a stimulation of the prostaglandin-E2 production in the POAH, which, as the main control center of thermoregulation in humans, would induce an adjustment of the core temperature setpoint (Figure 5.22) [6].

In addition to this classical model, others have postulated there might be alternative models. These pathways are summarized in Figure 5.23 [81].

However, if fever is induced, three phases can be distinguished: (i) increase of fever, (ii) plateau phase of fever, and (iii) decrease of fever (Figure 5.24). Usually, fever follows the normal fluctuation pattern of body temperature, only on a distinctly higher level. That is why fever, just like the normal body temperature, is higher in the evening (evening "peaks") than in the morning. After the setpoint has been shifted to a higher level (first phase), for example, a core

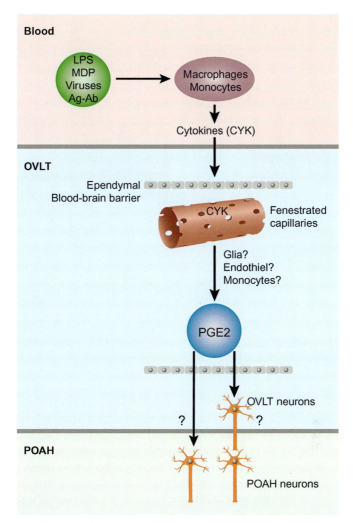

FIGURE 5.22 Hypothetical scheme showing the anatomical structures and physiological trans-mitters involved in the induction of fever in humans according to Jessen (Jessen 2001). Exogenous pyrogens (lipopolysaccharides [LPS], muramyl dipeptide [MDP], viruses, antigen-antibody complexes [Ag-Ab]) in the blood come in contact with immune-competent cells (macrophages, monocytes). These immune cells release thereafter cytokines such as interleukin-1 alpha (IL-1 alpha), interleukin-1 beta (IL-1 beta), and/or interleukin-6 (upper panel). Through the fenestrated capillaries of the organum vasculosum laminae terminalis (OVLT) in the vicinity of the POAH, these relative large cytokines can pass into the perivascular space. There, the cytocines are activating resident monocytes, endothiel, or glia cells that trigger the production of prostaglandin-E2 (PGE2). It is currently not fully understood how PGE2 acts on the POAH. Either the PGE reaches the POAH via diffusion and/or special neurons, which then induce the upward change of the setpoint of core body temperature. *Adapted from Ref. [6].*

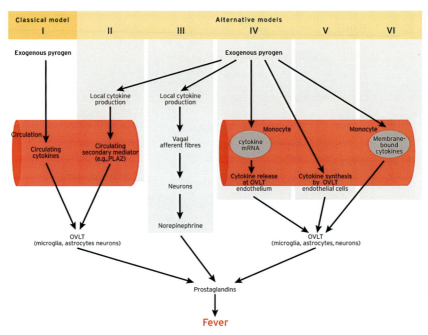

FIGURE 5.23 Additional pathways might be able to trigger OVLT (organum vasculosum laminae terminales). *Adapted from Ref. [81].*

body temperature of 40 °C, the body reacts in the same way as if it were exposed to a cold environment. This leads to an increased cold sensation, a strong vasoconstriction in the skin vessels (reduction of heat loss), and the occurrence of muscle shivering. In this phase, the person concerned looks pale, has cool acrae and warm or hot regions that are located close to the body core (neck, forehead). When the new setpoint temperature is reached, a plateau phase follows, the length of which depends on the disease (second phase). Actually, the plateau phase is characterized by the fact that no activation of heat loss mechanisms can be observed in the subject such as profuse sweating or a massive increase in peripheral skin blood flow or vasodilation. In short, the body is handling the increased temperature as if it were normal. This lack of increased heat loss mechanisms at a core temperature distinctly above normal temperature distinguishes fever from hyperthermia, for example, by physical exertion or strong external heat supply. During the decrease of fever (defervescence), the setpoint temperature is shifted back to a normal temperature around 36.5-37.0 °C. As a result the setpoint temperature lies now distinctly below the actual core body temperature, which as a rule leads to an increased activation of the heat loss mechanisms (evaporation, vasodilation, increased skin blood flow). Due to the rapid vasodilation and the relatively too low blood volume in consequence, a febrile circulatory collapse with warm, moist skin, tachycardia, and low diastolic blood pressure can follow during the critical defervescence phase.

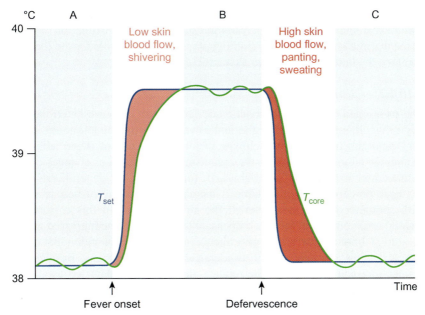

FIGURE 5.24 Typical time course of fever. First of all, an adjustment setpoint (blue line) of the core body temperature is initiated by the anterior hypothalamus (thermoregulatory center) (A). The actual core temperature (green line) follows with a delay (time lag) (increase of fever). In the plateau phase, actual core temperature measured and setpoint value are equal (B). The decrease of fever is initiated by an adjustment of the setpoint value to a lower body temperature, which is again followed by the actual core temperature with a delay (C). The typical physiological accessory symptoms during the increase and decrease of fever are shown in the illustration. *Adapted from Ref. [6].*

5.10   THERMOREGULATION AND CLIMATE

Environmental conditions have a significant influence on the human thermo-regulation and the feeling of comfort.

5.10.1   Definitions

5.10.1.1   *Atmosphere, Climate, Weather, Microclimate*

The layer of air held by the gravity of the Earth consists of different gases (nitrogen, oxygen, CO_2, inert gases) as mentioned before. In meteorology, the specific qualities of the atmosphere above a defined region of the Earth are summarized under the term climate. These qualities are decisively determined by the position above the surface of the Earth. An observation period of at least 20 years is prerequisite for the characterization of a climate (altitude, tropic, desert, and polar climate). The recorded meteorological factors are, among others, air temperature, relative air humidity, wind velocity, as well as amount and temperatures of radiation. A season is a term given to the state of the atmosphere over a period of three to four months (seasons spring, summer, autumn, winter). If the state of the atmosphere can be predicted for 48 h, this is referred to as weather

[35]. A special bioclimate for humans is the indoor climate, because, besides the climatic parameters, nonclimatic variables such as protective function (housing) or the thermic resistance of the clothing play a significant role for the feeling of comfort; the latter is also known as microclimate.

5.10.1.2 Climate Indices

The mentioned climatic and nonclimatic factors can be represented individually or as climate indices. For the understanding of the climate index curves, it is important that individual climate factors can be compensated, alleviated, or augmented in their effect on the organism by simultaneous changes of other climatic and/or nonclimatic variables. If, for example, an air temperature that has been perceived as comfortable is increased, no feeling of heat originates with simultaneous augmentation of the air movement. In this case, the increased convective heat losses compensate the effects of the increased air temperature. Climate indices thus give numerical values for the individual climate values leading to a synopsis of an identical effect on humans. Besides room temperature and air velocity, the relative air humidity is depicted as a further important climatic variable. It becomes evident that with low relative humidity (10%) and high wind velocity (3.0 m/s) an actual air temperature of 37 °C is perceived by the organism as 25 °C (comfort). However, a combination of high relative humidity (95%), low wind velocity (0.1 m/s), and a room temperature of 29 °C is perceived as a disagreeable, damp-warm indoor-climate (discomfort). Relevant detailed studies with various climatic combinations have led to the term *effective temperatures*. A distinction is made (i) between normal effective temperature that applies to persons with usual streetwear (clothing) and (ii) basal effective temperature, which applies to persons with unclothed upper part of the body. However, the same room temperature can be perceived quite differently by different persons with the same clothing and activity. Wearing light summer clothing with a relative air humidity of 50% and a wind velocity of 0.1 m/s, most test persons regard an indoor climate of 25-27 °C as comfortable. For some, however, this room climate might already be "too cool". The reason for this subjectively different thermic perception lies in the individually different balance of heat production and heat loss depending, among other things, on numerous factors like age, body size, body composition, and hormone levels [35].

5.11 OUTLOOK: GLOBAL WARMING AND HUMAN HEALTH

The WHO report by Campbell-Lendrum et al. [82] as well as more recently the Intergovernmental Panel on Climate Change (IPCC) [83] stated that climate change is an emerging risk factor for human health due to the fact that almost the entire planet has experienced surface warming (Figure 5.25).

According to Cubasch [85], we have to consider that each of the last three decades in particular has been successively warmer at the Earth's surface than any preceding decade since 1850 (Figure 5.26).

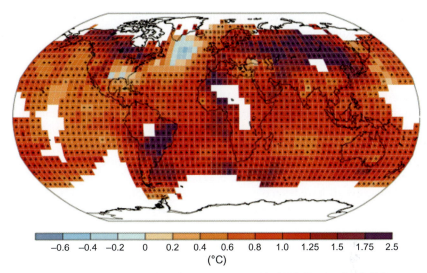

FIGURE 5.25 Observed change in surface temperature 1901-2012 (Stocker, 2013) [98].

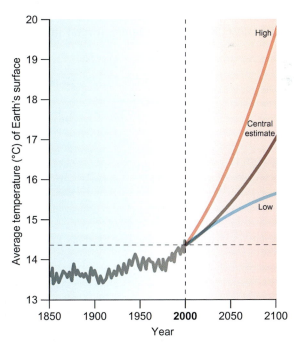

FIGURE 5.26 Global temperature record since instrumental recording began in 1860 and projection to 2100 according to the IPCC (Stocker, 2013).

In addition to that, the Northern Hemisphere during the time span of 1983-2012 likely experienced the warmest 30-year period of the last 1400 years. Also, the globally averaged combined land and ocean surface temperature data as calculated by a linear trend showed a warming of 0.85 °C over the period 1880-2012. When multiple independently produced datasets exist, the total increase between the average of the 1850-1900 period and the 2003-2012 period is 0.78 °C. The continental-scale surface temperature reconstructions show multi-decadal periods during the Medieval Climate Anomaly (year 950-1250) that were in some regions as warm as in the late twentieth century. These regional warm periods did not occur as coherently across regions as the warming in the late twentieth century (Figure 5.27). Changes in many extreme weather and climate events have been observed since about 1950. Finally, it is likely that the number of these heat waves has increased in large parts of Europe, Asia, and Australia.

Global warming, along with the growing incidence of extreme climatic conditions and the rising population densities in metropolitan areas lead to considerable increases in risk for human health [83,84]. Furthermore, this could lead to a substantial reduction in economical productivity [85]. Thus, there should be increasing interest from the scientific and political communities to explore, understand, and answer the emerging questions of the impact of global warming and heat waves on the health of the different populations around the world (rural and urban) and its impact on economy, with the ultimate aim to develop multifactorial preventive strategies for reducing climate-related issues such as morbidity and mortality [86,87]. For example, several studies have shown that

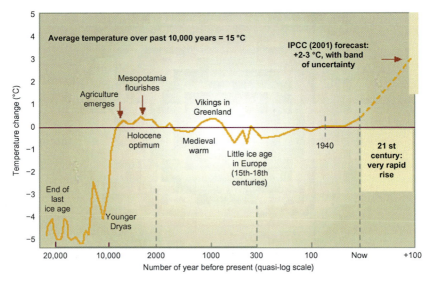

FIGURE 5.27   Average global temperature changes in the last 20,000 years.

increased rates of morbidity and mortality during heat waves can be found especially in risk groups such as children in their first year of life, the elderly, bedridden patients, drug addicts, alcoholics, diabetics, immobile and homeless persons [67,68,88]. Preventive measures demand an awareness of the problems, not only in the population, but also in many responsible governmental institutions, and especially in the group of medical professionals and social services [8]. More than 70,000 additional deaths occurred in Europe during the summer of 2003 [84]. Major distortions occurred in the age distribution of the deaths (older population), but no harvesting effect was observed in the months following August 2003 [84,86,87]. Thus, warming constitutes a new health threat in an aged Europe that may be difficult to detect at the country level. Depending on its size, for example, centralizing the count of daily deaths on an operational geographical scale seems to be a clear priority to Public Health Service in Europe. Furthermore, a change in lifestyle is required in the group of people mentioned above, including dietary measures, sufficient fluid intake, and regulation of room climate. To gain a better understanding of heat waves on human health, a new generation of web-based, autonomous, noninvasive, light, and easy-to-handle methodologies have to be developed for collecting significant physiologic information on the impact of heat stress on humans in rural and urban areas compared with people in metropolitan areas. Such technology could be used to build a database that will support the understanding of the relationship between extreme climate conditions during sleep, rest, and work and the potential health risks. Moreover, this data could be used to develop a computer model as an early warning system for predicting urban population- and region-specific human health risks of future extreme climate condition situations, and provide architectural assistance and support, for example, in future city developments to optimize the protection of the population from climate hazards [83]. Especially, in recent years, heat waves were found to commence earlier in the year and to last longer, locally accompanied by heavy rain and varying levels of humidity. In the near future, we have to face even worse global weather conditions with increasing global temperatures [83]. The three different kinds of estimations concerning the global temperature development for the twenty-first century given in Figure 5.26 exceed by far the human comfort interval and must be looked upon as an extreme strain, because the body's heat-defensive mechanisms have to be permanently active. Full recovery during sleep is almost impossible under heat wave conditions, because night temperatures are kept high, so that the cardiovascular system is permanently stressed during day and night, causing significant increases in morbidity and mortality. Furthermore, since 2007 more people are living in cities than in rural areas, which means that in the near future these extreme climates will hit a population that is growing older, has impaired levels of fitness, and suffers increasingly from diseases such as diabetes, cardiovascular deconditioning, and dementia. Another severe factor is the exponentially increasing level of obesity and adiposity in the population. In addition, due to, for example, cardiovascular diseases, patients are forced to take medications that constrain the body's defensive mechanisms against heat. The aging, currently about 30-40% of the population in German cities, has to be considered as a risk group [84]. These people are frequently not considering themselves at risk, and consequently do not take any precautions and are mostly unable to maintain appropriate behavior to extreme climates such as adequate fluid intake. This consideration is of utmost importance, because the behavior of humans determines 90% of their thermoregulation. As outlined before, the body's own defensive mechanisms are of high importance, although they cannot compensate for the substantial consequences owed to deficits in behavior. While the statistical link between heat stress and health risks has been well documented, a better understanding of the relationship between climate conditions and health risks is clearly needed. Therefore, from a technological perspective, to achieve this goal specifically, new methodologies have to be available that can simultaneously record, for example, core and body shell temperatures, heart rate, activity and resting periods, sleep stages, as well as environmental conditions (ambient temperature, humidity, and atmospheric pressure), some of them already described in chapter 2. Furthermore, research studies have to be initiated (i) in a larger and longer scale, (ii) in different age groups, (iii) in healthy and unhealthy subjects, (iv) under various environmental conditions (seasons), and (v) under different physical activities. Once the feasibility of such technology is confirmed, it could be used to provide the basis for the development of a prediction system for determining the impact of heat stress on humans. This would promote, for example, the generation of human heat stress data in representative samples in diverse urban areas during different environmental conditions, and provide the basis to link research on climate change and health to other research disciplines such as meteorology, informatics, architecture, and engineering. These collaborative efforts could be used to develop computer models for predicting urban population- and region-specific human health risks of future extreme climate conditions, and to provide continuous support regarding preventive strategies for reducing climate-related morbidity and mortality. Furthermore, the development of individual mobile health risk monitoring systems could be used to assist local, regional, and global rural/urban infrastructure and architectural development planning [85].

These new methodologies and datasets could be used to assess in the impact of global warming on occupational health issues. According to the Kjellstrom study on "Global Assessment of the Health Impacts of Climate Change" [89], it is quite clear that occupational heat stress is already a significant problem in East Asia, South Asia, Southeast Asia, Central America, tropical Latin America, North Africa, East Africa, West Africa, and the Middle East. The study estimates that the global number of occupational heat stress fatalities due to climate change (additional workplace deaths) may amount to 12,000-30,000 cases in 2030 and 26,000-54,000 in 2050 [86,87,89]. For nonfatal heat stroke cases (in addition to cases in 1975), study calculations indicate 35,000-65,000 in 2030 and 40,000-73,000 in 2050. According to the study, there could also be more than 20 million heat exhaustion cases globally in 2030 due to climate change and possibly 40 million cases in 2050 (each case assumed to be seriously affected for 1 day). This would lead to a loss of work capacity globally of 1.0-1.7% (depending on climate model used) in 2030 and 1.7-2.4% in 2050. As these authors state, this may look like small changes, but in the worst affected regions (South Asia and West Africa), the estimated annual work capacity losses at population level are at least twice as high, not to mention that a healthy workplace climate can also be considered as a component of human rights to health. They also calculated the direct economic impact from sustained reductions in work output and gave an interesting example. If the economic growth per person of an economic unit (country, province, locality, company) is assumed to be 4% per year, in a 30-year period the income per person would increase from $2000 per year to $6500. If this growth was undermined by a 1% or 2% annual productivity loss, the resulting income after 30 years would be $4900 and $3500 respectively, which indicates losses of 36% or 67% of the additional economic growth. In this way, the climate change impact on work productivity clearly undermines the efforts to achieve the economic improvement targets in the Millennium Development Goals [87,89]. If the work capacity losses are expressed as Disability Adjusted Work Years lost (= lost years of fully healthy life in the age range 15-64 years) called DAWYs, the impact is even more apprehensible: The "health" losses in the hottest regions are similar to those caused by all cases of tuberculosis or all injuries in this age range. In their most interesting study, Kjellstrom et al. concluded that increasing occupational heat stress due to climate change is a very significant health and welfare challenge in tropical and other hot parts of the world [87,89,90].

In conclusion, (i) there are several proofs that the planet is warming; (ii) this is mainly due to human activities; (iii) this trend will continue for several decades or longer [91]; so that (iv) weather and climate will exert an increasing major influence on human health by such means as heat waves, floods, and storms, and of more indirect influences (v) on the distribution and transmission intensity of infectious diseases, as well as on the availability of freshwater and food [91–93]; and (vi) climate change has now been recognized as a global dimension issue that will increasingly affect human health and well-being on a wide scale.

5.12   SUMMARY

Man is counted among the endothermic organisms (mammals, birds) whose body temperature (36-40 °C) lies distinctly above the average temperature of their living environment. This high temperature gradient can be maintained only if heat development and heat loss are in balance. This heat balance is enabled by a high basal energy rate (tachymetabolism), insulating layers for the reduction of heat losses (subcutaneous fat tissue, hair coat), as well as complex mechanisms of temperature regulation (e.g., blood flow regulation, sweating). This enables endothermic organisms to keep their body temperature constant during a wide range of different states of activity and under wide ranges of different varying environmental conditions. Ectothermic organisms (reptiles, fishes), however, have a metabolism three to four times lower (bradymetabolism) and are less adaptive to changes in environmental temperatures. The high body temperatures of endothermic creatures (animals, animate beings) permit an active, all-year-round way of life, largely independent of environmental conditions. The temperature within the body, however, varies at different anatomical sites and depends on the respective metabolic activity of the organs (heat production). In the body core (brain, heart, liver) the tissue temperature is higher than in the periphery (legs, arms) under resting conditions. Furthermore, the subjective perception of temperature differs individually. It is influenced not only by air temperature, but also by air humidity and wind velocity. Temperature that is perceived as neither too hot nor too cold is called the indifference temperature. Under these conditions, heat production and heat loss are in balance.

Different transport mechanisms are available for heat transport from the body core to the periphery (internal heat transport) and from the periphery to the environment (external heat transport). The internal heat transport is affected mainly by means of blood circulation in a convective manner (convection = heat transport by means of a moving medium). Internal heat is also transported from the vascularized subcutaneous tissue to the body surface via vasoconstriction and vasodilatation. This is apparent especially in the extremities and acrae (e.g., hands, fingers, feet, ears) where the internal heat flow to the periphery can be dramatically increased or decreased. External heat transport in humans is enabled additionally by means of evaporation (evaporation of sweat) and radiation (long-wave, infrared radiation). Under resting conditions and ambient indifference temperature (27-31 °C), heat loss by radiation prevails. With physical work and/or warm environmental conditions, the organism depends more and more on evaporative heat loss. Physical and chemical processes provoking heat production or loss are regulated by the hypothalamus (preoptic area) by comparing the afferent information with an intrinsic reference value. This reference value is generated in the hypothalamus and shows cyclic changes throughout the course of the day (circadian rhythm). Hyperthermia (body core temperature >37.5 °C) is characterized by a disproportion between heat loss and heat production. However, the reference value, the so-called core temperature setpoint, which is created in the preoptic area of the hypothalamus, is unchanged in hyper- and hypothermia. During an increased metabolism, for example, strenuous physical exercise or heavy external heat gain (sauna), body core temperatures may rise rapidly to >40 °C. As a consequence, heat cramps, heat collapse, heat exhaustion, and even a life-threatening heat stroke may result. Hypothermia exists when the body temperature lies at or <35.5 °C. In cold water (5-10 °C), this value can be reached after just 10-20 min. Elderly people with a reduced metabolism, and infants (unfavorable surface-volume relationship) are particularly prone to suffering from hyper- or hypothermias. During a classic fever, in contrast to hyper- and hypothermia, the core temperature setpoint is up-regulated and mediated by exogenous and endogenous pyrogenes. A rising fever leads to a cold sensation and—among other features—is associated with increased heat production via muscle shivering. With declining fever, an adjustment back to normal core temperature takes place, which leads to promotion of heat loss mechanisms such as sweating and increased blood circulation of the skin. If heat losses exceed the production of heat in the organism for a longer period of time, the body core temperature continuously decreases.

With global warming, it can be foreseen that there will be a growing incidence of extreme climatic conditions. This will lead to considerable health problems and a substantial reduction in economical productivity. Thus, there should be increasing interest from the scientific and political communities to explore, understand, and answer the emerging questions of the impact of global warming on the public health of different populations around the world (rural and urban) and its impact on economies, with the ultimate aim to develop multifactorial preventive strategies for reducing climate-related issues such as morbidity and mortality.

REFERENCES

[1] Piantadosi CA. The biology of human survival life and death in extreme environments. 2003. Available from: http://search.ebscohost.com/login.aspx?direct=true&scope=site&db=nlebk&db=nlabk&AN=161222.

[2] McKnight T, Hess D. Climate zones and types: the Köppen system. Physical geography: a landscape appreciation. Upper Saddle River, NJ: Prentice Hall; 2000.

[3] Chapman F. The jungle is neutral. London: The Reprint Society; 1949.

[4] Edholm OG, Bacharach AL. The physiology of human survival. London: Academic Press; 1965.

[5] Sloan A. Man in extreme environments. Springfield, IL: Charles C. Thomas; 1979.

[6] Jessen C. Temperature regulation in humans and other mammals. Berlin, Heidelberg: Springer; 2001.

[7] Kleiber M. The fire of life: an introduction to animal energetics. New York: Wiley; 1961.

[8] Parsons K. Human thermal environments. London: Taylor & Francis; 2003.

[9] Schmidt-Nielsen K. Animal physiology. New York: Cambridge University Press; 1990.

[10] Gunga H. Wärmehaushalt und Temperaturregulation. In: Speckmann H, Hescheler J, Köhling R, editors. Physiologie. Elsevier; 2013. p. 601–28.

[11] Cooper KE, Cranston WI, Snell ES. Temperature in the external auditory meatus as an index of central temperature changes. J Appl Physiol 1964;19(5):1032–4.

[12] Shiraki K, Konda N, Sagawa S. Esophageal and tympanic temperature responses to core blood temperature changes during hyperthermia. J Appl Physiol 1986;61(1):98–102.

[13] Clark WG, Lipton JM. Drug-related heatstroke. Pharmacol Ther 1984;26(3):345–88.

[14] Nielsen B, Nielsen M. Body temperature during work at different environmental temperatures. Acta Physiol Scand 1962;56(2):120–9.

[15] Aikas E, Karvonen MJ, Piironen P, Rousteenoja R. Intramuscular, rectal and oesophageal temperature during exercise. Acta Physiol Scand 1962;54(3–4):336–70.

[16] Cooper KE, Kenyon JR. A comparison of temperatures measured in the rectum, oesophagus, and on the surface of the aorta during hypothermia in man. Br J Surg 1957;44(188):616–9.

[17] Cranston WI, Gerbrandy J, Snell ES. Oral, rectal and oesophageal temperatures and some factors affecting them in man. J Physiol 1954;126(2):347–58.

[18] Mairiaux P, Sagot JC, Candas V. Oral temperature as an index of core temperature during heat transients. Eur J Appl Physiol Occup Physiol 1983;50(3):331–41.

[19] Saltin B, Gagge AP, Stolwijk JA. Body temperatures and sweating during thermal transients caused by exercise. J Appl Physiol 1970;28(3):318–27.

[20] Pandolf KB, Sawka MN, Gonzalez RR. Human performance physiology and environmental medicine at terrestrial extremes. Brown & Benchmark; 1988.

[21] Molnar GW, Read RC. Studies during open-heart surgery on the special characteristics of rectal temperature. J Appl Physiol 1974;36(3):333–6.

[22] Greenleaf JE, Castle BL. External auditory canal temperature as an estimate of core temperature. J Appl Physiol 1972;32(2):194–8.

[23] Nadel ER, Horvath SM. Comparison of tympanic membrane and deep body temperatures in man. Life Sci I 1970;9(15):869–75.

[24] Saltin B, Hermansen L. Esophageal, rectal, and muscle temperature during exercise. J Appl Physiol 1966;21(6):1757–62.

[25] Stolwijk JA, Saltin B, Gagge AP. Physiological factors associated with sweating during exercise. Aerosp Med 1968;39(10):1101–5.

[26] Bräuer A, Weyland W, Fritz U, Schuhmann MU, Schmidt JH, Braun U. Determination of core body temperature. A comparison of esophageal, bladder, and rectal temperature during postoperative rewarming. Anaesthesist 1997;46(8):683–8.

[27] Aulick LH, Robinson S, Tzankoff SP. Arm and leg intravascular temperatures of men during submaximal exercise. J Appl Physiol Respir Environ Exerc Physiol 1981;51(5):1092–7.

[28] Dickey WT, Ahlgren EW, Stephen CR. Body temperature monitoring via the tympanic membrane. Surgery 1970;67(6):981–4.

[29] Tabor MW, Blaho DM, Schriver WR. Tympanic membrane perforation: complication of tympanic thermometry during general anesthesia. Oral Surg Oral Med Oral Pathol 1981;51(6):581–3.

[30] Wallace CT, Marks WE, Adkins WY, Mahaffey JE. Perforation of the tympanic membrane, a complication of tympanic thermometry during anesthesia. Anesthesiology 1974;41(3):290–1.

[31] Marcus P. Some effects of radiant heating of the head on body temperature measurements at the ear. Aerosp Med 1973;44(4):403–6.

[32] Marcus P. Some effects of cooling and heating areas of the head and neck on body temperature measurement at the ear. Aerosp Med 1973;44(4):397–402.

[33] McCaffrey TV, McCook RD, Wurster RD. Effect of head skin temperature on tympanic and oral temperature in man. J Appl Physiol 1975;39(1):114–8.

[34] Byrne C, Lim CL. The ingestible telemetric body core temperature sensor: a review of validity and exercise applications. Br J Sports Med 2007;41(3):126–33.

[35] Gunga H, Sandsund M, Reinertsen R, Sattler F, Koch J. A non-invasive device to continuously determine heat strain in humans. J Therm Biol 2008;33(5):297–307.

[36] Opatz O, Trippel T, Lochner A, Werner A, Stahn A, Steinach M, et al. Temporal and spatial dispersion of human body temperature during deep hypothermia. Br J Anaesth 2013;111(5):768–75.

[37] Fox RH, Solman AJ, Isaacs R, Fry AJ, MacDonald IC. A new method for monitoring deep body temperature from the skin surface. Clin Sci 1973;44(1):81–6.

[38] Smith P, Davies G, Christie MJ. Continuous field monitoring of deep body temperature from the skin surface using subject-borne portable equipment: some preliminary observations. Ergonomics 1980;23(1):85–6.

[39] Folk GE, Semken HA. The evolution of sweat glands. Int J Biometeorol 1991;35(3):180–6.

[40] Milan FA. The human biology of circumpolar populations. Cambridge University Press; 1980.

[41] Schaefer O, Hildes JA, Greidanus P, Leung D. Regional sweating in Eskimos compared to Caucasians. Can J Physiol Pharmacol 1974;52(5):960–5.

[42] Montain SJ, Cheuvront SN, Lukaski HC. Sweat mineral-element responses during 7h of exercise-heat stress. Int J Sport Nutr Exerc Metab 2007;17(6):574–82.

[43] Höfler W. Changes in regional distribution of sweating during acclimatization to heat. J Appl Physiol 1968;25(5):503–6.

[44] Adolph EF. Physiology of man in the desert. New York: Interscience Publishers; 1947.

[45] Allan JR, Wilson CG. Influence of acclimatization on sweat sodium concentration. J Appl Physiol 1971;30(5):708–12.

[46] Buono MJ, Ball KD, Kolkhorst FW. Sodium ion concentration vs sweat rate relationship in humans. J Appl Physiol 2007;103(3):990–4.

[47] Montain SJ, Sawka MN, Wenger CB. Hyponatremia associated with exercise: risk factors and pathogenesis. Exerc Sport Sci Rev 2001;29(3):113–7.

[48] Avellini BA, Shapiro Y, Pandolf KB, Pimental NA, Goldman RF. Physiological responses of men and women to prolonged dry heat exposure. Aviat Space Environ Med 1980; 51(10):1081–5.

[49] Avellini BA, Shapiro Y, Fortney SM, Wenger CB, Pandolf KB. Effects on heat tolerance of physical training in water and on land. J Appl Physiol Respir Environ Exerc Physiol 1982;53(5):1291–8.

[50] Avellini BA, Kamon E, Krajewski JT. Physiological responses of physically fit men and women to acclimation to humid heat. J Appl Physiol Respir Environ Exerc Physiol 1980;49(2):254–61.

[51] Bar-Or O. Effects of age and gender on sweating pattern during exercise. Int J Sports Med 1998;19(Suppl. 2):S106–7.

[52] Gagnon D, Jay O, Lemire B, Kenny GP. Sex-related differences in evaporative heat loss: the importance of metabolic heat production. Eur J Appl Physiol 2008;104(5):821–9.

[53] Gagnon D, Kenny GP. Sex modulates whole-body sudomotor thermosensitivity during exercise. J Physiol 2011;589(Pt 24):6205–17.

[54] Machado-Moreira CA, Smith FM, van den Heuvel AM, Mekjavic IB, Taylor NA. Sweat secretion from the torso during passively-induced and exercise-related hyperthermia. Eur J Appl Physiol 2008;104(2):265–70.

[55] Machado-Moreira CA, Caldwell JN, Mekjavic IB, Taylor NA. Sweat secretion from palmar and dorsal surfaces of the hands during passive and active heating. Aviat Space Environ Med 2008;79(11):1034–40.

[56] Machado-Moreira CA, Wilmink F, Meijer A, Mekjavic IB, Taylor NA. Local differences in sweat secretion from the head during rest and exercise in the heat. Eur J Appl Physiol 2008;104(2):257–64.

[57] Madeira LG, da Fonseca MA, Fonseca IA, de Oliveira KP, Passos RL, Machado-Moreira CA, et al. Sex-related differences in sweat gland cholinergic sensitivity exist irrespective of differences in aerobic capacity. Eur J Appl Physiol 2010;109(1):93–100.

[58] Shapiro Y, Pandolf KB, Avellini BA, Pimental NA, Goldman RF. Physiological responses of men and women to humid and dry heat. J Appl Physiol Respir Environ Exerc Physiol 1980;49(1):1–8.

[59] Shapiro Y, Pandolf KB, Avellini BA, Pimental NA, Goldman RF. Heat balance and transfer in men and women exercising in hot-dry and hot-wet conditions. Ergonomics 1981;24(5):375–86.

[60] Taylor NA, Machado-Moreira CA. Regional variations in transepidermal water loss, eccrine sweat gland density, sweat secretion rates and electrolyte composition in resting and exercising humans. Extreme Physiol Med 2013;2(1):4.

[61] Havenith G, Fogarty A, Bartlett R, Smith CJ, Ventenat V. Male and female upper body sweat distribution during running measured with technical absorbents. Eur J Appl Physiol 2008;104(2):245–55.

[62] Smith CJ, Havenith G. Body mapping of sweating patterns in athletes: a sex comparison. Med Sci Sports Exerc 2012;44(12):2350–61.

[63] Smith CJ, Havenith G. Body mapping of sweating patterns in male athletes in mild exercise-induced hyperthermia. Eur J Appl Physiol 2011;111:1391–404.

[64] Rowell LB. Human cardiovascular control. Oxford University Press; 1993.

[65] Gauer OH, Henry JP. Neurohormonal control of plasma volume. Int Rev Physiol 1976;9:145–90.

[66] Gathiram P, Wells MT, Raidoo D, Brock-Utne JG, Gaffin SL. Portal and systemic plasma lipopolysaccharide concentrations in heat-stressed primates. Circ Shock 1988;25(3):223–30.

[67] Falk B. Effects of thermal stress during rest and exercise in the paediatric population. Sports Med 1998;25(4):221–40.

[68] Falk B, Dotan R. Children's thermoregulation during exercise in the heat: a revisit. Appl Physiol Nutr Metab 2008;33(2):420–7.

[69] Sinclair WH, Crowe MJ, Spinks WL, Leicht AS. Pre-pubertal children and exercise in hot and humid environments: a brief review. J Sports Sci Med 2007;6(4):385–92.

[70] Larose J, Boulay P, Sigal RJ, Wright HE, Kenny GP. Age-related decrements in heat dissipation during physical activity occur as early as the age of 40. PLoS One 2013;8(12):e83148.

[71] Popkin BM, D'Anci KE, Rosenberg IH. Water, hydration, and health. Nutr Rev 2010;68(8): 439–58.

[72] Kenefick RW, Leon LR. Pathophysiology of heat-related illnesses. In: Auerbach PS, editor. Wilderness medicine. Philadelphia: Elsevier; 2012. p. 215–31.

[73] Shkolnik A, Taylor CR, Finch V, Borut A. Why do Bedouins wear black robes in hot deserts? Nature 1980;283(5745):373–5.

[74] Schmidt-Nielsen K. Osmotic regulation in higher vertebrates. Harvey Lect 1963;58:53–93.

[75] Schroter RC, Robertshaw D, Baker MA, Shoemaker VH, Holmes R, Schmidt-Nielsen K. Respiration in heat stressed camels. Respir Physiol 1987;70(1):97–112.

[76] Gomez CR. Disorders of body temperature. In: Handbook of clinical neurology. Elsevier; 2014. p. 947–57.

[77] Gisolfi CV, Spranger KJ, Summers RW, Schedl HP, Bleiler TL. Effects of cycle exercise on intestinal absorption in humans. J Appl Physiol 1991;71(6):2518–27.

[78] Maughan RJ. Distance running in hot environments: a thermal challenge to the elite runner. Scand J Med Sci Sports 2010;20(Suppl. 3):95–102.

[79] Maughan RJ, Shirreffs SM. Development of individual hydration strategies for athletes. Int J Sport Nutr Exerc Metab 2008;18(5):457–72.

[80] Dinarello CA. Infection, fever, and exogenous and endogenous pyrogens: some concepts have changed. J Endotoxin Res 2004;10(4):201–22.

[81] Netea MG, Kullberg BJ, Van der Meer JW. Circulating cytokines as mediators of fever. Clin Infect Dis 2000;31(Suppl. 5):S178–84.

[82] Campbell-Lendrum D, Corvalán C. Climate change and developing-country cities: implications for environmental health and equity. J Urban Health 2007;84(Suppl. 1):109–117.

[83] Cubasch U, Wuebbles D, Chen D, Facchini MC, Frame D, Mahowald N, et al. In: Climate change 2013: the physical science basis. Contribution of working group I to the fifth assessment

report of the intergovernmental panel on climate change 2013. Cambridge, United Kingdom and New York, NY, USA: Cambridge University Press; 2013.

[84] Robine JM, Cheung SL, Le Roy S, Van Oyen H, Griffiths C, Michel JP, et al. Death toll exceeded 70,000 in Europe during the summer of 2003. C R Biol 2008;331(2):171–8.

[85] Dunne JP, Stouffer RJ, John JG. Reductions in labour capacity from heat stress under climate warming. Nat Clim Change 2013;3(6):563–6.

[86] Berry HL, Bowen K, Kjellstrom T. Climate change and mental health: a causal pathways framework. Int J Public Health 2010;55(2):123–32.

[87] Kjellstrom T, Friel S, Dixon J, Corvalan C, Rehfuess E, Campbell-Lendrum D, et al. Urban environmental health hazards and health equity. J Urban Health 2007;84(1):86–97.

[88] Falk B, Bar-Or O, MacDougall D, Goldsmith H, McGillis L. Longitudinal analysis of the sweating response of pre-, mid-, and late-pubertal boys during exercise in the heat. Am J Hum Biol 1992;4(4):527–35.

[89] Kjellstrom T editor. Climate change exposures, chronic diseases and mental health in urban populations: a threat to health security, particularly for the poor and disadvantaged. Kobe, Japan: WHO Centre for Health Development; 2012.

[90] Hollowell DR. Perceptions of, and reactions to, environmental heat: a brief note on issues of concern in relation to occupational health. Global Health Action 2010;3:5632.

[91] Pachauri R, Reisinger A. Contribution of working groups I, II and III to the fourth assessment report of the intergovernmental panel on climate change. 2007, IPCC, Geneva, Switzerland. pp 104.

[92] McMichael AJ, Campbell-Lendrum D, Kovats S, Edwards S, Wilkinson P, Wilson T, et al. Global Climate Change. In: Ezzati M, Lopez A, Rodgers A, Murray CJL, editors. Comparative quantification of health risks: global and regional burden of disease attributable to selected major risk factors. Geneva: World Health Organization; 2004. p. 1543–650.

[93] Oreskes N. Beyond the ivory tower. The scientific consensus on climate change. Science 2004;306(5702):1686.

[94] Aschoff J. Exogenous and endogenous components in circadian rhythms. Cold Spring Harb Symp Quant Biol 1960;25:11–28.

[95] Kunsch K, Kunsch S. Der Mensch in Zahlen: eine Datensammlung in Tabellen mit über 20000 Einzelwerten. München: Spektrum, Akademischer Verlag; 2007.

[96] Leon LR, Kenefick RW. Pathophysiology of heat-related illnesses. In: Auerbach PS, editor. Wilderness medicine. 6th ed. Philadelphia: Elsevier; 2012. p. 215-231.

[97] Folk GE. Textbook of environmental physiology. Philadelphia: Lea & Febiger; 1974.

[98] Stocker TF, Qin D, Plattner GK, Tignor M, Allen SK, Boschung J, et al., editors of IPCC: Summary for Policymakers. Cambridge: Cambridge University Press; 2013.

[99] Kamler K. Extreme survival. London: Robinson; 2004.

Chapter 6

Cold Environments

Mathias Steinach* and Hanns-Christian Gunga†
*Postdoctoral Research Associate, Center for Space Medicine and Extreme Environments, Institute of Physiology, CharitéCrossOver (CCO), Charité University Medicine Berlin, Berlin, Germany
†Professor, Center for Space Medicine and Extreme Environments, Institute of Physiology, CharitéCrossOver (CCO), Charité University Medicine Berlin, Berlin, Germany

PREFACE

This chapter differs distinctly from the previous and the following ones. We decided to use this chapter on Cold Environments to report on research of our own working group on extreme environments, which is currently ongoing in the Arctic and Antarctic regions. We intended to illustrate on one hand the extraordinary human capabilities to live and work in extreme environments, and on the other hand, we thought it might be of interest to the reader to participate, at least in one chapter, in such ongoing research studies.

6.1 INTRODUCTION

Cold environments are environments in which the ambient temperature of the atmosphere is close to or below 0 °C. There are four main types of cold environment: polar, high mountain, glacial, and periglacial, the latter mainly represented by the Arctic and Antarctic regions that receive less intensive solar radiation because the sun's light hits the earth at an oblique angle, that is, spreading over a larger area and passing through the earth's atmosphere in a longer distance. Thus the energy gets absorbed, scattered, and reflected. The present-day global distribution of these four environments is shown in Figure 6.1 [1].

Usually the Arctic region is defined as north of 60° north latitude, or the region from the north pole south of the timberline; the Antarctic is defined as south of 60° south latitude. About 10% of the earth's land surface is covered by ice today. Antarctica accounts for 85% of the total and Greenland for 10%; the rest is comprised of vast periglacial regions in Russia and Canada. These regions are characterized by low-growing tundra vegetation and permanently

Human Physiology in Extreme Environments. http://dx.doi.org/10.1016/B978-0-12-386947-0.00006-X

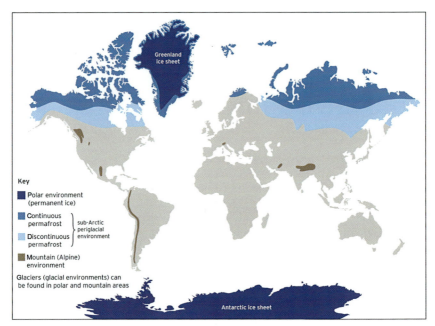

FIGURE 6.1   The different polar regions of the earth, *created from [1]*.

frozen soils. Extremely cold conditions can also be found in the mountain regions, such as the Himalayas, Andes, and Rocky Mountains. In the past, as mentioned in chapter 1, large parts of the Earth were covered with ice, and only about 20,000 years ago, the United States and northern Europe were covered with an extremely large ice shield. Today, in Antarctica and Siberia, extremely low temperatures can be observed in winter, sometimes reaching below −70 °C. Antarctica and northern polar regions are very dry environments with relatively low amounts of precipitation (i.e. snow accumulation). *Periglacial* literally means edge of glacial. These environments are located on the fringes of polar or glacial environments, mainly in parts of Siberia, Canada, and the Antarctic ice sheet. Glacial environments are specifically associated with glaciers. While some enormous glaciers are found in polar environments, most of the world's actively moving glaciers are found high up in high mountain regions (alpine environment). The heavy winter snowfall in these areas provides the ice to feed the glaciers. Then, in the summer, meltwater lubricates the glaciers, allowing them to move like giant conveyor belts down the valleys.

High mountain areas experience very cold winters with heavy snow. Because of the high altitude, the temperature can drop far below −30 °C, and strong winds worsen the environmental conditions for humans. The extreme winter cold is replaced with warmer weather in the summer. Then the ambient temperature can be far above 0 °C. As human physiological adaptations are very limited—such as involuntary muscle contractions to produce heat, non-shivering thermogenesis by beta-adrenergic activation of brown-fat tissue—behavior, technological

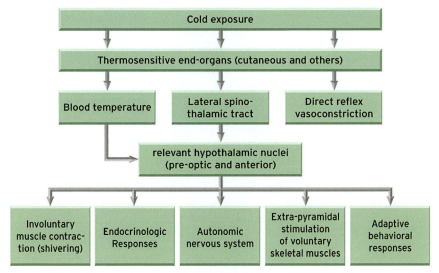

FIGURE 6.2   Physiological pathways activated by cold exposure in humans, *adapted from [2].*

inventions, and logistical support play a central role in surviving in such a cold environment (Figure 6.2) [2].

Technological inventions and logistical support especially enable humans to maintain a permanent yearlong stay in cold environments. In the arctic polar regions, the Inuit have developed such extraordinary skills and technical solutions over several thousand years to withstand such a harsh environmental climate. One key technical solution is shown in Figure 6.3, typical Inuit clothing.

This design ensures flexible thermal management in the cold. During rest the fur inside the suit and around the face reduces heat losses to the environment, and during heavy work, the bands around the arms and extremities can be opened to allow heat transfer from the body to the environment; these parts act then as thermal windows. Even sweat can be partly removed by these avenues, so that it does not accumulate inside the suit. The other technological invention is the snowhouse, called *igloo* or *iglu*, a type of shelter built of snow, originally built by the Inuit at the northeastern shores of Greenland [3]. A typical cross section through an iglu is shown in Figure 6.4.

But aside from the challenging thermal demands of the cold, the extreme metabolic aspects of humans living and even exercising in cold environments will be illustrated in the following paragraph dealing with the Yukon Arctic Ultra (YAU), a long-distance race in the northern part of Canada—taken as an example of the challenges in the Arctic region before we will describe the Antarctic regions in chapter 6.3.

6.2   ARCTIC REGIONS

The Yukon Territory lies in the northwest of Canada bordering the Northwest Territories to the east, British Columbia to the south, and the U.S.-Alaska to the

FIGURE 6.3   Typical Inuit suit—at the arms and legs the suit can be opened and used as a "thermal window".

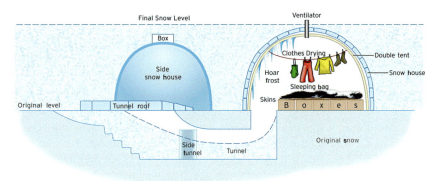

FIGURE 6.4   Cross cut view of an igloo—the two snow house parts are connected by a tunnel.

west [4]. Its name is derived from the Yukon river because most of its territory lies in the watershed of this river. The landscape is characterized by many glacier-fed lakes, rugged mountains—the highest point is also Canada's highest peak, Mount Logan (5959 m). A considerable amount of the Yukon landscape is comprised of national parks (Kluane National Park, Ivvavik National Park, and Vuntut National Park), some of which have also been declared World Heritage Sites by the UNESCO to preserve the unspoiled pristine nature of the landscape. The largest city is the capital Whitehorse, with a population of about 20,000, which is two-thirds of the entire Yukon population; the second largest city is Dawson City with a population of only about 1300 inhabitants.

During the nineteenth century fur trade, and later the gold rush, brought European settlers to the Yukon Territory where the inhabitants of the native First Nations had already established settlements and trading routes. Recent politics try to preserve the heritage of the First Nation descendants such as the Vutnut-Gwitchin, Selkirk, or Trondëk-Hwëchin First Nation and their cultures and languages. According to the 2006 Canada census, 57% of the Yukon population are of multiple ethnicity, and 13% are of aboriginal or First Nation ethnicity [5]. The gold rush and the subsequent mining industry heavily influenced today's economy, while other fields such as hydropower and tourism have gained much importance [4].

The climate of the Yukon is characterized by its global location, leading to a continental arctic climate [6]. The average temperatures during the summer can become quite hot, with peaks reaching over 30 °C, while the lows during the winter can become very cold, reaching ambient daytime temperatures of −25 and −55 °C during the night. Figure 6.5 illustrates the measured ambient temperatures during the YAU 2013. As can be seen, the temperatures, especially during the beginning of the YAU 2013, were relatively mild.

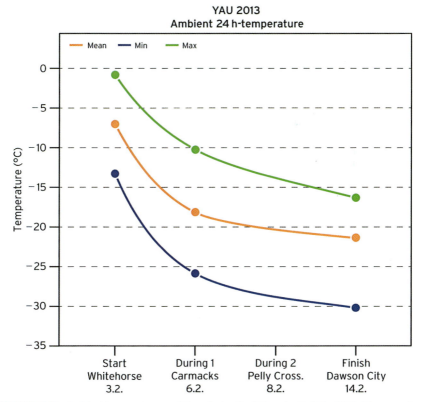

FIGURE 6.5 Ambient temperature conditions during the Yukon Arctic Ultra 2013. The respective locations and dates of measurement, e.g. "Start in Whitehorse on the 3rd of February", are given on the x-axis.

6.2.1   The Yukon Arctic Ultra

The Yukon Arctic Ultra (YAU) is an ultramarathon in the extreme environment of the arctic northwestern Yukon Territory of Canada [7,8]. The race takes place during February each year and offers extreme athletes the opportunity to enter competition in the disciplines of running, cross-country-skiing, or mountain biking. Athletes from these disciplines can enter the different distance categories. The distances covered are standard marathon (42 km), 100 miles (160 km), 300 miles (480 km), and 430 miles (690 km), where the longest distance of 430 miles takes place biannually. Thus the race combines high levels of physical activity in an extremely cold environment.

Participants of the YAU need to be adults (both sexes can participate) and are required to prove their skills (e.g., outdoor orientation, capability to make fire and prepare food, capability to bivouac) in order to be allowed to enter the race. A survival training course covering topics like first aid, shelter, gear-check and food and water preparation in the cold is mandatory for those athletes who enter the YAU for the first time. Usually, the entering athletes have a long history of participation in marathons and other ultra-endurance-events, some even in extreme environments, such as the Marathon des Sables in Morocco [9]. In addition, participants must cover all their own costs (travel to the Yukon, stay, equipment, insurance) and also disburse a fee (up to €2000) to enter the YAU, which covers the organization and realization of the race (e.g., logistics, trail preparation, safety, medical supervision, insurance). No prize money is awarded.

The YAU follows the traditional path of the Yukon Quest dog sled race and starts in the Yukon's capital Whitehorse with several checkpoints until the finish in Dawson City (430 miles) or Pelly Crossing (300 miles). The trail's elevation changes over the course of the race from 610 meter at the start to the highest elevation of 975 meter until the finish in Dawson with 305 meter. Figure 6.6 shows the route of the YAU.

The route is prepared by support staff using snowmobiles; however, elevations as well as unforeseen snowfall and overflow (water flooding the trail) can make the trail cumbersome and difficult to travel on. In addition, the trail is marked with wooden sticks with fluorescent tops (from the Yukon Quest dog sled race); however, heavy snowfall can obliterate these markers.

The athletes are required to carry their own mandatory equipment (e.g., bivvy bag, sleeping bag, thermo-protective clothes, food and water, head torch, flare matches and lighter, first aid kit, stove, small saw. In addition, a snow shovel, crampons and a GPS device are mandatory for the 430-mile participants) in a sledge behind them while moving. Non-mandatory gear can be prepared as drop bags to be deposited at the checkpoints, limited to three drop bags for the 300 mile and four for the 430 mile participants. Except for the few checkpoints, participants rest, sleep, prepare their food, and perform all other activities outdoors.

The YAU was established in 2003 and has taken place annually ever since. Thus the 2013 YAU was the 10th anniversary with a high participation rate—overall 76 participants entered the YAU in its various categories and distances;

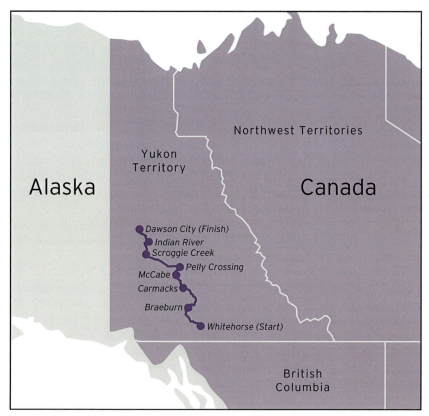

FIGURE 6.6   The course of the Yukon Arctic Ultra in North-West Canada.

from those only 40 participants were able to finish the race (52%), which indicates the difficult conditions of the YAU. The race started February 3rd 2013 at 10:30 a.m. As an orientation, the finish times for the 430-mile race by foot ranged from 186 h 50 min (= 7 days 18 h 50 min) to 300 h 35 min (= 12 days 12 h 35 min); 12 participants were able to finish from 24 athletes who entered this category (50%).

The presented study was conducted with participants in the 430-mile foot race category who volunteered to particpate in the study measurements. Conditions, circumstances, and results of the study will be described in the subsequent paragraphs.

6.2.2   Study Design

The participants of the YAU 2013 were recruited through the direct approach with the support of the YAU 2013 organizers; adults of both sexes were admitted. All had taken part in previous YAU races as well as many other endurance events—this was seen as beneficial in order to increase the chance for the participants to reach the finish in Dawson City, but also to have participants who were already familiar with the race in order to decrease their perceived burden of wearing the measurement devices. However, none of the participants were professional runners—all had regular occupations outside their interest in ultra-endurance events. The participants were informed of the implications of the study and gave their written consent to participate as volunteers.

In detail, a total of six volunteer runners (five male, one female) of the 430-mile race did take part from which four eventually reached the finish in Dawson City (66%)—two participants had to scratch from the race in its early stage due to knee problems (one male) and general problems (one female—feelings of dizziness and illness) respectively. Further details are given in the section about anthropometric data of the participants.

The measurements conducted in this study were split in two distinct sets: one set of measurements before (Pre), during, and after (Post) the YAU 2013 as well as continuous measurements throughout the race.

The measurements were taken two nights before the start in Whitehorse (Pre), then at the checkpoints Carmacks (after 277 km = "During 1"), Pelly Crossing (after 382 km = "During 2"), and finally in Dawson City (after 690 km = "Post"). The measurements consisted of body weight, body composition and energy intake, sleep parameters, and heart rate variability. The measurements were combined in order to impose the least strain upon the participants: Measurements started with taking the sleep measurement during the night, followed by the measurements of body weight and composition and heart rate variability the next morning. Sleep was measured using the actimeter SenseWear device and the Zeo-Sleep device, which is a validated system for sleep analysis [10] made up of a headband measuring EEG signals sending data to a device closeby where the analysis regarding sleep phases is performed. Body weight and body composition were measured using a calibrated scale and the validated bio-impedance procedure (as described in chapter 2 of this book; manufacturer Akern); energy intake was estimated using food protocols, which was possible because the participants took thorough notes of all food they consumed. Heart rate variability was evaluated after eight-minute-long measurements were taken after awakening using the Polar Heart Rate Monitor (RS800CX) and subsequent beat-to-beat analysis performing power spectral analysis (due to benign arrhythmic events in one participant, the data from only three participants could be used for analysis of heart rate variability; technical problems did not allow measurements at "During 1").

Continuous measurements were conducted using the actimeter SenseWear for evaluation of daily energy expenditure and sleep parameters (laytime—the time at rest in bed but not asleep, sleeptime, and sleep efficiency) as well as the heart rate monitor to measure heart rate (as average values per minute).

The resulting data were analyzed using respective analysis and statistical software evaluating for statistical significance. Due to the pilot nature of this study with the small number of four participants (three for heart rate variability), only descriptive statistics were applied.

6.2.3   Results and Discussion

6.2.3.1   Anthropometric Data

As mentioned, a total of six participants entered the 430-mile footrace of the YAU 2013 as volunteers of this study. However, only four were able to reach the finish in Dawson City, while two had to abandon the race at an early stage. For this reason only the data of the four individuals reaching Dawson City were considered in the study evaluation. These participants were all male and by median 44 years old (25th and 75th percentile: 37.5 and 47.5 years), 179.0 cm in height (25th and 75th percentile: 178 and 182 cm), and 78.0 kg in weight (25th and 75th percentile: 73 and 87.7 kg) with a median BMI of 23.7 kg/m² (25th and 75th percentile: 22.7 and 27.6 kg/m²).

6.2.3.2   Energy Expenditure

The evaluation of the total daily energy expenditure revealed very high rates of daily expenses by the participants compared to general values. Especially high were the rates at the beginning of the race and again at the end when the participants got closer to the finish. The highest values reached a peak of 38,000 kJ per day, and average daily expenditures rated in metabolic equivalents reached up to 4.5 MET per day. Figure 6.7 shows the daily energy expenditure of the four participants.

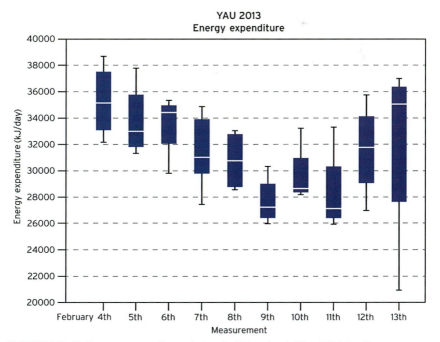

FIGURE 6.7   Daily energy expenditures during the Yukon Arctic Ultra 2013 ($n=4$).

With regard to statistical analysis over the course of the race, no results attained statistical significance (RM ANOVA); however, the probability value is very close to the threshold value ($p=0.066$). When considered for the entire race, each participant expended an average amount of energy of approximately 350 MJ.

When comparing energy intake and expenditure, a considerable energy deficit emerges. As Figure 6.8 reveals, there is a substantial difference between energy intake and expenditure; this difference is of statistical significance (*t*-test, $p=0.0007$).

The analysis also reveals that in average only about 44% of the energy expenses accumulated over the course of the race were covered by energy intake during the same time.

The participants of the YAU 2013 exercised daily on a high endurance level. Exercise—depending on mode and level of activity—can yield high increases in energy expenditure [11] (see chapter 3 regarding exercise in this book). In addition, energy expenditure can be increased during exposure to a cold environment [12,13] because its increase is one form of human adaptation to a cold

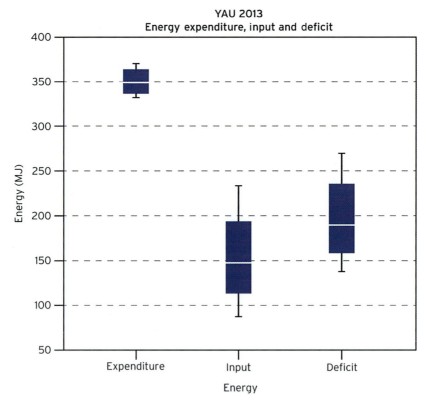

FIGURE 6.8   Total energy expenditure, input and deficit during the Yukon Arctic Ultra 2013 ($n=4$).

environment [14–16]. Increases in thyroid activity and brown adipose tissue activation seem to play an important role [17–19] as the function of brown adipose tissue was recently discussed in human adults [20].

The measurements regarding energy expenditure revealed rather high values in the participants (4.5 MET) with an average total daily energy expenditure of around 35,000 kJ, which surpasses the total daily energy expenditure of R.F. Scott and his men when calculating their daily energy expenditure during their race for the South Pole, which was around 26,000 kJ [21]. Other ultra-endurance events have also shown a high energy expenditure in the past as well; however, the daily expenditures during the YAU seem to be very high in comparison [22,23].

It is interesting to note that the values were higher at the start and the end of the race, which can be attributed to the motivation and excitement of the participants. This has been shown before as measured by oxygen uptake in endurance athletes [24].

With regard to energy intake and deficit, there appears to be a rather great deficit because only about 44% of the expended energy was being fueled by the energy intake. Personal interviews with the participants revealed that they simply had no appetite to consume more food than they did. This was also the case in another study where participants of an endurance event had a reduced energy intake [25], which was identified in yet another study of this subject to be due to low appetite and gastrointestinal problems [26].

6.2.3.3 Body Composition

As previously described, only a fraction of the energy expenditure was covered by energy intake. As a result, changes in body weight and composition occurred. In average, the volunteer participants lost 6.2 kg of body mass over the course of the race with a maximum loss of 7.9 kg in one participant, which equals an average loss of 7.7% of the starting body weight. The changes in body mass led to significant results (RM ANOVA, $p < 0.001$). Figure 6.9 illustrates the changes in body mass measured at the four measurement points.

With regard to the changes in body composition, there was initially a loss in fat mass that persisted throughout the race. As the race progressed, there was also a considerable loss in fat-free mass. The change in fat mass is statistically significant (RM ANOVA, $p < 0.001$), but not in fat-free mass, while the probability value for the changes in fat-free mass is very close to the threshold value (RM ANOVA, $p = 0.052$). On average, there was a loss of 3.9 kg in fat mass and 2.3 kg in fat-free mass over the course of the race, which equals a loss of 27.9% from the starting fat mass and 3.4% from the starting fat-free mass respectively. Figure 6.10 depicts the changes in fat mass and fat free mass.

The loss of body mass that occurred in the fat mass and later on also in the fat-free mass can most likely be attributed to the rather high energy deficit resulting from the high-energy expenditure and relatively low energy intake. Investigations with other ultra-endurance events also showed decreases in body mass—mostly attributed to the loss in fat mass but also in fat-free mass [27–30].

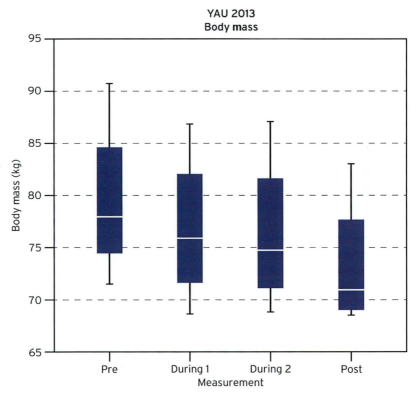

FIGURE 6.9 Changes in body mass during the Yukon Arctic Ultra 2013 ($n=4$). The times of measurement are given on the x-axis (Pre=Feb 3rd, During 1=Feb 6th, During 2=Feb 8th and Post=Feb 14th).

In addition, dehydration may also play a role in body mass loss in participants of ultra-endurance events [31]. As mentioned before, the activation of brown adipose tissue in response to the external cold stimuli might also have contributed to an increased energy expenditure and thus to the loss in body mass, especially in fat mass [32,33], therefore it might be possible that an ultra-endurance event like the Yukon Arctic Ultra leads to a high energy expenditure through a combination of physical exertion of the participants but also the cold ambient temperatures.

6.2.3.4 Sleep Parameters

The analysis of the measurements conducted with the actimeter SenseWear revealed that there was a considerable deficit in laytime and sleeptime when comparing the data with general values. The average laytime was 329 min (5 h, 29 min), and the average sleeptime was 253 min (4 h, 13 min). In addition, there was a pronounced sleep deficit at the start and the end of the race (median sleeptimes 125 and 191 min for the first and last night respectively). Also it is of interest to note that there were no increases in these sleep parameters when

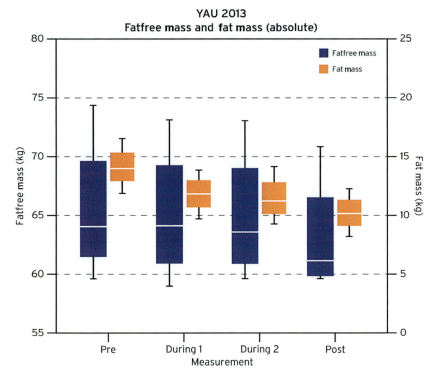

FIGURE 6.10 Changes in fat mass and fat free mass during the Yukon Arctic Ultra 2013 ($n=4$).

participants slept indoors at the checkpoints (nights four and six). The average calculated sleep efficiency was about 75%. The changes in laytime were of statistical significance (RM ANOVA, $p=0.023$); the changes in sleeptime were not of statistical significance, although the probability parameters were very close to the threshold value (RM ANOVA, $p=0.05$). Figure 6.11 illustrates the changes in sleeptime over the course of the race.

With regard to the changes measured with the ZEO-Sleep device, there were noticeable changes in the sleep phases of the participants: a pronounced decrease in REM-sleep directly after the beginning of the race and an increase in deep-sleep directly after the end of the race, with a great variation in some of the participants. The change in REM-sleep was of statistical significance (RM ANOVA, $p=0.005$), the change in deep-sleep was not (RM ANOVA, $p=0.386$). Figures 6.12 and 6.13 illustrate these changes.

The findings of this study seem to be in accordance with findings of past research: Cold exposure leads to a disrupted sleep and changes in sleep phases in humans [34] as well as in the animal model [35]. Cold-induced shivering occurring during sleep with an increase in oxygen consumption and thus energy expenditure may play a role leading to disrupted sleep [36]. In addition, it was found that stress hormone levels increase in response to cold stimuli, which in

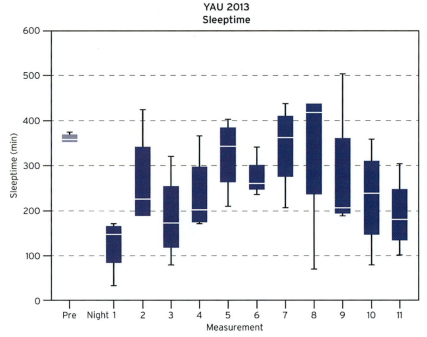

FIGURE 6.11   Changes in sleeptime during the Yukon Arctic Ultra 2013 ($n=4$).

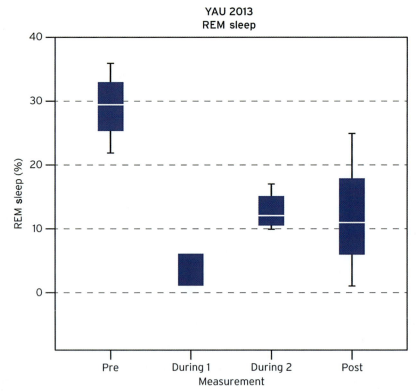

FIGURE 6.12   Changes in REM-sleep (Rapid Eye Movement) during the Yukon Arctic Ultra 2013 ($n=4$).

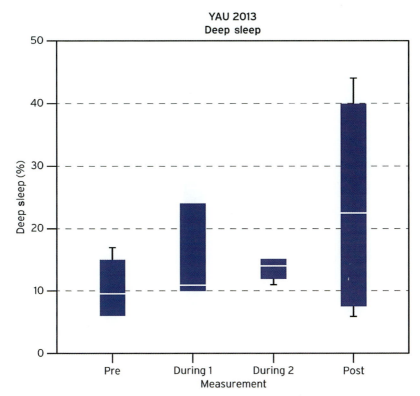

FIGURE 6.13   Changes in deep sleep during the Yukon Arctic Ultra 2013 ($n=4$).

turn may lead to poor sleep quality [37] and a reduction in REM-sleep [38], which can reactively increase again during subsequent sleep periods [39].

In turn, sleep deprivation—as it was found with participants of this study—may itself impair thermoregulation [40], which in turn could decrease the body's capability for sleep homeostasis due to the complex interaction through hypothalamic neuropeptides [41,42]. This also might increase the risk for cold injuries [43]. Therefore, athletes' sleep in a cold environment such as the Yukon Territory during the YAU with a high level of physical exertion may be impaired due to a combination of the influencing factors, cold environment and the increased level of stress hormones in response to exercise [44–46].

6.2.3.5   Heart Rate Variability

The analysis of the changes in vegetative control as expressed and measured through the parameter of heart rate variability revealed interesting results. Due to benign arrhythmic events (ventricular extrasystoles) in one participant, the data of only three participants could be used in this analysis. The changes within the time domain revealed a decrease in beat-to-beat-variability

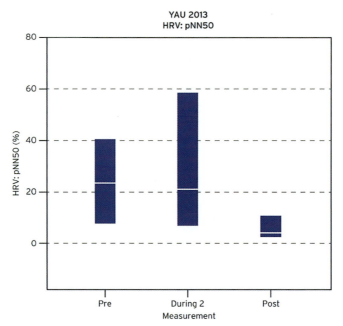

FIGURE 6.14 Changes in heart rate variability: pNN50 during the Yukon Arctic Ultra 2013 (HRV, Heart Rate Variability; $n=3$).

as expressed in the pNN50[1]; these changes were not of statistical significance, however (RM ANOVA, $p=0.429$). The changes within the frequency domain revealed a decrease in the low frequency to high frequency ratio (LF/HF-ratio) as an expression of the sympathetic-vagal-balance with an increase of rhythms predominantly influenced by the parasympathetic nervous system; these changes were of statistical significance (RM ANOVA, $p=0.035$). Figures 6.14 and 6.15 illustrate these changes in heart rate variability.

The analysis of the heart rate variability is a reliable tool used in several clinical and scientific applications to evaluate the heart's autonomic function and the state of the vegetative nervous system in general [47–49].

It is known that regular endurance exercise for health purposes can lead to an increased heart rate variability in both younger and older human subjects [50], often with a higher parasympathetic component in endurance-trained individuals compared to controls [51].

However, as was the case in this study with an ultra-endurance exercise, heart rate variability may also decrease post exercise [52], which was attributed to a downregulation of beta-adrenergic receptors with a subsequent increase in parasympathetic influence reflected by a decrease in the LF/HF ratio [53]. Fatigue might also play a role in this context as was discussed in another

---

1. The pNN50 denotes the percentage of consecutive RR-intervals that differ from each other by a time interval of 50 ms or more.

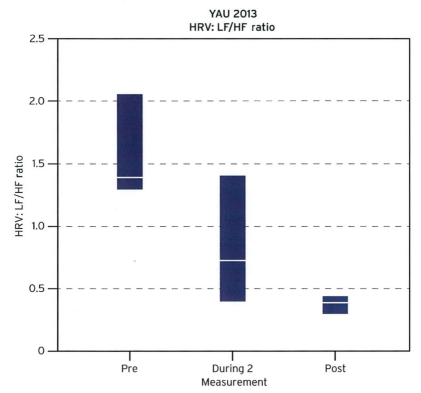

FIGURE 6.15   Changes in heart rate variability: LF/HF-ratio during the Yukon Arctic Ultra 2013 (HRV, Heart Rate Variability; LF, Low Frequency portion; HF, High Frequency portion; $n=3$).

investigation where a lower heart rate variability was found in fatigued athletes versus non-fatigued ones [54].

The statistically significant results from this study—despite the very low number of evaluated subjects—seem to be in accordance with these previous findings, although changes in heart rate variability are not always educible in respective studies [55].

6.2.4   Limitations

It is in the nature of a pilot study that many of the parameters associated with such a study have not yet been established. This undertaking was therefore also conducted as a feasibility study to assess whether such a study can be conducted with regard to work efforts, logistics, financial expenses, and other parameters. As a result only a very small number of volunteer participants were recruited to participate ($n=6$). In addition, due to the nature of this very demanding race, a rather high dropout rate occurred, leading to eventually only four participants reaching the finish and entering study evaluation. Furthermore, due to benign arrhythmic events in one participant the data of only three participants could be used for analysis of heart rate variability. With regard to statistical analysis, a very small number leads to a low power, which makes it difficult to detect statistically significant changes in the study cohort even if they do exist.

Also, not all study parameters that we would have wanted to evaluate did get analyzed. Some of the additional parameters would have been hormone changes (erythropoiesis, metabolism, thyroid hormones), blood parameters (blood counts, hemostasis), stress parameters, and others. It is planned to analyze these parameters in future studies of the YAU to allow for a more complex analysis.

With regard to energy expenditure, it has been argued that actimeter such as the SenseWear device used in this study do not correctly reproduce high energy expenditure values and thus underestimate this parameter [56]. Although the newest algorithm available for the analysis has been applied, it is possible that the energy expenditures—and thus the energy deficit—were even higher than described. In addition, for the estimation of energy intake, all foods brought and consumed by the participants were accounted for; however, some foods such as soups and other meals and snacks such as nuts at the checkpoints had to be estimated using food protocols. Rather conservative ratios using the upper end of quantity and energy content were applied.

6.3 ANTARCTIC REGIONS

6.3.1 Introduction

6.3.1.1 Antarctica: Geological, Climatic, and Historical Background

Today's Antarctica used to be a part of the ancient supercontinent Gondwana 200 million years ago. Due to the continental drift, Antarctica reached its current position in the south [57,58]. Antarctica contains all areas of land, ice, and sea around the South Pole, extends to the southern polar circle (66° 33′ S), and consists of an area of 13.6 million km^2 [59,60]. Figure 6.16 illustrates Antarcticas dimensions and continental ice [61]. Antarctica's ice shield holds approximately 30 million km^3 ice ($=65\%$ of all freshwater resources) [61], which leads to extensive reflection of sunlight (known as "Albedo") [62]. Antarctica's climate is dominated by seasonal changes. Depending on latitude, months of complete darkness in the Antarctic winter alternate with months of 24 h bright daylight in the Antarctic summer [63]. Figure 6.17 illustrates the duration of solar radiation at the location of the German research stations Georg-von-Neumayer I and II (1982-2004) as illustrated in Figure 6.20; as can be seen, a period of approximately 80 days of complete darkness results from the stations' location [64].

Very low temperatures are the outcome of the stations' local conditions. As Figure 6.18 illustrates, maximum temperatures measured at the station barely reach 0 °C during the summer (December-January) and can be as low as −50 °C in the Antarctic wintertime's darkness period (2008-2011) [60].

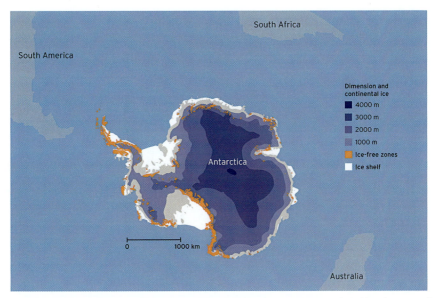

FIGURE 6.16 Antarctica—dimension and continental ice (light blue to dark blue in steps of 1000 m), ice-free zones (orange), ice shelf (white), *adapted from [61]*.

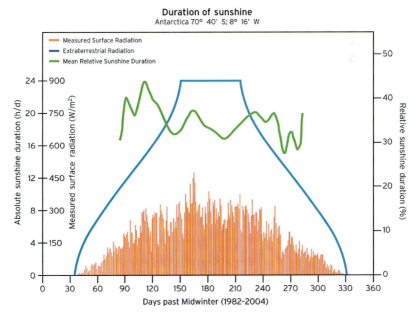

FIGURE 6.17 Annual duration of sunshine (absolute and relative) at the location of the Georg-von-Neumayer-Station during the period of 1982-2004, *adapted from [64]*.

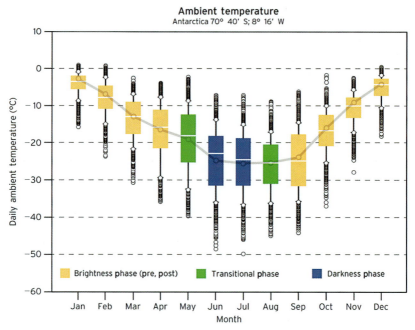

FIGURE 6.18   Annual ambient temperatures at the location of the Georg-Neumayer-Station II and III 2008-2011 (boxes: ambient temperatures over course of the day, circles and curve: temperatures at noon), *created from AWI weather data [60].*

6.3.1.2   Antarctica: Discovery and Settlement

It is still subject of debate who first discovered Antarctica in 1820: Captain Fabian von Bellinghausen (Baltic-Germany), Captain Edward Bransfield (Ireland), or seal hunter Nathaniel Palmer (United States). Seal hunter John Davis (United States) was the first to set foot on the Antarctic continent one year later on February 7, 1821 [65]. In 1895 the 6th International Geographical Congress in London released a note that was to encourage the exploration of Antarctica: "*that this congress record its opinion that the exploration of the Antarctic regions is the greatest piece of geographical exploration still to be undertaken*" [66]. In the famous race to the South Pole, Norwegian Roald Amundsen reached the South Pole December 14, 1911, about one month before his second-place opponent Sir Robert Falcon Scott arrived there. Scott later died in a storm with his men only 18 kilometers away from the next depot [67].

Later expeditions were conducted by Richard E. Byrd (UK), who was the first to fly over Antarctica in 1929 [68,69] and Vivian Fuchs (UK) who managed the first successful Antarctic transit from November 24, 1957 to May 2, 1958. There he covered a distance of 3440 km in 99 days [70]. Other renowned explorers of the Antarctic were Ernest Shackleton [66,71], Jean-Louis Etienne (France), Will Steger (United States), Reinhold Messner (Italy), and Arved Fuchs

(Germany) [72,73]. Recently, added to this list is the White Mars Project, a polar traverse to cross the Antarctic in winter—the first ever attempt to do so [74]. The participants face very adverse conditions on their 2000-mile (3200 km) journey with virtually complete darkness, temperatures as low as −90 °C, hypobaric conditions while travelling on the 3200 m a.s.l. Antarctic plateau with complete dependency on technology. Using two modified caterpillars towing the crew's cabooses, fuel tanks, and equipment container, the crew of five men started their traverse on March 21. 2013, at Crown Bay. Their route takes them via the South Pole to Ross Island, which they were scheduled to reach on September 21, 2013. However, as this is being written (end of August 2013), the crew is stuck after a journey of 313 km due to very adverse weather conditions [74].

It should be clear that these field studies show in a remarkable fashion the analogies of such ventures to long-term space flights. Therefore, scientific investigation on the crew will encompass monitoring of nutritional status, hormone levels, changes in human circadian rhythm and sleep, vitamin D status, immunological changes, hemostasis, ophthalmological impact, as well as microbiological changes in relation to isolation, and cardiovascular and neurocognitive deconditioning.

During the International Geophysical Year from July 1, 1957 to December 31, 1958, utilization of the Antarctic was restricted to scientific research in the areas of seismology, glaciology, meteorology, oceanography, astronomy, and geomagnetism [75,76]. It was then that the Amundsen-Scott Polar Station was constructed [77]. Currently, 82 research stations exist in Antarctica, of which about half are operated year round. Many are situated along the coastline, which makes provision and maintenance much easier [77].

Stays in the Antarctic can be divided into three categories [78]:

- Polar Trek: A small group of explorers often use traditional means of transportation, thus leading to high loads of physical and psychological stress among the participants.
- Summer Camp: Scientists and workers reside in Antarctic research stations for a time limited to the Antarctic summer using the favorable temperature and light conditions. The number of summer inhabitants in the Antarctic can reach up to 2000 persons [79].
- Overwintering: About 1000-3000 persons reside in Antarctica during the winter (March-September). They are mostly the crew of year-round inhabited research stations; that number is currently 47 from 20 nations.

Depending on the nature of stay in Antarctica, inhabitants may have to face great exhaustion [80], deep cold and dry air [81], and a different circadian cycle [82]. It has been shown that the environmental changes can lead to hypothermia [83,84], changes in circadian rhythm [85], changes in the immune system [86], and changes in the hormonal system [87–89] in humans. It has been discussed that the number of inhabitants in polar regions is likely to increase, for example, for the extraction of natural resources [90], therefore knowledge gained in stays in the Antarctic could be used for practical applications.

6.3.1.3 Antarctica: German Research Stations

Germany currently runs the year-round inhabited research station Georg-von-Neumayer III as well as the Kohnen station, which is only used during the Antarctic summer. In addition, Germany uses the so-called Dallmann Laboratories on the Antarctic peninsula close to the Argentine Station Carlini [60].

6.3.1.3.1 Georg-von-Neumayer Stations

It was decided in 1977 to continue the already started scientific research on a year-round manned Antarctic station. The planning was carried out in the 1980s by the Alfred Wegener Institute (AWI) for Polar and Marine research [60]. On February 24, 1981, the first year-round manned Antarctic research station, Georg-von-Neumayer-Station, was officially inaugurated at Atka Bay [91]. It is named after German geophysicist and polar scientist Georg von Neumayer (1826-1909) [92].

The first station was replaced in 1992 because its structural integrity could no longer be maintained. This station also had to be abandoned and replaced 17 years later in 2009 by the research station Georg-von-Neumayer III (70° 40′ S, 08° 16′ W), which was specifically built to enable the station to remain above ground indefinitely due to a hydraulically controlled basement because its predecessors were eventually covered in precipitation and lost in the ice. Figure 6.19 shows the Georg-von-Neumayer-Station III along with a view of the basement with hydraulics attached to the supporting foot. Note the new installed telecommunication unit which increased the stations communication abilities but could also lead to challenges within the groups social coherence regarding communication times and social isolation.

FIGURE 6.19 German Antarctic research station Georg-von-Neumayer III, exterior view with the new telecommunication system and view of basement (Photographs taken by Prof. Gunga).

The crew of Georg-von-Neumayer III in the Antarctic winter consists of approximately nine employees from different fields, while during the summer months the station offers room for up to 50 guest scientists. Under the patronage of the AWI, the station gathers data in aerial-chemical, geophysical, and meteorological investigations and, since the beginning of 2000, also medical and physiological studies. The station contains cubicles, culinary rooms, a mess hall that serves as a day room and common room, a sickbay with an operation unit and dental-medical facilities, workshops as well as other units for the technical and logistical management of the station. New on Neumayer III is a unit located on the roof for the start of weather balloons, which improves the logistics for the realization of these measurements [60].

The difficult weather conditions during the dark phase make it nearly impossible to reach the station by airplane or even snowmobile. Hence, the scientists and engineers are virtually completely isolated during the Antarctic winter. Merely an Internet connection and a satellite phone allow for contact with the outside world. A rescue in emergency situations, which is practically impossible during the Antarctic winter, can last several weeks during the summer months. The provisioning of the station takes place once a year through the research vessel *Polarstern* (*Pole Star*) [60].

6.3.1.4 Environments like the Antarctic as Space Analogues

It is the explicit aim of the European Space Agency (ESA) and National Aeronautics and Space Agency (NASA) that a manned flight to Mars be carried out within the next decades [93,94]. A stay in space confronts the individual with considerable physiological and psychosocial challenges with effects on most bodily functions. This requires sound preparation to guarantee the health and well-being of the astronauts and to guarantee the success of such a mission [95]. A manned mission to Mars could last up to 1000 days depending on mission objectives [96,97]. Before the International Space Station ISS was in function, the majority of real space missions were limited to short term stays of less than 21 days [98], now the number of long term missions are increasing.

It is this limited horizon of experience that necessitates terrestrial studies to simulate a long-term stay in space. Several methods to achieve this exist: bed rest and immersion studies, parabolic flights, and studies with volunteers in isolation and confinement—such as in Antarctic research stations. While other settings of isolation and confinement offer a fast termination in case of an emergency, a stay in the Antarctic does not and is therefore a realistic model for a long-term stay in space, also with regard to unforeseen events, seasonal changes, hostile environment, and hence high technical dependency [76,99–101].

From these investigations arise a multitude of results relevant for medicinal applications such as altitude medicine, exercise physiology, travel and occupational medicine with a gain in knowledge in cardiovascular and musculoskeletal diseases, and psychological disorders [98,102].

To summarize: Life and work of humans in extreme environments such as the Antarctic can be characterized by the following parameters: They are of high technical dependency (life support, water, climate); there exists great physical and psychosocial isolation from the usual environment; inhabitants are living and working in a confined space with a small number of unknown individuals and they have to face challenging tasks on site (physical, psychosocial, cognitive). There also exists a risk for failure with high economic and individual loss. There is virtually no possibility for rescue and there is a high necessity to communicate and cooperate within the crew and to the outside (Mission Control) with postponement of personal interests [76,99].

Long-term stays in extreme environments such as the Antarctic offer therefore—because of their analogy to many conditions encountered in space—the possibility to examine these conditions and their impact on humans as a model for long-term space flights [76,103,104], which is why the space agencies have incorporated these models into their study program [94,105,106].

Norwegian explorer Fridtjof Nansen (1861-1930) answers the question why humans should risk setting foot in hostile extreme environments at all: *"What is the point in fiddling in the ice up there? There is nothing but ice, of no importance for mankind."* His answer: *"In order to say that, we already have to know these areas, which we don't,"* and he continued: *"A larger scientific gain can only be had by long term studies on the ground."* [107], that is, by humans themselves.

6.3.1.4.1 Investigation of Physiological Changes in Overwinterers in the Antarctic 2008-2011

As described, there are major parallels between long-term overwinterings in Antarctic research stations and long-term flights in space. Several studies aimed to better understand the implications of long-term isolation and confinement with regard to space flights, such as HYDREMSI-89 [108], ISEMSI-90 [109], EXEMSI-92 [110], Biosphere-2 [111], Hubes-94 [112], SFINCSS-99 [113], BIOS-3 [114], and MARS500 [115] to provide improved crew selection, training, as well as countermeasures in space in order to maintain crew health and to ensure mission success. The results presented here are based on a study conducted to investigate long-term changes in the physiology of overwinterers of the German Antarctic research stations Georg-von-Neumayer II and III with regard to the parameters of body weight, body composition, sleep and activity, and several hormonal and metabolic parameters in the years of 2008 (Neumayer II) and 2009-2011 (Neumayer III).

6.3.2 General Descriptions

6.3.2.1 Georg-von-Neumayer-Stations

6.3.2.1.1 Geographical Location

German Antarctic research stations Georg-von-Neumayer II and Georg-von-Neumayer III, operated by the AWI, were the location of the investigation. Station II

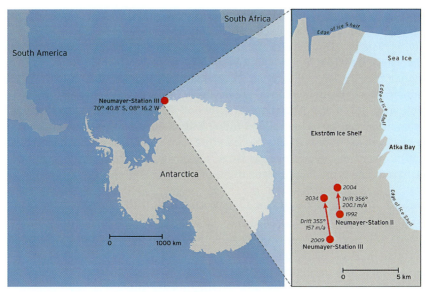

FIGURE 6.20 Location of the German Antarctic research station Georg-von-Neumayer III, and location in the ice shelf with movement of stations Georg-von-Neumayer II and III, *created from [60]*.

was and Station III is located in Atka bay 70° 40′ S, 8° 16′ W on the Ekström ice shelf. The shelf ice is about 200 m in depth around the station's base and moves at a rate of approximately 160 m per year (i.e. 43 cm per day). The location of Neumayer III and the ice shelf movement can be seen in Figure 6.20 [60].

6.3.2.1.2 Description of Georg-von-Neumayer-Stations II and III

The station Georg-von-Neumayer II consisted of a tube construction with two parallel steel tubes (east tube and west tube) 90-m in length and about 6 m in diameter connected to each other. Ship containers served as space surfaces inside of the tubes. The main tubes were air conditioned in sections and contained the containers for the cubicles, the laboratories, the kitchen, the day room, the workshops, the sickbay, as well as for the radio room and the thaw room. In the third tube chill camps, waste disposal, and garbage storage as well as the vehicle hall were accommodated [60].

Since February 2009 the new Georg-von-Neumayer-Station III of the AWI has been in operation. This is the first Antarctic station that combines human stay and research on a platform above the snowy surface with a garage built within the snow [60].

Within the platform a total of 100 containers accommodate the living and day rooms, the kitchen, the mess hall, the sickbay with operating room, the labs, the radio room, the energy supply, and the thaw room. The building consists of two stories and is 24 m wide and 68 m long; the platform height lies 6 m above the snowy surface. Situated in the underground is the garage (depth approximately 8.2 m), containers for the stock storage and waste storage, and the fuel for workshops as well as for the vehicles (engine sledge). From the ground of the garage up to the highest point of the roof structure (balloon filling hall) the station rises 29.2 m. All together the station offers more than 4400 m^2 of usable area from which about 1850 m^2 are available in the form of air-conditioned usable areas in labs and other rooms. A stairwell with a lift connects the four floors from the garage up to the balloon filling hall on the roof. Power is provided via a block power station, an emergency power generator, and a wind energy plant [60]. Approximately 16 km away from Neumayer III is the shelf ice from where delivered goods are received and transported to the station.

6.3.2.2 Participants of the Overwintering in the Antarctic

The members of each overwintering crew resided at the Antarctic station for 14-15 months (Neumayer II in 2008 and Neumayer III in 2009, 10 and 11) performing their respective duties and research work. The recruitment to take part in the study was carried out by the AWI. There were no special prerequisites involved so all overwinterers were eligible for the study. The AWI conducted several weeks of training and exercise testing as well as mountaineering safety courses with the volunteers. After self- and team assessment, the final crew composition for each overwintering campaign was set.

An overwintering team was composed of nine participants. In 2008, this consisted of seven men and two women, the same in 2009 and 2010. In 2011 there were six men and three women. Each team was headed by a station manager, which was usually the medical doctor, who was a specialist trained in surgery and internal medicine as well as emergency dental procedures. In addition to the medical profession, the manager was also responsible for the logistical and human resources organization of the station management and carried out the measurement procedures associated with the study. The other crewmembers were scientists (meteorologists, biologists, geologists), technicians, electricians, computer technicians, and a cook. Aside from their respective fields of occupation, all crew members had to tend to duties associated with station maintenance.

Traditionally, each new team was welcomed by the previous overwintering crew and instructed within two months of their duties at the station. As of February, each overwintering crew was then on its own, and the actual isolation began. It was not until November that the first visits of guest scientists and members of the new team were to be expected. During the time of isolation the Internet and a satellite phone were the only means to contact the outside world. With the arrival of the new crew, the isolation period ended and a new cycle of overwintering commenced.

Aside from their daily duties (station maintenance, satellite surveys, evaluation of scientific data, operation of weather balloon ascent, etc.), the crew also participated in group activities such as eating meals, small celebrations, and the "melting service" to produce fresh water. Crew members could use their spare time in the station's gym, sauna, library, workshops, or TV with Internet access. They also engaged in dance courses, music, and photography and published the journal "Atka-Express" on a monthly basis [60,116].

Because of the risks involved (injuries due to fall, frostbite, and hypothermia), outside activities during the Antarctic winter was reduced to a minimum. Nevertheless, each crewmember was equipped with appropriate cold-protection clothing and emergency equipment. Also each participant was allowed to bring a box with personal items.

The supply of food to the Neumayer Station II and III was conducted in accordance with the requirements of increased energy consumption and because of months of inaccessibility. The food was composed primarily of frozen products in which their storage in the Antarctic was expectedly unproblematic. Fresh fruit and vegetables were available once per year after delivery by the research vessel *Polarstern* so these products could not be provided from June to November. The composition of the mixed diet consisted of vegetables, meat, fish, starchy side dishes, and fruits. There was no rationing, so an excess supply of food was available. Moreover, there was no regulation of caloric intake [60].

The overwintering participants in the respective winter campaigns could voluntarily participate in the study measurements and also at any time refuse to participate further without any consequences. Some participants refused to participate in the whole or parts from the study. Of the 36 participants in the four observation periods from 2008 to 2011, a total of 31 subjects took part in the measurements (=86.1%). The volunteers were informed orally and in writing of the measurements and the background of the study and gave their written consent after adequate time for reflection. The study was approved by the ethics committee of the Charité University Medicine Berlin. There were no separately formulated exclusion or inclusion criteria aside from the selection process of the AWI: adults with the ability to realize the scope of the study in the Antarctic and on this basis to participate of his or her own free will. The participants were healthy and were not taking any medication (apart from oral contraceptives by the female participants).

The participants were encouraged to independently measure the values for body composition (body weight and BIA measurements) as well as sleep and activity parameters (SenseWear measurement) in 14-day intervals. In about eight-week intervals blood samples were taken for later analysis. However, different temporal and operational requirements as well as interindividually different demands led to somewhat different measurement time points between the campaigns. For statistical analysis, Analysis of Variance of Repeated Mesaures (RM ANOVA) was applied with a two-sided p-value of 0.05 as an indicator for statistical significance.

6.3.2.3 Body Composition

As described in other parts of this book, the determination of body composition plays an important role in internal medicine, exercise and training physiology [117], and many other adjacent fields, but especially that of extreme environments. Data regarding body composition can be obtained using several different technological

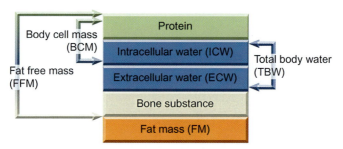

FIGURE 6.21   Human body composition, *adapted from [118]*. The bone structure in this model consists only of mineral components.

approaches, some of which are very precise but costly, while others offer low cost and portability but not always acceptable precision and accuracy. Information about body composition also differs with regard to the inherent model used because different models divide the human body composition into a different number of compartments—please refer to the respective chapter in this book for more detail. Figure 6.21 offers a simple summarizing outline of human body composition and the relationship of the different compartments toward each other [118].

For the determination of body composition in the overwinterers, Bio Impedance Analysis (BIA) was applied. The BIA uses the fact that electric resistance is proportional to the length of a conductor and inversely proportional to its cross-sectional area. Furthermore, two types of resistance build up in the human body: ohmic resistance (Resistance R) and capacitor resistance (Reactance Xc). Resistance R is mainly determined by the amount of ions solved in the body water of the fat-free mass, which is (in healthy adults) very precisely hydrated at a constant 73%. The reactance is mainly determined by the amount and constitution of cell membranes—which act as a capacitor—and thus increase with an increase in body cell mass. The total resistance (Impedance Z) results of combination of R and Xc. Several different kinds of BIA-units exist, applying different techniques such as single-frequency versus multifrequency technology. In general, BIA can be considered a safe, easy-to-use, cost-effective, portable, fast, and reliable method to determine human body composition as long as no electrolyte imbalance or fluid shifts are present in the measured individual, the body fat percentage is in a normal range, and the BIA equation used to determine the parameters is adequate for the measured person [118–122].

6.3.2.3.1   Determination in Overwinterers

The resistance values of the overwintering crews in Antarctica were measured with a single-frequency (50 kHz) BIA device (Akern BIA 101, Bioresearch, Florence, Italy). This apparatus operates on a resistive resistance 0-1000 Ω and a reactance of 200 Ω. Two self-adhesive disposable electrodes (Classic Tabs Ag/AgCl, Medical Healthcare GmbH, Karlsruhe, Germany) were each attached to the right extensor side of the distal wrist and hand and at the right extensor side of the ankle and the metatarsus and were connected to the cables of the device according to the information provided by the manufacturer. After switching on, the device automatically started the measurement and gave out the resistance values. Body weight was determined by a mechanical scale (manufacturer Seca) with the volunteers wearing only underwear. The measurements were performed by the overwinterers independently and according to their respective timetables and requirements. The data generated were transmitted via the Internet to Berlin for further processing and analysis. Total body water (TBW) and fat-free mass (FFM) were calculated using the formulas by Sun et al. [123]. Based on these calculations, the respective relative proportions of these compartments with regard to the measured body mass were calculated and used for further analysis.

6.3.2.4   Energy Expenditure

The metabolism of an organism consists of all biochemical events of anabolic and catabolic nature. The daily energy requirement necessary to fuel them is known as the Daily Energy Expenditure and is comprised of the Basal Metabolic Rate, Thermic Effects of Activity, and Thermic Effects of Food [11,124].

The basal metabolic rate (BMR) is closely correlated with body mass and fat-free mass in particular, and the proportion of FFM explains up to 85% of the variance of the BMR [125]. It is therefore clear why the BMR for men is higher by an average of 5-10% than for women with the same body mass, because women have a higher percentage of fat mass which is comparatively hypometabolic (the FFM is seven times more metabolically active than fat mass) [126]. Another influencing factor is age: The BMR of children and adolescents is higher, while it gradually decreases with age, which is explained by the increased metabolic processes of growth during childhood and adolescence. The BMR makes up to 60-75% of total daily energy expenditure for a moderately active individual. Within the fat-free mass, especially the metabolically active organs as heart, kidneys, liver, and brain, which only represent 5% of the body mass, are responsible for about 65% or the BMR [127]. During exercise conditions, it is particularly the musculature that is responsible for the increase in $\dot{V}O_2$ and thus energy expenditure [128–130]. For a moderately active person, the thermic effects of activity make up about 15-30% of the daily energy requirement, but this number can vary greatly depending on the activity or inactivity of the individual, as well as training status, age, and gender. Several technical approaches exist to measure energy expenditure, such as indirect calorimetry, the doubly labelled water method or the—in comparison—cost-effective and portable method of actimetry, which was used in this study.

6.3.2.4.1   Determination in Overwinterers

Actimeters are mobile devices designed to measure acceleration data of their wearer and thus allow for assessment of activity, energy expenditure, and sleep parameters [131,132].

Because of the benefits, such as outpatient use, less effort, low cost, and relatively high acceptance and compliance on the part of the subjects, actimeters are used for assessment of energy expenditure and activity for different questions [133–135].

The actimeter SenseWear (Pro3 version) used in this study in Antarctica to determine the energy expenditure is a device made by BodyMedia, Pittsburgh, PA, USA. It is worn on the back of the right upper arm and detects physiological data, including motion by measurement of the acceleration in two axes ($\pm2.0\,g$), heat flow (0.0-$300.0\,W/m^2$), the skin temperature, near-body temperature ($20.0\,°C$ to $40.0\,°C$), and skin conductance (Galvanic Skin Response GSR; $50.0\,nSiemens$ to $17.0\,\mu Siemens$) according to the manufacturer [136]. The data are averaged and recorded with a frequency of $1\,min^{-1}$. After completion of the measurement, the data are transferred to a computer via a USB interface so that they can be exported and analyzed using the software of the manufacturer [136]. The device measures $89\times57\times22\,mm$ and weighs approximately $82\,g$ (including belt and battery). The device has received a high level of acceptance in terms of comfort and skin tolerance [137].

The 24-h measurements using the SenseWear device were conducted by the overwintering crews independently and according to their time requirements. The overwinterers transferred the data onto a separate computer from which the data were sent to Berlin for further analysis.

6.3.2.5   Circadian Rhythm and Sleep

The circadian rhythm—the variation of various body functions over the time course of a day—is a ubiquitous phenomenon [138]. The body core temperature, which was an early subject of studies, is closely linked to the regulation of circadian rhythms, as well as the plasma concentrations of various neurotransmitters. From the first studies that showed that the rhythm is maintained even in isolation from external influences such as the light-dark cycle, it was concluded that it had to be of endogenous origin. Later studies showed that the hypothalamus and in particular the area of the suprachiasmatic nucleus (SCN) are key locations in the regulation of circadian rhythm (the so-called "internal clock"). Through an endogenous oscillator two proteins are synthesized: Clock and Cycle. Within the nucleus, they cause the creation of two other proteins: PER (period) and CRY (cryptochrome), which form a complex and act as negative feedback on the synthesis of Clock and Cycle. They eventually form a feedback mechanism of synthesis and degradation that takes about 24 h. Glutamate from the retinohypothalamic tract carrying the signal "light" can alter the rate of synthesis and therefore synchronize this internal mechanism with the external light-dark cycle, one of the most important time donors ("zeitgeber"). Figure 6.22 illustrates the mechanisms of the internal clock [139].

The SCN then acts on the core body temperature as well as various hormone levels and thus affects all other cells of the body and their molecular clocks in the sense of synchronization. The close proximity of the SCN to core

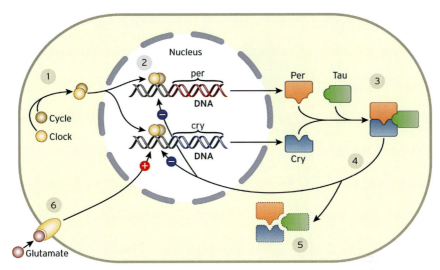

FIGURE 6.22 Cellular clock proteins and interaction within the SCN, Proteins Cycle and Clock (1) facilitate the production of proteins Period and Cryptochrome (2) which form a complex with protein Tau (3) and act as a negative feedback on their own creation (4). The complex degrades within 24 hours (5); Glutamate carrying the information "light" can influence this cycle (6). *Adapted from [139].*

areas of food and energy regulation in the hypothalamus and the simultaneous importance of respective neurotransmitters VIP and neuropeptide Y in the regulation of appetite show the link between the sleep function, energy balance and body composition [140,141].

Besides the known photoreceptors, cones and rods, other types of melanopsin-containing retinal ganglion cells have been discovered that are considered to be primary photoreceptors for the circadian rhythm [142,143]. These receptors have their highest sensitivity for light of wavelengths around 460 nm, the blue portion of the light that shows the greatest ability to suppress melatonin secretion and affect the circadian rhythm. Melatonin is secreted by the pineal gland, which has afferent inputs and efferent outputs to other parts of the circadian system. The pineal gland receives information from the superior cervical ganglion, which receives afferent information from the SCN itself. Melatonin from the pineal gland in turn affects the SCN and thus the circadian rhythm by inducing a sleep-promoting state via an inhibition of neurons in the SCN. Figure 6.23 provides a schematic overview of these basic regulatory centers and their interconnection [138].

A functional model of the circadian rhythm has an endogenous circadian process (process C), a homeostatic process (process S), and influences from the environment. The endogenous process C promotes wakefulness during the day and the initiation and consolidation of sleep during the night along with the synchronization with the external conditions. In addition, the homeostatic process S accumulates as a function of previous wakefulness so that the propensity

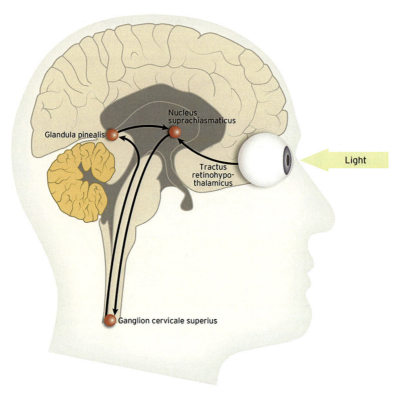

FIGURE 6.23 Regulatory centers and pathways of the human circadian system, *adapted from [138].*

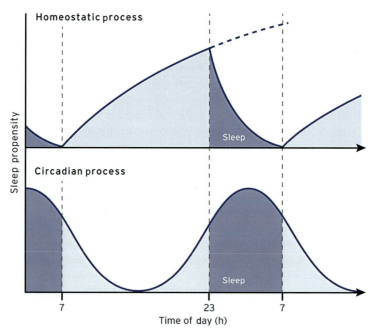

FIGURE 6.24 Processes governing sleep propensity, *adapted from [144].*

for sleep (fatigue) steadily increases until this drive is again reduced by sleep. These relationships are modeled in Figure 6.24 [144].

Sleep and its regulation, function, lack thereof, and other associated phenomena are yet not fully understood. However, it is known that continuous lack of sleep leads to a reduction in cognitive and exercise performance, is detrimental to one's health, and can eventually be fatal. Changes in circadian rhythm and sleep can be assessed through different means such as polysomnography, actigraphy, and sleep protocols.

6.3.2.5.1 Determination in Overwinterers

As for the determination of energy expenditure in the overwinterers, the actimeter SenseWear was also used to determine sleep parameters. For more general information about actimeter function, see the description in the previous section and the respective literature.

Since the 1980s, actimeters have been used for the assessment of sleep in humans [145]. Studies comparing actimetry and polysomnography (PSG) to validate the data showed that in young healthy subjects an agreement of 91-93% could be achieved [145]. Advantages lie in the possibility of measuring with the actimeter on an outpatient basis, such as in the familiar living environment of volunteers and patients, and thus achieving high acceptance and compliance of these measurements over several days. Further, the use of the actimeter is relatively inexpensive and requires less effort [146], compared to the gold standard—the measurement by means of polysomnography—which is complicated and expensive [147].

Actimetry allows the distinction between sleep and wake state [148], and it also offers sufficiently reliable information regarding sleep habits and sleep disorders of examined patients. However, the interindividual differences are very high, which does not allow individual prediction of the sleep stage [147]. Further disadvantages are a lower precision of measurements taken and the dependence on the subject compliance [149].

The 24-h measurements using the SenseWear device were conducted by the overwintering crews independently and according to their time requirements.

The following participant parameters were evaluated using the evaluation software SenseWear (Professional version 7.0): laytime (how long the participants were lying at rest but not sleeping), sleeptime (the participants were asleep as measured by the SenseWear device), sleep efficiency (ratio between laytime and sleeptime), and sleep fragmentation (number of wakes per night).

Data on laytime and sleeptime were calculated through a proprietary software algorithm. The parameter sleep efficiency was calculated from the laytime and sleeptime data. The sleep fragmentation was determined manually and includes all detected wake episodes during a period of sleep.

6.3.2.6 Hormones

6.3.2.6.1 25-OH-Vitamin D

The collective term *vitamin D* (calciferol) combines the substances vitamin D3 (cholecalciferol) and vitamin D2 (ergocalciferol), which are hormonally active after conversion in the metabolism. The formation of vitamin D in the human skin—depending on sun exposure due to geographical location, altitude, clothing, and work—makes up to 95% of the vitamin D requirement [150].

A photochemical conversion of the pro-vitamin D3 (7-dihydrocholesterol 7-DHC) by UV light of wavelengths 280-320 nm causes the formation of pre-vitamin D3, and finally of vitamin D3 (cholecalciferol). The wavelength of 298 nm causes the maximum production rate [151]. The resulting vitamin D3 is bound to transport proteins (D-Binding Protein DBP and albumin) and transported to the liver. A small portion is stored in adipose tissue and skeletal muscle, from where it can be released in times of deficiency; the half-life is approximately two months. In the liver, vitamins D3 and D2 are converted into 25-hydroxy-vitamin D. From there this metabolite passes through the bloodstream bound to DBP, to the kidneys, where the second activation step to 1,25-dihydroxy-vitamin D (calcitriol) is catalyzed. The reaction step is stimulated by parathyroid hormone and inhibited by the resulting product. Its target organs are the intestine, bone, kidney, adrenal gland, and others [150]. Figure 6.25 illustrates the formation of vitamin D in humans [150].

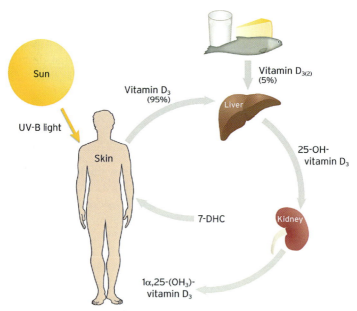

FIGURE 6.25 Formation of calcitriol in humans, *adapted from [150]*. Vitamin D_3 and also D_2 can be supplied through nutrition. However, the predominant proportion of Vitamin D is formed in reaction to UV-light exposure.

The effects of calcitriol are divided into calcemic and non-calcemic effects. The former serves to maintain the calcium and phosphate homeostasis. The latter serves to:

i) Regulate cell growth and differentiation, ii) Modulate the immune response (monocytes/macrophages, B-and T-lymphocytes), iii) Stimulate microbial defense, iv) Control the renin-angiotensin system, v) Control muscular function, vi) control brain development and mood regulation [150].

The main effect on calcium and bone metabolism consists of an increased intestinal calcium absorption via an increased formation of the calcium-binding protein calbindin and maintaining the length of intestinal villi. In the kidneys, calcitriol causes increased reabsorption of calcium, and in the bone tissue calcification of the bone matrix; in addition, it causes an inhibition of the parathyroid hormone so that in sum it leads to a buildup of bone structure. At very high doses, however, calcitriol also acts on osteoclasts and may cause bone loss [152,153].

Furthermore, there is an association with functional processes of the immune system: regulation or inhibition of specific and nonspecific stimulation of immune defense and control of chemotaxis and vascular permeability [154]. Other positive effects of vitamin D could be shown on the function of the nervous system, the inhibition of diseases like the metabolic syndrome, lung diseases (COPD, asthma), and several types of cancer [155,156].

In central European latitudes there is a lack of vitamin D due to low amounts of sunlight in the winter months [157], and up to two-thirds of the population exhibit blood concentrations of less than 20 ng/ml (specification for 25-OH-vitamin D). Values between 40 and 60 ng/ml are currently seen as optimal values, and 30 ng/ml is considered as a temporary lowest acceptable limit [158]. But even this seems hard to reach with a normal mixed diet during the winter months, because it would mean taking up to 2000 IE vitamin D3 per day. This would correspond to 1 kg mackerel, 1.5 kg avocado, 10 kg of pork, or 50 l of whole milk. Even the daily intake of 200 IU, as recommended for adults, is rarely achievable, because significant vitamin D levels are only to be found in sea fish (as in cod liver oil: 10,000 IE/100 g) [158].

In this context, because 95% of the vitamin D originates from endogenous formation, sun exposure is the main factor for the body's vitamin D supply. In general, a daily exposure of the face and the upper and lower extremities for a minimum of 15 min is sufficient for a person with skin type 2 to achieve a production of about 3000-5000 IU of vitamin D [159].

These relationships are of interest especially for individuals like astronauts or those living in a manner analogous to overwinterers in the Antarctic due to the prolonged dark phases and the corresponding reduction of the endogenous vitamin D synthesis.

6.3.2.6.2 Thyroid-Stimulating Hormone

The follicular cells of the thyroid gland mainly secrete thyroxine (T4) (80%), the low active precursor for the higher active triiodothyronine (T3), which is released directly to a lesser extent from the thyroid gland (about 20%). A large part of the active form T3 is formed in the periphery through deiodination of T4 [160]. Synthesis and secretion in the thyroid are under the control of the thyroid-stimulating hormone (TSH) from the anterior pituitary, which causes increased production and release of thyroid hormones and in turn is controlled by the thyreotropine-releasing hormone (TRH) originating from the hypothalamus. In addition to these control mechanisms, the thyroid hormones act themselves via feedback loops on the release of TRH and TSH as well as the thyroid gland in terms of a negative feedback [160].

The thyroid hormones have important regulatory functions for growth [161,162], as well as for the development of the brain before and after birth [163]. A congenital deficiency of thyroid hormones can lead to diminished mental capacity [163,164]. As well as in the periphery, such as skeletal muscle, conversion to the active form T3 is required for the central effects of thyroid hormone [165]. Deiodinase enzymes (isoforms D2 and D3) were detected in different cells of the brain [166]. Several more studies have investigated thyroid hormone brain receptor interaction in humans and animal models.[2]

The thyroid hormones also play an important role in thermoregulation and cold adaptation. High levels of triiodothyronine (T3) are necessary to allow for the up-regulation of brown adipocytes UCP1 (Uncoupling Protein 1 of the brown adipose tissue). It was recently possible to also show the existence of brown adipose tissue in adult humans [20,171]. T3 leads to an increased gene expression of the UCP1 after its binding to the receptor TR-β [172]. As has been shown, brown adipose tissue contains high levels of type II iodothyronine 5′ deiodinase, an enzyme that transforms tetraiodothyronine (T4) into the functional form of T3 [19,173]. Sympathetic activation leads to an increased conversion rate of this enzyme, which in turn causes an increased tissue concentration of T3 [173]. The relationship between cold exposure and T3 production in brown adipose tissue was only hypothesized [174] but is now—through the above mechanisms (exposure to cold—sympathetic activation—increased deiodinase activity—resulting higher T3 formation) well documented [19,175]. In addition, TSH seems to have a direct influence on the type II deiodinase and the UCP1 [176]. In the light of the extremely cold environment the Antarctic overwinterers have to work in, the investigation of changes in thyroid hormone levels was of high interest.

---

2. A study reported decreased T3 concentrations in the brains of D2 knockout rats, while there were normal concentrations in the serum of T3 and increased systemic TSH and T4 levels [167]. In humans, a study showed particularly high concentration of the isoform D3 deiodinase in the hippocampus and the temporal lobe [168]. To overcome the blood-brain barrier, the thyroid hormones require a transport mechanism. OATP1C1 is a transporter, in particular for T4 and rT3 [169]. For transporter MCT8, it was shown that loss of function may result in psychomotor retardation [170]. Thus, several factors in the maintenance of T3 concentration are involved in the brain, such as the concentrations of T3 and T4 in the blood, transport activities for the introduction into the cells and the deiodinase activity.

6.3.2.6.3   Analysis of Hormone Levels

Blood was taken bimonthly as part of a routine health check by the station's physician. It was then centrifuged (15 min at 2500 rpm), the serum aliquoted, stored at −20 °C, and transported, observing the cooling chain, to Germany for analysis using the ELISA-method. 25-OH-vitamin D was analyzed using the test kit by manufacturer IDS, Frankfurt Main, Germany (Ref AC57F1), standard values 47.7-144 nmol/l (male and female, 5th to 95th percentile); TSH was analyzed using the test kit by manufacturer IBL, Hamburg, Germany (Ref RE55221), standard values 2.32 ± 0.07 mlU/ml (male and female, mean ± standard deviation). The resulting data were analyzed using Microsoft Excel Version 2007 (12.0.4518), Systat SigmaPlot Version 12 (12.2.0.45) and MedCalc Software Version 12 (12.2.1.0).

6.3.3   Results and Discussion

6.3.3.1   Anthropometric Data

The data show that more than two-thirds of the participants were male and one-third female. In median the participants were 32 years old and predominantly of normal weight (BMI median 25.9). As to be expected, the male participants were of greater weight and height. Regarding age, the female participants were about 10 years younger than their male colleagues (median 30 versus 40.7 years).

6.3.3.2   Body Composition

The analysis for fat mass showed no significant change in fat mass of all participants over the course of three overwintering campaigns. It showed no significant difference between the sexes. Within the two sexes there was no significant difference over the course of the three phases (Figure 6.26).

The median values showed a decrease in body weight for 2008, 2010, and 2011 and an increase for 2009; combined for all campaigns there was no change over the course of the overwintering. Combined for all seasons, there was a decrease in fat-free mass. In analogy, combined for all seasons there was an increase in fat mass because body weight remained relatively the same.

Both during a space flight [177] as well as while living in an isolated and confined environment [110,178] changes in body composition and body weight can be found, losses as well as increases [179]. Furthermore, body weight may stay the same with changes in individual body compartments, as demonstrated in the isolation study EXEMSI-92 [110]. A review of body weight changes of 60 overwinterers showed that 17% of participants lost weight, 6% showed no change, and 77% gained weight. The average increase was 2 kg [180], so there seems to be an overall tendency toward an increase. Altogether, however, changes are corresponding to a tendency for groups only, while the interindividual variation is very large.

Results of clinical studies on the changes of body composition in the Antarctic are not uniform: The fat mass showed no increase in one study [181],

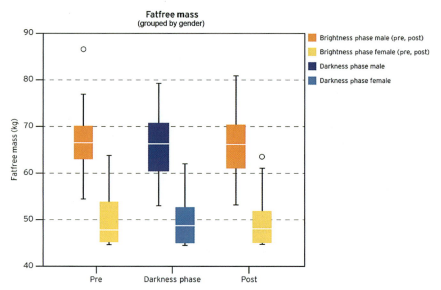

FIGURE 6.26   Boxplots of changes of fatfree mass of overwintering campaigns 2008-2011 by gender (n = 30, 21 male, 9 female).

and fat-free mass showed an increase in another study [182]. In other studies there was an increase in fat mass [183] and an increase in fat mass and a decrease in fat-free mass [184].

The presented results with the high inter- and intraindividual variations in body weight and body composition with—for all seasons combined—reproducible increases in fat mass and decreases in fat-free mass—although without reaching statistical significance—appear to be in accordance with the available studies of this topic.

6.3.3.3   Sleep and Activity Parameters

A significant difference between the sexes was found for the parameter of lay-time associated with a statistically significant increase for the female participants (Figure 6.27). The differences between males and females might be due to different activities which were not evaluated up until now.

As could be expected, there was a statistically significant difference in the energy expenditures between the sexes. The values of daily energy expenditure decreased during the darkness period for all participants combined, while at the same time there was an observed decrease in fat-free mass—as mentioned in the previous paragraph—which might have influenced each parameter's change (Figure 6.28).

In the analysis of the mean values of the combined campaigns, laytime, sleeptime, and sleep efficiency showed a tendency for a decrease and sleep fragmentation a tendency for an increase (Figures 6.29–6.32).

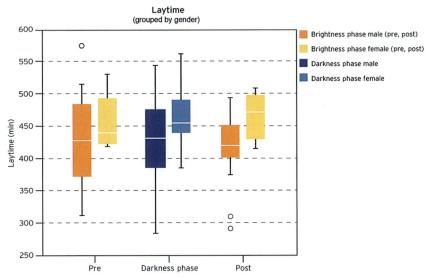

FIGURE 6.27   Boxplots of changes of laytime of overwintering campaigns 2008-2011 by gender (n = 29, 20 male, 9 female).

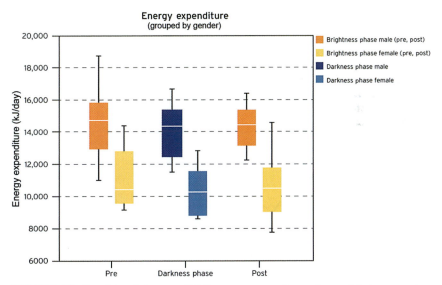

FIGURE 6.28   Boxplots of changes of daily energy expenditure of overwintering campaigns 2008-2011 by gender (n = 29, 20 male, 9 female).

A significant difference between the sexes was found for the parameters of the daily energy expenditure with significantly lower values for the female participants. In the analysis of the mean values of the combined campaigns, there was a decrease in daily energy expenditure over the course of the overwintering, although without reaching statistical significance (Figure 6.33).

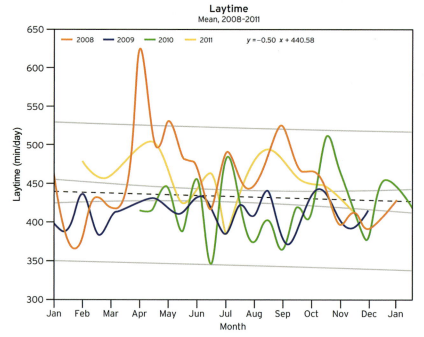

FIGURE 6.29 Curves of the values of the laytime for the campaigns 2008-2011 as mean values of the respective seasons (n = 29, 20 male, 9 female). Also shown are the regression line (dashed line) and the 95% confidence and 95% prediction intervals.

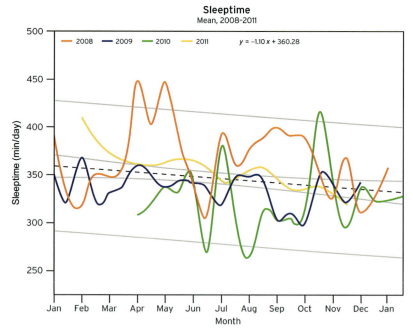

FIGURE 6.30 Curves of the values of the sleeptime for the campaigns 2008-2011 as mean values of the respective seasons (n = 29, 20 male, 9 female). Also shown are the regression line (dashed line) and the 95% confidence and 95% prediction intervals.

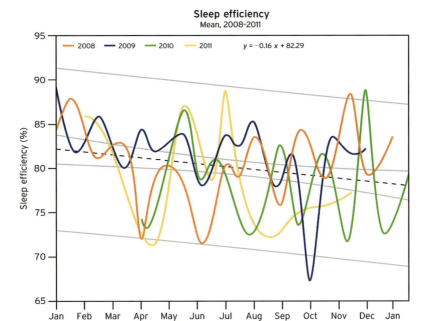

FIGURE 6.31 Curves of the values of the sleep efficiency for the campaigns 2008-2011 as mean values of the respective seasons (n = 29, 20 male, 9 female). Also shown are the regression line (dashed line) and the 95% confidence and 95% prediction intervals.

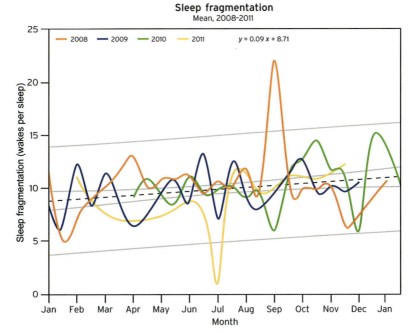

FIGURE 6.32 Curves of the values of the sleep fragmentation for the campaigns 2008-2011 as mean values of the respective seasons (n = 29, 20 male, 9 female). Also shown are the regression line (dashed line) and the 95% confidence and 95% prediction intervals.

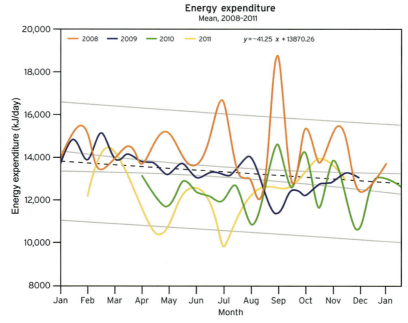

FIGURE 6.33   Curves of the values of daily energy expenditure for the campaigns 2008-2011 as mean values of the respective seasons (n = 29, 20 male, 9 female). Also shown are the regression line (dashed line) and the 95% confidence and 95% prediction intervals.

In summary, the results represent a change over the course of the campaigns in sleep and activity among the participants. Laytime, sleeptime, and sleep efficiency decrease, while laytimes of female participants are significantly higher than those of men. For both sexes, the sleep fragmentation increases continuously.

Isolation and confinement in polar regions usually lead to changes in the psychosocial context by isolation from family, friends, and familiar environment; the duration of the winter darkness period; and adverse weather conditions [78,185]. The restriction of privacy, the non-separation of work and leisure, and the barely existing possibility to avoid other team members may lead to social conflicts [185–187]. Exhaustion, negative affect, decreased communication, and sleep disturbance were frequently shown in previous studies in this context [78]. Past studies regarding sleep and energy expenditure in isolation produced mixed results: In an isolation study (ISEMSI-90) that also utilized an actimeter, no deterioration of the duration and quality of sleep could be detected [188].

With regard to findings from Antarctic overwinterings, studies showed a decrease in sleep duration and quality [189] and an increase in sleep during the winter [190]. A study using polysomnographic measurements in the Antarctic found decreases in the depth of sleep (sleep stages 3 and 4) during the darkness phase, but with high interindividual differences [190].

In particular, during the time period of mid-winter (June 21) sleep disorders can be increasingly present [191,192]. A study on the McMurdo Station showed that up to 64% of the overwinterers had trouble sleeping [186]. In addition to the changes in the circadian rhythm [193], influences of the cold environment [194] and psychosocial stress [189] were discussed as being crucial.

The results collected in the submitted work on sleep in Antarctica largely corroborate the results obtained by different authors with a decrease in sleep duration, sleep quality (sleep efficiency), and an increase in sleep disorders. It seems conceivable that a decrease in physical activity, looming on the basis of diminishing daily energy expenditure, could be a possible cause, as well as the changes of illumination with different spectra and the external change of the light-dark cycle.

6.3.3.4   Hormones

6.3.3.4.1   25-OH-Vitamin D

The analysis of 25-OH-vitamin D showed a significant change in the 25-OH-vitamin D concentration of all participants over the course of three overwintering campaigns. It showed no significant difference between the sexes. Within the two sexes there was a significant difference over the course of the three phases with a decrease toward the darkness phase and an increase for the subsequent brightness phase. The median values of the female participants were below those of the male participants in all three phases of overwintering (Figure 6.34).

These changes in the mean values in all campaigns also showed a decrease with a minimum in the darkness phase (June-August) and for campaigns 2009 and 2010 a recovery of the values at the end of the measurement period (Figure 6.35).

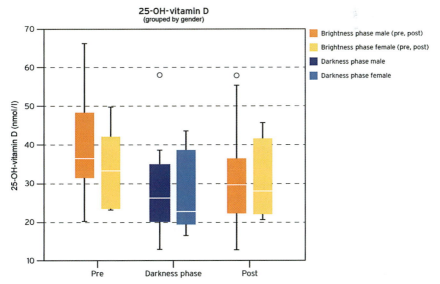

FIGURE 6.34   Boxplots of changes of 25-OH-vitamin D concentration of overwintering campaigns 2008-2010 by gender (n = 23, 17 male, 6 female).

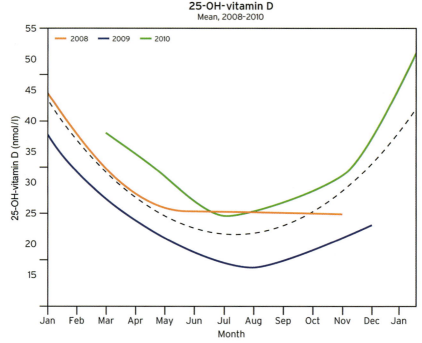

FIGURE 6.35   Curves of the values of 25-OH-vitamin D for the campaigns 2008-2010 as mean values of the respective seasons (a total of n = 23, 17 male, 6 female). Also shown is the regression curve (dashed curve, second order equation).

It is important to note that the concentrations achieved reached very low values in the majority of the participants and were partly below the general recommendations [195].

As already stated in the methodology section, people living in northern latitudes need to ensure an adequate intake of vitamin D during the winter months to prevent various diseases and to ensure optimal functioning of the body [150,196]. Regarding the overwinterers in Antarctica, previous studies found a decrease of 25-OH-vitamin D but no change in bone mass (bone mineral density BMD) [197,198]. Other results, however, did show a decrease in BMD (mean 1% of the proximal femur, $p < 0.05$) [199]; other authors pointed out that the level of physical activity might have a greater overall impact on bone density in the overwintering participants [200].

The documented option to compensate for vitamin D deficiency through supplementation [198,201] should be used in polar regions, but also in people with low UV radiation during the year (miners, astronauts) to ensure the health of these individuals. This recommendation could be made for the overwinterers at the station Neumayer III as well. Likewise, an adequate physical activity should be maintained in order to counteract loss of bone density [200].

In summary, it can be stated that there is a statistically significant decrease in 25-OH-vitamin D concentration in overwinterers in Antarctica, which does not recover to the initial values for some of the crew, which confirms the results of previous studies.

6.3.3.4.2   Thyroid-Stimulating Hormone

The profiles of the mean values of the changes in TSH concentrations in 2009 and 2010 showed a decrease with a subsequent increase (maximum around June), a decline (minimum around September-October) and then a rise again. The values for 2008 showed a similar behavior with an advancement of approximately two months and an overall decline of the values (Figure 6.36).

Analysis of the mean values of the combined campaigns showed a periodic pattern with an increase toward the dark phase with a subsequent decrease of the values; these changes seem to be analogous to those shown in previous studies. This "Polar T3 Syndrome" is characterized by an unstimulated increase of TSH secretion, which is stimulatable by TRH and without feedback resistance

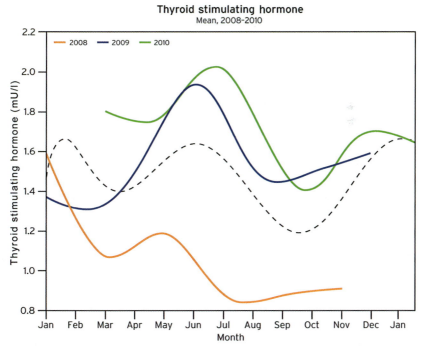

FIGURE 6.36   Curves of the values of TSH for the campaigns 2008-2010 as mean values of the respective seasons (a total of n = 23, 17 male, 6 female). Also shown is the regression curve (dashed curve, seventh order equation).

[202]. Furthermore, there is a slight decrease in the serum concentrations of free T3 and T4 (fT3 and fT4) with simultaneously increasing production and clearance rate of T3 [18]. Along with the polar T3 syndrome go decrements in cognitive function and worsening of mood [203,204] of the overwintering personnel. This functional hypothyroidism is interpreted in the context of a hypothermic cold adaptation [16], with a decrease in core body temperature and up to a 40% increase in daily energy needs [89]. A recent study interprets the finding as follows [18]: Chronic cold exposure in the Antarctic environment causes an increase in thyroid activity, which can be detected via the increased serum concentration of thyreoglobulin and increased iodine excretion in the urine and which is interpreted as an increased T3 production and clearance. The T3 production occurs thereby mainly in peripheral tissues [89], in particular the skeletal muscles and brown adipose tissue, through conversion by deiodination of T4 without release of T3 into the bloodstream, so there is no increase of T3 that can be measured in the blood. This could help to explain why, depending on the study conditions, different studies found conflicting results that could not reflect the T3 production and clearance [205,206]. The model of increased peripheral T3 consumption is supported by studies that have shown that the mood changes could be improved by administration of thyroxine [202]. Various studies also showed that the changes in the polar T3 syndrome during long-term stays in modern polar stations apparently continue and that this condition is not being counteracted through modern climatic control within the station [202,207]. One might speculate that light could be of more importance than ambient temperature.

A circannual rhythm can be shown for the TSH secretion both for temperate latitudes [208] as well as for overwinterers in Antarctica [209], which is probably determined by environmental parameters (cold, photoperiod) [210]. The periodicity of the stay in Antarctica shows a bimodal behavior with an increase in November and June and a decrease in March. Results of this study show a periodicity of similar behavior with a decrease toward March and an increase toward June/July (darkness phase) and another (delayed) increase toward November/December. Figure 6.37 shows data from previous studies [202] superimposed with results from this investigation.

Various studies have aimed at the effect of cold on the thyroid hormones, but no, little, or partly inconsistent results were found [211–213]. Especially in studies on thyroid changes in humans during long-term polar stays, factors such as energy intake, iodine deficiency, changes in the circadian day-night rhythm, isolation, physical activity, and the storage of plasma samples can influence the analysis and the study outcome [211].

The results of the present study corroborate the results presented in many previous investigations, although there is a high interindividual variation. With reference to the research for long-term stays in space, these results should be taken into account in order to maintain the health and performance of the astronauts.

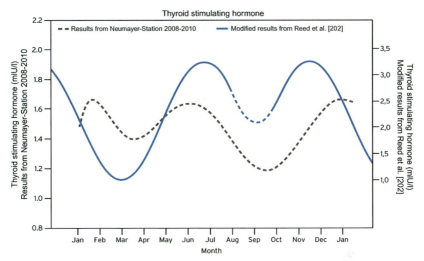

FIGURE 6.37   TSH serum-concentrations in overwinterers, *adapted from Reed et al. 2001 [202]*, superimposed with the regression curve of results TSH for the campaigns 2008-2010 at Neumayer III.

6.3.4   Limitations

This study regarding changes in various parameters of overwinterers in the Antarctic is subject to various restrictions. It was not possible to employ a uniform plan for the implementation of the measurements due to various temporal and operational requirements of the individual winter campaigns. In this study—rather than in a controlled study with paid subjects—the subjects performed the measurements as an add-on to their regular station duties. This limitation was compensated for in the statistical analyses and graphical illustration by taking into account the real time points when the measurements took place. However, a standardized measurement plan would be beneficial in future investigations. Further, there were no blood samples for campaign 2011 because extreme weather conditions prevented their safe transport, so the blood samples had to be analyzed later, but this is not unusual for a field study of this kind.

  Another limitation is the isolation itself because the measurement and transmission of the data was performed by the subjects of the study, which may result in inconsistent measurement methods (e.g., attaching the electrodes to the BIA measurements, attaching of the SenseWear device, and duration of the measurements). And finally, due to the distant location and limited possibility of control, there is an increased opportunity for noncompliance of individual measurements.

6.4   SUMMARY

This chapter on Cold Environments dealt with reports on research of our own working group on extreme environments, which is currently ongoing in Arctic and Antarctic regions.

First, the results of a pilot study in the frame of the YAU, an ultra-endurance event in the extreme environment of northwest Canada's Yukon Territory over a 430 mile (690 km) distance were reported. Data analysis revealed pronounced changes in several parameters with a high energy expenditure and a considerable energy deficit, a decrease in body mass, fat mass, and fat-free mass, changes in laytime and sleeptime with low values (where laytime is the time at rest in bed but while awake and sleeptime the time asleep), especially at the start and toward the end of the race, as well as changes in sleep phases with decreases in REM-sleep at the start and increases in deep-sleep after the end of the race. Furthermore, toward the end of the YAU there were noticeable decreases in heart rate variability with an increase of parasympathetic influence.

Results from Antarctic overwintering showed that from 2008 to 2011 changes in body composition occurred, with a decrease in fat-free mass and an increase in fat mass in the 31 participating subjects during the overwintering periods without reaching statistical significance, while the body weight remained relatively constant. There was a decrease in laytime and sleeptime in the overwinterers. Because the decline in sleeptime was greater than the one in laytime, a decrease in sleep efficiency took place. Furthermore, there was an increase in sleep fragmentation. These changes did not reach statistical significance; however, there seemed to be a deterioration in sleep quality. The total daily energy expenditure showed a decline but without reaching statistical significance. A decrease reaching statistical significance was shown for the parameter 25-OH-vitamin D for all participants, some of which did not reach the initial pre-darkness values. In some participants very low values were detected (below 30 nmol/l), which is why substitution should be considered. For TSH a similar periodicity was demonstrated, as was the case in previous studies, with an increase during the darkness phase which seems to be in accordance with the interpretation of the polar T3 syndrome. Between the sexes there were textbook value differences in body weight, body composition, and energy expenditure, with lower values for women. Of interest was the finding in which the laytime for women was significantly higher than for men. In general, the findings in the Neumayer III research station substantiate the findings of previous studies regarding the increase in fat mass, decrease in fat-free mass, changes in sleep and activity parameters, as well as the changes in hormones. This is not trivial because Neumayer III is a new station by comparison and its location is relatively "northern" compared to other Antarctic stations. The findings from this study may give rise to information to improve future overwinterings in Antarctica—such as to physical activity, sleep hygiene, and vitamin D supplementation—so that the welfare and health of the participants can be maintained.

REFERENCES

[1] Peel MC, Finlayson BL, McMahon TA. Updated world map of the Köppen-Geiger climate classification. Hydrol Earth Syst Sci 2007;11:1633–44.

[2] Danzl D. Accidental hypothermia. In: Auerbach PS., editor. Wilderness Medicine. 6th ed. Philadelphia: Elsevier Mosby; 2012. p. 116–42.

[3] Giese P. Igloo – the traditional Arctic Snow Dome. http://www.kstrom.net/isk/maps/houses/igloo.html; 2013.

[4] Yukon Government. The Yukon. www.gov.yk.ca; 2013.

[5] Statistics Canada. 2006 census. http://www23.statcan.gc.ca/imdb-bmdi/pub/instrument/3901_Q2_V3-eng.pdf; 2013.

[6] Climate Weather Office Canada. Environment Canada. http://climate.weatheroffice.gc.ca/; 2013.

[7] Pollhammer RTGO. http://www.arcticultra.de/; 2013.

[8] Hines M. The Yukon Arctic Ultra. Healthy Body Publishing; 2010.

[9] Hines M. The Marathon des Sables. Healthy Body Publishing; 2010.

[10] Shambroom JR, Fabregas SE, Johnstone J. Validation of an automated wireless system to monitor sleep in healthy adults. J Sleep Res 2012, April;21(2):221–30.

[11] McArdle WD, Katch FI, Katch VL. Human energy expenditure during rest and physical activty. In: Exercise physiology: energy, nutrition and human performace. 6th ed. Baltimore/Philadelphia: Lippincott Williams & Wilkins; 2007. p. 195–208.

[12] Rintamaki H. Performance and energy expenditure in cold environments. Alaska Med 2007;49(Suppl. 2):245–6.

[13] Ouellet V, Labbe SM, Blondin DP, Phoenix S, Guerin B, Haman F, et al. Brown adipose tissue oxidative metabolism contributes to energy expenditure during acute cold exposure in humans. J Clin Invest 2012, February 1;122(2):545–52.

[14] Launay JC, Savourey G. Cold adaptations. Ind Health 2009, July;47(3):221–7.

[15] Yakimenko MA. Thermoregulation in man during cold adaptation. Arctic Med Res 1991;(Suppl.)534–6.

[16] Savourey G, Barnavol B, Caravel JP, Feuerstein C, Bittel JH. Hypothermic general cold adaptation induced by local cold acclimation. Eur J Appl Physiol Occup Physiol 1996;73(3–4):237–44.

[17] Solter M, Brkic K, Petek M, Posavec L, Sekso M. Thyroid hormone economy in response to extreme cold exposure in healthy factory workers. J Clin Endocrinol Metab 1989, January;68(1):168–72.

[18] Andersen S, Kleinschmidt K, Hvingel B, Laurberg P. Thyroid hyperactivity with high thyroglobulin in serum despite sufficient iodine intake in chronic cold adaptation in an Arctic Inuit hunter population. Eur J Endocrinol 2012, March;166(3):433–40.

[19] Laurberg P, Andersen S, Karmisholt J. Cold adaptation and thyroid hormone metabolism. Horm Metab Res 2005, September;37(9):545–9.

[20] Virtanen KA, Lidell ME, Orava J, Heglind M, Westergren R, Niemi T, et al. Functional brown adipose tissue in healthy adults. N Engl J Med 2009, April 9;360(15):1518–25.

[21] Noakes TD. The limits of endurance exercise. Basic Res Cardiol 2006, September;101(5):408–17.

[22] Hill RJ, Davies PS. Energy expenditure during 2 wk of an ultra-endurance run around Australia. Med Sci Sports Exerc 2001, January;33(1):148–51.

[23] Knechtle B, Knechtle P, Schuck R, Andonie JL, Kohler G. Effects of a Deca Iron Triathlon on body composition: a case study. Int J Sports Med 2008, April;29(4):343–51.

[24] Viru M, Hackney AC, Karelson K, Janson T, Kuus M, Viru A. Competition effects on physiological responses to exercise: performance, cardiorespiratory and hormonal factors. Acta Physiol Hung 2010, March;97(1):22–30.

[25] Bescos R, Rodriguez FA, Iglesias X, Knechtle B, Benitez A, Marina M, et al. Nutritional behavior of cyclists during a 24-hour team relay race: a field study report. J Int Soc Sports Nutr 2012;9(1):3.

[26] Enqvist JK, Mattsson CM, Johansson PH, Brink-Elfegoun T, Bakkman L, Ekblom BT. Energy turnover during 24 hours and 6 days of adventure racing. J Sports Sci 2010, July;28(9):947–55.

[27] Schutz UH, Billich C, Konig K, Wurslin C, Wiedelbach H, Brambs HJ, et al. Characteristics, changes and influence of body composition during a 4486 km transcontinental ultramarathon: results from the Transeurope Footrace mobile whole body MRI-project. BMC Med 2013;11:122.

[28] Hoffman MD, Lebus DK, Ganong AC, Casazza GA, Van LM. Body composition of 161-km ultramarathoners. Int J Sports Med 2010, February;31(2):106–9.

[29] Knechtle B, Salas FO, Andonie JL, Kohler G. Effect of a multistage ultra-endurance triathlon on body composition: World Challenge Deca Iron Triathlon 2006. Br J Sports Med 2008, February;42(2):121–5.

[30] Bischof M, Knechtle B, Rust A, Knechtle P, Rosemann T. Changes in skinfold thicknesses and body fat in ultra-endurance cyclists. Asian J Sports Med 2013, March;4(1):15–22.

[31] Rust CA, Knechtle B, Knechtle P, Wirth A, Rosemann T. Body mass change and ultraendurance performance: a decrease in body mass is associated with an increased running speed in male 100-km ultramarathoners. J Strength Cond Res 2012, June;26(6):1505–16.

[32] Yoneshiro T, Aita S, Matsushita M, Kameya T, Nakada K, Kawai Y, et al. Brown adipose tissue, whole-body energy expenditure, and thermogenesis in healthy adult men. Obesity (Silver Spring) 2011, January;19(1):13–6.

[33] Saito M. Brown adipose tissue as a regulator of energy expenditure and body fat in humans. Diabetes Metab J 2013, February;37(1):22–9.

[34] Palca JW, Walker JM, Berger RJ. Thermoregulation, metabolism, and stages of sleep in cold-exposed men. J Appl Physiol (1985) 1986, September;61(3):940–7.

[35] Mahapatra AP, Mallick HN, Kumar VM. Changes in sleep on chronic exposure to warm and cold ambient temperatures. Physiol Behav 2005, February 15;84(2):287–94.

[36] Haskell EH, Palca JW, Walker JM, Berger RJ, Heller HC. Metabolism and thermoregulation during stages of sleep in humans exposed to heat and cold. J Appl Physiol Respir Environ Exerc Physiol 1981, October;51(4):948–54.

[37] Goodin BR, Smith MT, Quinn NB, King CD, McGuire L. Poor sleep quality and exaggerated salivary cortisol reactivity to the cold pressor task predict greater acute pain severity in a non-clinical sample. Biol Psychol 2012, September;91(1):36–41.

[38] Buguet A. Sleep under extreme environments: effects of heat and cold exposure, altitude, hyperbaric pressure and microgravity in space. J Neurol Sci 2007, November 15;262(1–2):145–52.

[39] Dewasmes G, Loos N, Candas V, Muzet A. Effects of a moderate nocturnal cold stress on daytime sleep in humans. Eur J Appl Physiol 2003, June;89(5):483–8.

[40] Romeijn N, Verweij IM, Koeleman A, Mooij A, Steimke R, Virkkala J, et al. Cold hands, warm feet: sleep deprivation disrupts thermoregulation and its association with vigilance. Sleep 2012, December;35(12):1673–83.

[41] Nunez A, Rodrigo-Angulo ML, Andres ID, Garzon M. Hypocretin/Orexin neuropeptides: participation in the control of sleep-wakefulness cycle and energy homeostasis. Curr Neuropharmacol 2009, March;7(1):50–9.

[42] Krauchi K, Deboer T. The interrelationship between sleep regulation and thermoregulation. Front Biosci (Landmark Ed) 2010;15:604–25.

[43] Sauvet F, Bourrilhon C, Besnard Y, Alonso A, Cottet-Emard JM, Savourey G, et al. Effects of 29-h total sleep deprivation on local cold tolerance in humans. Eur J Appl Physiol 2012, September;112(9):3239–50.

[44] Schwarz L, Kindermann W. Beta-endorphin, catecholamines, and cortisol during exhaustive endurance exercise. Int J Sports Med 1989, October;10(5):324–8.

[45] Yoshizumi M, Nakaya Y, Hibino T, Nomura M, Minakuchi K, Kitagawa T, et al. Changes in plasma free and sulfoconjugated catecholamines before and after acute physical exercise: experimental and clinical studies. Life Sci 1992;51(3):227–34.

[46] Mattsson CM, Enqvist JK, Brink-Elfegoun T, Johansson PH, Bakkman L, Ekblom B. Reversed drift in heart rate but increased oxygen uptake at fixed work rate during 24 h ultra-endurance exercise. Scand J Med Sci Sports 2010, April;20(2):298–304.

[47] Rajendra AU, Paul JK, Kannathal N, Lim CM, Suri JS. Heart rate variability: a review. Med Biol Eng Comput 2006, December;44(12):1031–51.

[48] Togo F, Takahashi M. Heart rate variability in occupational health –a systematic review. Ind Health 2009, December;47(6):589–602.

[49] Wheat AL, Larkin KT. Biofeedback of heart rate variability and related physiology: a critical review. Appl Psychophysiol Biofeedback 2010, September;35(3):229–42.

[50] Levy WC, Cerqueira MD, Harp GD, Johannessen KA, Abrass IB, Schwartz RS, et al. Effect of endurance exercise training on heart rate variability at rest in healthy young and older men. Am J Cardiol 1998, November 15;82(10):1236–41.

[51] Dixon EM, Kamath MV, McCartney N, Fallen EL. Neural regulation of heart rate variability in endurance athletes and sedentary controls. Cardiovasc Res 1992, July;26(7):713–9.

[52] Kaikkonen P, Rusko H, Martinmaki K. Post-exercise heart rate variability of endurance athletes after different high-intensity exercise interventions. Scand J Med Sci Sports 2008, August;18(4):511–9.

[53] Sztajzel J, Atchou G, Adamec R, Bayes de LA. Effects of extreme endurance running on cardiac autonomic nervous modulation in healthy trained subjects. Am J Cardiol 2006, January 15;97(2):276–8.

[54] Schmitt L, Regnard J, Desmarets M, Mauny F, Mourot L, Fouillot JP, et al. Fatigue shifts and scatters heart rate variability in elite endurance athletes. PLoS One 2013;8(8):e71588.

[55] Scott JM, Esch BT, Shave R, Warburton DE, Gaze D, George K. Cardiovascular consequences of completing a 160-km ultramarathon. Med Sci Sports Exerc 2009, January;41(1):26–34.

[56] Drenowatz C, Eisenmann JC. Validation of the SenseWear Armband at high intensity exercise. Eur J Appl Physiol 2011, May;111(5):883–7.

[57] Ehrmann WU, Mackensen A. Sedimentological evidence for the formation of an East Antarctic ice sheet in Eocene/Oligocene time. Palaeogeogr Palaeoclimatol Palaeoecol 1992;93(1–2):85–112.

[58] Vaughan APM, Livermore RA. Episodicity of Mesozoic terrane accretion along the Pacific margin of Gondwana: implications for superplume-plate interactions. Illustrated edition 2005 ed. Geological Society; 2005 p. 143–78.

[59] Faure G, Mensing M. Antarctica – the continent. In: The transantarctic mountains. 1st ed. Netherlands: Springer; 2010. p. 36–62.

[60] AWI. Alfred Wegener Institute for Polar and Marine Research, Bremerhaven. www.awi.de; 2012.

[61] Faure G, Mensing M. The East Antarctic ice sheet. In: The transantarctic mountains. 1st ed. Netherlands: Springer; 2010. p. 573–624.

[62] King JC, Turner J. Physical climatology. In: Antarctic climatology. Cambridge atmospheric and space science series, Cambridge University Press; 1997. p. 68–140.

[63] Straßl H. Ein Nomogramm für die Dauer von Polartag und Polarnacht bei beliebiger Ekliptikschiefe. Polarforschung 1951;21(1):40–3.

[64] König-Langlo G. Annual sunshine duration at the location of the Georg-von-Neumayer-Station. Bremerhaven: Alfred Wegener Institute for Polar and Marine Research; 2006. Personal communication.

[65] Gurney A. Der weiße Kontinent – die Geschichte der Antarktis und ihre Entdecker. München: Sierra Taschenbuch; 2002.

[66] Shackleton SE. Mit der Endurance ins ewige Eis: Meine Antarktisexpedition 1914-1917. 4th ed. München: Piper Taschenbuch; 2006.

[67] Langner RK. Duell im ewigen Eis: Scott und Amundsen oder Die Eroberung des Südpols. Frankfurt/Main: Fischer Verlag; 2011.

[68] Byrd RE. Little America: aerial exploration in the antarctic, the flight to the South Pole. 1st ed. New York: G. P. Putnam's Sons; 1930.

[69] Byrd RE. Aufbruch ins Eis. München: National Geographic Taschenbuch; 2007.

[70] Fuchs V, Hillary E. The crossing of Antarctica: the commonwealth transantarctic expedition, 1955-1958. Whitefish, MT, USA: Literary Licensing, Llc; 2011.

[71] Lansing A. 635 Tage im Eis. München: Bertelsmann Verlag; 2002.

[72] Etienne JL. Transantarctica. München: Sierra Taschenbuch; 2000.

[73] Messner R. Antarktis: Himmel und Hölle zugleich. Frankfurt/Main: Fischer; 2004.

[74] The White Mars Project. http://www.thecoldestjourney.org; 2013.

[75] Russel RCJ. International geophysical year. VSD; 2012.

[76] Suedfeld P, Weiss K. Antarctica natural laboratory and space analogue for psychological research. Environ Behav 2000, January;32(1):7–17.

[77] Surhone LM, Timpledon MT. Earth's rotation, North Pole, Antarctic, Amundsen-Scott South Pole station, rotation, geographic coordinate system. South Pole: Betascript Publishing; 2010.

[78] Palinkas LA, Suedfeld P. Psychological effects of polar expeditions. Lancet 2008, January 12;371(9607):153–63.

[79] Rothblum ED. Psychological factors in the antarctic. J Psychol 1990, May;124(3):253–73.

[80] Sproule J, Jette M, Rode A. Medical observations of members of the Soviet/Canadian 1988 Polar Bridge Expedition. Arctic Med Res 1991;(Suppl.)489–90.

[81] De Freitas CR, Symon LV. A bioclimatic index of human survival times in the Antarctic. Polar Record 1987;23:651–9.

[82] Broadway JW, Arendt J. Seasonal and bright light changes of the phase position of the human melatonin rhythm in Antarctica. Arctic Med Res 1988;47(Suppl. 1):201–3.

[83] Cattermole TJ. The epidemiology of cold injury in Antarctica. Aviat Space Environ Med 1999, February;70(2):135–40.

[84] Ikaheimo TM, Hassi J. Frostbites in circumpolar areas. Glob Health Action 2011;4.

[85] Kennaway DJ, Van Dorp CF. Free-running rhythms of melatonin, cortisol, electrolytes, and sleep in humans in Antarctica. Am J Physiol 1991, June;260(6 Pt 2):R1137–44.

[86] Muller HK, Lugg DJ, Quinn D. Cell mediated immunity in Antarctic wintering personnel; 1984-1992. Immunol Cell Biol 1995, August;73(4):316–20.

[87] Leppaluoto J, Paakkonen T, Korhonen I, Hassi J. Pituitary and autonomic responses to cold exposures in man. Acta Physiol Scand 2005, August;184(4):255–64.

[88] Palinkas LA, Reedy KR, Shepanek M, Smith M, Anghel M, Steel GD, et al. Environmental influences on hypothalamic-pituitary-thyroid function and behavior in Antarctica. Physiol Behav 2007, December 5;92(5):790–9.

[89] Reed HL, Silverman ED, Shakir KM, Dons R, Burman KD, O'Brian JT. Changes in serum triiodothyronine (T3) kinetics after prolonged Antarctic residence: the polar T3 syndrome. J Clin Endocrinol Metab 1990, April;70(4):965–74.

[90] Schiermeier Q. The great Arctic oil race begins. Nature 2012, February 2;482(7383):13–4.

[91] Gunga H-C. Forscher am Ende der Welt – Die Deutsche Antarktisstation. In: Das Neue Universum 99. Erstausgabe ed. München: Südwest-Verlag; 1982. p. 39–42.

[92] Priesner C. Georg von Neumayer Neue Deutsche Biographie. Berlin: Duncker & Humblot; 1999 p. 166–8 [Band 19].

[93] Committee on assessing the solar system exploration program, National Research Council. Grading NASA's solar system exploration program: a midterm review. National Academic Press; 2008.

[94] International Business Publication – USA. European space agency handbook (world business, investment and government library). International Business Publications; 2009.

[95] Gunga H-C, Steinach M, Kirsch K. Weltraummedizin- und biologie. In: Ley W, Wittmann K, Hallmann W, editors. Handbuch der Raumfahrttechnik. 3rd ed. München: Carl Hanser Verlag; 2008. p. 575–87.

[96] Andersen DT, McKay CP, Wharton RA, Rummel JD. Testing a Mars science outpost in the Antarctic dry valleys. Adv Space Res 1992;12(5):205–9.

[97] Horneck G. Astrobiological aspects of Mars and human presence: pros and cons. Hippokratia 2008, August;12(Suppl. 1):49–52.

[98] Gunga H-C, Steinach M, Kirsch K. Erfahrungshorizont. In: Ley W, Wittmann K, Hallmann W, editors. Handbuch der Raumfahrttechnik. 3rd ed. München: Carl Hanser Verlag; 2008. p. 577–8.

[99] Bishop SL. Evaluating teams in extreme environments: from issues to answers. Aviat Space Environ Med 2004, July;75(Suppl. 7):C14–21.

[100] Tanaka M, Watanabe S. Overwintering in the Antarctica as an analog for long term manned spaceflight. Adv Space Res 1994;14(8):423–30.

[101] Stuster J. Analogue prototypes for Lunar and Mars exploration. Aviat Space Environ Med 2005, June;76(Suppl. 6):B78–83.

[102] Gunga H-C, Steinach M, Kirsch K. Ausblick. In: Ley W, Wittmann K, Hallmann W, editors. Handbuch der Raumfahrttechnik. 3rd ed. München: Carl Hanser Verlag; 2008. p. 587.

[103] Harrison AA, Clearwater YA, McKay CP. The human experience in Antarctica: applications to life in space. Behav Sci 1989, October;34(4):253–71.

[104] Palinkas LA, Gunderson EK, Johnson JC, Holland AW. Behavior and performance on long-duration spaceflights: evidence from analogue environments. Aviat Space Environ Med 2000, September;71(Suppl. 9):A29–36.

[105] ESA. The future of European space exploration. Towards a European long-term strategy. Executive summary. Paris: The European Space Agency; 2005.

[106] NASA. The vision for space exploration. Washington, D.C.: National Aeronautics and Space Administration; 2004.

[107] Brubakk AO. Man in extreme environments. Aviat Space Environ Med 2000, September;71(Suppl. 9):A126–30.

[108] Collet J, Novara M. Global approach to simulation: a gateway to long-term human presence in space. Adv Space Res 1992;12(1):285–99.

[109] Maillet A, Gunga HC, Gauquelin G, Fortrat JO, Hope A, Rocker L, et al. Effects of 28-day isolation (ESA-ISEMSI'90) on blood pressure and blood volume regulating hormones. Aviat Space Environ Med 1993, April;64(4):287–94.

[110] Gunga HC, Kirsch KA, Rocker L, Maillet A, Gharib C. Body weight and body composition during sixty days of isolation. Adv Space Biol Med 1996;5:39–53.

[111] Nelson M, Allen JP, Dempster WF. Biosphere 2: a prototype project for a permanent and evolving life system for Mars base. Adv Space Res 1992;12(5):211–7.

[112] Larina IM, Bystritzkaya AF, Smirnova TM. Psycho-physiological monitoring in real and simulated space flight conditions. J Gravit Physiol 1997, July;4(2):113–4.

[113] Sandal GM. Culture and tension during an international space station simulation: results from SFINCSS '99. Aviat Space Environ Med 2004, July;75(Suppl. 7):C44–51.

[114] Wheeler RM. Bios-3 project in Krasnoyarsk. Russia Life Support Biosph Sci 1994;1(2):83–4.

[115] Vigo DE, Ogrinz B, Wan L, Bersenev E, Tuerlinckx F, Van Den Bergh O, et al. Sleep-wake differences in heart rate variability during a 105-day simulated mission to Mars. Aviat Space Environ Med 2012, February;83(2):125–30.

[116] Graser N. Kalte Füße inklusive: Mein Jahr in der Antarktis. 1st ed. München: Knaur Taschenbuch; 2008.

[117] McArdle WD, Katch FI, Katch VL. Body composition assessment. In: Exercise physiology: energy, nutrition and human performace. 6th ed. Baltimore/Philadelphia: Lippincott Williams & Wilkins; 2007. p. 773–809.

[118] Kyle UG, Bosaeus I, De Lorenzo AD, Deurenberg P, Elia M, Gomez JM, et al. Bioelectrical impedance analysis—part I: review of principles and methods. Clin Nutr 2004, October;23(5):1226–43.

[119] Kotler DP, Burastero S, Wang J, Pierson Jr. RN. Prediction of body cell mass, fat-free mass, and total body water with bioelectrical impedance analysis: effects of race, sex, and disease. Am J Clin Nutr 1996, September;64(Suppl. 3):489S–97S.

[120] Sun G, French CR, Martin GR, Younghusband B, Green RC, Xie YG, et al. Comparison of multifrequency bioelectrical impedance analysis with dual-energy X-ray absorptiometry for assessment of percentage body fat in a large, healthy population. Am J Clin Nutr 2005, January;81(1):74–8.

[121] Stahn A, Terblanche E, Gunga H-C. Selected applications of bioelectrical impedance analysis: body fluids, blood volume, cell and fat mass. In: Preedy V, editor. The handbook of anthropometry: physical measures of human form in health and disease. New York: Springer; 2012. p. 415–40.

[122] Stahn A, Terblanche E, Gunga H-C. Use of bioelectrical impedance: general overview and principles. In: Preedy V, editor. The handbook of anthropometry: physical measures of human form in health and disease. New York: Springer; 2012. p. 49–90.

[123] Sun SS, Chumlea WC, Heymsfield SB, Lukaski HC, Schoeller D, Friedl K, et al. Development of bioelectrical impedance analysis prediction equations for body composition with the use of a multicomponent model for use in epidemiologic surveys. Am J Clin Nutr 2003, February;77(2):331–40.

[124] Müller MJ, Bosy-Westphal A. Energieverbrauch. In: Speckmann E-J, Hescheler J, Köhling R, editors. Physiologie. 5th ed. München: Elsevier GmbH; 2008. p. 587–93.

[125] Illner K, Brinkmann G, Heller M, Bosy-Westphal A, Muller MJ. Metabolically active components of fat free mass and resting energy expenditure in nonobese adults. Am J Physiol Endocrinol Metab 2000, February;278(2):E308–15.

[126] Nelson KM, Weinsier RL, Long CL, Schutz Y. Prediction of resting energy expenditure from fat-free mass and fat mass. Am J Clin Nutr 1992, November;56(5):848–56.

[127] Muller MJ, Bosy-Westphal A, Kutzner D, Heller M. Metabolically active components of fat free mass (FFM) and resting energy expenditure (REE) in humans. Forum Nutr 2003;56:301–3.

[128] McArdle WD, Katch FI, Katch VL. Functional capacity of the cardiovascular system. In: Exercise physiology: energy, nutrition and human performance. 6th ed. Baltimore/Philadelphia: Lippincott Williams & Wilkins; 2007. p. 356–7.

[129] McArdle WD, Katch FI, Katch VL. Special aids to exercise training and performance. In: Exercise physiology: energy, nutrition and human performance. 6th ed. Baltimore/Philadelphia: Lippincott Williams & Wilkins; 2007. p. 590–1.

[130] Richardson RS, Knight DR, Poole DC, Kurdak SS, Hogan MC, Grassi B, et al. Determinants of maximal exercise VO_2 during single leg knee-extensor exercise in humans. Am J Physiol 1995, April;268(4 Pt 2):H1453–61.

[131] Welk GJ, Schaben JA, Morrow Jr. JR. Reliability of accelerometry-based activity monitors: a generalizability study. Med Sci Sports Exerc 2004, September;36(9):1637–45.

[132] Taraldsen K, Chastin SF, Riphagen II, Vereijken B, Helbostad JL. Physical activity monitoring by use of accelerometer-based body-worn sensors in older adults: a systematic literature review of current knowledge and applications. Maturitas 2012, January;71(1):13–9.

[133] Hart TL, McClain JJ, Tudor-Locke C. Controlled and free-living evaluation of objective measures of sedentary and active behaviors. J Phys Act Health 2011, August;8(6):848–57.

[134] Cupisti A, Capitanini A, Betti G, D'Alessandro C, Barsotti G. Assessment of habitual physical activity and energy expenditure in dialysis patients and relationships to nutritional parameters. Clin Nephrol 2011, March;75(3):218–25.

[135] Rawson ES, Walsh TM. Estimation of resistance exercise energy expenditure using accelerometry. Med Sci Sports Exerc 2010, March;42(3):622–8.

[136] BodyMedia I. SenseWear Manual. SW-H, Rev. 1st ed. Pittsburgh, PA, USA: BodyMedia, Inc.; 2007.

[137] Jones V, Bults R, de WR, Widya I, Batista R, Hermens H. Experience with using the sensewear BMS sensor system in the context of a health and wellbeing application. Int J Telemed Appl 2011;2011:671040.

[138] Reid KJ, Zee PC. Circadian rhythm disorders. Semin Neurol 2009 September; 29(4):393–405.

[139] Birbaumer N, Schmidt RF. Wachen, Aufmerksamkeit und Schlafen. In: Schmidt RF, Lang F, editors. Physiologie des Menschen. 30th ed. Heidelberg: Springer Medizin Verlag; 2007. p. 202–22.

[140] Ahima RS, Flier JS. Leptin. Annu Rev Physiol 2000;62:413–37.

[141] Stephens TW, Basinski M, Bristow PK, Bue-Valleskey JM, Burgett SG, Craft L, et al. The role of neuropeptide Y in the antiobesity action of the obese gene product. Nature 1995, October 12;377(6549):530–2.

[142] Thapan K, Arendt J, Skene DJ. An action spectrum for melatonin suppression: evidence for a novel non-rod, non-cone photoreceptor system in humans. J Physiol 2001, August 15;535(Pt 1):261–7.

[143] Foster RG, Hankins MW. Non-rod, non-cone photoreception in the vertebrates. Prog Retin Eye Res 2002, November;21(6):507–27.

[144] Beersma D. Models of human sleep regulation. Sleep Med Rev 1998;2(1):31–43.

[145] Weeß H-G. Diagnostische Methoden. In: Stuck BA, Maurer JT, Schredl M, Weeß H-G, editors. Praxis der Schlafmedizin. Heidelberg: Springer Medizin Verlag; 2009. p. 23–78.

[146] Morgenthaler T, Alessi C, Friedman L, Owens J, Kapur V, Boehlecke B, et al. Practice parameters for the use of actigraphy in the assessment of sleep and sleep disorders: an update for 2007. Sleep 2007, April;30(4):519–29.

[147] Conradt R, Brandenburg U, Ploch T, Peter JH. Actigraphy: methodological limits for evaluation of sleep stages and sleep structure of healthy probands. Pneumologie 1997, August;51(Suppl. 3):721–4.

[148] Sadeh A, Hauri PJ, Kripke DF, Lavie P. The role of actigraphy in the evaluation of sleep disorders. Sleep 1995, May;18(4):288–302.

[149] Chae KY, Kripke DF, Poceta JS, Shadan F, Jamil SM, Cronin JW, et al. Evaluation of immobility time for sleep latency in actigraphy. Sleep Med 2009, June;10(6):621–5.

[150] Lehmann B, Grant WB, Worm N. Physiologie von Vitamin D und Epidemiologie des Mangels. In: Reichrath J, Lahmann B, Spitz J, editors. Vitamin D Update 2012. München-Deisenhofen: Dustri-Verlag; 2012. p. 1–32.

[151] CIE. Technical report action spectrum for the production of previtamin D3 in human skin. CIE 2006;174:1–12.

[152] van Leeuwen JP, van DM, van den Bemd GJ, Pols HA. Vitamin D control of osteoblast function and bone extracellular matrix mineralization. Crit Rev Eukaryot Gene Expr 2001;11 (1–3):199–226.

[153] Wolff AE, Jones AN, Hansen KE. Vitamin D and musculoskeletal health. Nat Clin Pract Rheumatol 2008, November;4(11):580–8.

[154] van Etten E, Mathieu C. Immunoregulation by 1,25-dihydroxyvitamin D3: basic concepts. J Steroid Biochem Mol Biol 2005, October;97(1–2):93–101.

[155] Holick MF. Biologic effects of light: historical and new perspectives. In: Holick MF, Jung EG, editors. Biologic effects of light 1998, proceedings of a symposium, Basel, Switzerland, November 1-3, 1998. Boston, MA: Kluwer Academic Publishers; 1999. p. 10–32.

[156] Porojnicu AC, Robsahm TE, Dahlback A, Berg JP, Christiani D, Bruland OS, et al. Seasonal and geographical variations in lung cancer prognosis in Norway. Does vitamin D from the sun play a role? Lung Cancer 2007, March;55(3):263–70.

[157] Hintzpeter B, Mensink GB, Thierfelder W, Muller MJ, Scheidt-Nave C. Vitamin D status and health correlates among German adults. Eur J Clin Nutr 2008, September;62(9):1079–89.

[158] Heaney RP, Holick MF. Why the IOM recommendations for vitamin D are deficient. J Bone Miner Res 2011, March;26(3):455–7.

[159] Holick MF. Vitamin D, deficiency. N Engl J Med 2007, July 19;357(3):266–81.

[160] Shupnik MA, Ridgway EC, Chin WW. Molecular biology of thyrotropin. Endocr Rev 1989, November;10(4):459–75.

[161] Miell JP, Taylor AM, Zini M, Maheshwari HG, Ross RJ, Valcavi R. Effects of hypothyroidism and hyperthyroidism on insulin-like growth factors (IGFs) and growth hormone- and IGF-binding proteins. J Clin Endocrinol Metab 1993, April;76(4):950–5.

[162] Svanberg E, Healey J, Mascarenhas D. Anabolic effects of rhIGF-I/IGFBP-3 in vivo are influenced by thyroid status. Eur J Clin Invest 2001, April;31(4):329–36.

[163] Di L. I. Thyroid hormones and the central nervous system of mammals (review). Mol Med Rep 2008, May;1(3):279–95.

[164] Bernal J, Nunez J. Thyroid hormones and brain development. Eur J Endocrinol 1995, October;133(4):390–8.

[165] Bernal J. Action of thyroid hormone in brain. J Endocrinol Invest 2002, March;25(3):268–88.

[166] Kohrle J. Local activation and inactivation of thyroid hormones: the deiodinase family. Mol Cell Endocrinol 1999, May 25;151(1–2):103–19.

[167] Robbins J, Lakshmanan M. The movement of thyroid hormones in the central nervous system. Acta Med Austriaca 1992;19(Suppl. 1):21–5.

[168] Santini F, Pinchera A, Ceccarini G, Castagna M, Rosellini V, Mammoli C, et al. Evidence for a role of the type III-iodothyronine deiodinase in the regulation of 3,5,3′-triiodothyronine content in the human central nervous system. Eur J Endocrinol 2001, June;144(6):577–83.

[169] Chu C, Li JY, Boado RJ, Pardridge WM. Blood-brain barrier genomics and cloning of a novel organic anion transporter. J Cereb Blood Flow Metab 2008, February;28(2):291–301.

[170] Friesema EC, Grueters A, Biebermann H, Krude H, von MA, Reeser M, et al. Association between mutations in a thyroid hormone transporter and severe X-linked psychomotor retardation. Lancet 2004, October 16;364(9443):1435–7.

[171] van Marken Lichtenbelt WD, Vanhommerig JW, Smulders NM, Drossaerts JM, Kemerink GJ, Bouvy ND, et al. Cold-activated brown adipose tissue in healthy men. N Engl J Med 2009, April 9;360(15):1500–8.

[172] Spencer CA, Takeuchi M, Kazarosyan M. Current status and performance goals for serum thyroglobulin assays. Clin Chem 1996, January;42(1):164–73.

[173] Silva JE, Larsen PR. Adrenergic activation of triiodothyronine production in brown adipose tissue. Nature 1983, October 20;305(5936):712–3.

[174] Silva JE, Larsen PR. Potential of brown adipose tissue type II thyroxine 5′-deiodinase as a local and systemic source of triiodothyronine in rats. J Clin Invest 1985, December;76(6):2296–305.

[175] de Jesus LA, Carvalho SD, Ribeiro MO, Schneider M, Kim SW, Harney JW, et al. The type 2 iodothyronine deiodinase is essential for adaptive thermogenesis in brown adipose tissue. J Clin Invest 2001, November;108(9):1379–85.

[176] Murakami M, Kamiya Y, Morimura T, Araki O, Imamura M, Ogiwara T, et al. Thyrotropin receptors in brown adipose tissue: thyrotropin stimulates type II iodothyronine deiodinase and uncoupling protein-1 in brown adipocytes. Endocrinology 2001, March;142(3):1195–201.

[177] Grigoriev AI, Egorov AD. General mechanisms of the effect of weightlessness on the human body. Adv Space Biol Med 1992;2:1–42.

[178] Maillet A, Normand S, Gunga HC, Allevard AM, Cottet-Emard JM, Kihm E, et al. Hormonal, water balance, and electrolyte changes during sixty-day confinement. Adv Space Biol Med 1996;5:55–78.

[179] Custaud MA, Belin de CE, Larina IM, Nichiporuk IA, Grigoriev A, Duvareille M, et al. Hormonal changes during long-term isolation. Eur J Appl Physiol 2004, May;91(5–6):508–15.

[180] Edholm OG, Goldsmith R. Food intakes and weight changes in climatic extremes. Proc Nutr Soc 1966;25(2):113–9.

[181] Brotherhood JR, Budd GM, Regnard J, Hendrie AL, Jeffery SE, Lincoln GJ. The physical characteristics of the members during the international biomedical expedition to the Antarctic. Eur J Appl Physiol Occup Physiol 1986;55(5):517–23.

[182] Parker RH. Physiological adaptations and activity recorded at a polar base. Eur J Appl Physiol Occup Physiol 1985;54(4):363–70.

[183] Vats P, Singh SN, Singh VK, Shyam R, Upadhyay TN, Singh SB, et al. Appetite regulatory peptides in Indian Antarctic expeditioners. Nutr Neurosci 2005, August;8(4):233–8.

[184] Belkin V, Karasik D. Anthropometric characteristics of men in Antarctica. Int J Circumpolar Health 1999, July;58(3):152–69.

[185] Palinkas LA. Psychosocial effects of adjustment in Antarctica: lessons for long-duration spaceflight. J Spacecr Rockets 1990, September;27(5):471–7.

[186] Palinkas LA. Going to extremes: the cultural context of stress, illness and coping in Antarctica. Soc Sci Med 1992, September;35(5):651–64.

[187] Seymour GE. The concurrent validity of unobtrusive measures of conflict in small isolated groups. J Clin Psychol 1971, October;27(4):431–5.

[188] Tobler I, Borbely AA. European isolation and confinement study. Twenty-four hour rhythm of rest/activity and sleep/wakefulness: comparison of subjective and objective measures. Adv Space Biol Med 1993;3:163–83.

[189] Palinkas LA, Houseal M, Miller C. Sleep and mood during a winter in Antarctica. Int J Circumpolar Health 2000, January;59(1):63–73.

[190] Bhattacharyya M, Pal MS, Sharma YK, Majumdar D. Changes in sleep patterns during prolonged stays in Antarctica. Int J Biometeorol 2008, November;52(8):869–79.

[191] Joern AT, Shurley JT, Brooks RE, Guenter CA, Pierce CM. Short-term changes in sleep patterns on arrival at the South Polar Plateau. Arch Intern Med 1970, April;125(4):649–54.

[192] Usui A, Obinata I, Ishizuka Y, Okado T, Fukuzawa H, Kanba S. Seasonal changes in human sleep-wake rhythm in Antarctica and Japan. Psychiatry Clin Neurosci 2000, June;54(3):361–2.

[193] Steel GD, Callaway M, Suedfeld P, Palinkas L. Human sleep-wake cycles in the high Arctic: effects of unusual photoperiodicity in a natural setting. Biol Rhythm Res 1995, November;26(5):582–92.

[194] Angus RG, Pearce DG, Buguet AG, Olsen L. Vigilance performance of men sleeping under arctic conditions. Aviat Space Environ Med 1979, July;50(7):692–6.

[195] Biesalski HK. Vitamin D, recommendations: beyond deficiency. Ann Nutr Metab 2011;59(1):10–6.

[196] Huotari A, Herzig KH. Vitamin D and living in northern latitudes—an endemic risk area for vitamin D deficiency. Int J Circumpolar Health 2008, June;67(2–3):164–78.

[197] Yonei T, Hagino H, Katagiri H, Kishimoto H. Bone metabolic changes in Antarctic wintering team members. Bone 1999, February;24(2):145–50.

[198] Iuliano-Burns S, Ayton J, Hillam S, Jones G, King K, Macleod S, et al. Skeletal and hormonal responses to vitamin D supplementation during sunlight deprivation in Antarctic expeditioners. Osteoporos Int 2012 January 4;.

[199] Iuliano-Burns S, Wang XF, Ayton J, Jones G, Seeman E. Skeletal and hormonal responses to sunlight deprivation in Antarctic expeditioners. Osteoporos Int 2009, September;20(9):1523–8.

[200] Oliveri B, Zeni S, Lorenzetti MP, Aguilar G, Mautalen C. Effect of one year residence in Antarctica on bone mineral metabolism and body composition. Eur J Clin Nutr 1999, February;53(2):88–91.

[201] Smith SM, Gardner KK, Locke J, Zwart SR. Vitamin D supplementation during Antarctic winter. Am J Clin Nutr 2009, April;89(4):1092–8.

[202] Reed HL, Reedy KR, Palinkas LA, Van DN, Finney NS, Case HS, et al. Impairment in cognitive and exercise performance during prolonged antarctic residence: effect of thyroxine supplementation in the polar triiodothyronine syndrome. J Clin Endocrinol Metab 2001, January;86(1):110–6.

[203] Shurtleff D, Thomas JR, Schrot J, Kowalski K, Harford R. Tyrosine reverses a cold-induced working memory deficit in humans. Pharmacol Biochem Behav 1994, April;47(4):935–41.

[204] Palinkas LA, Cravalho M, Browner D. Seasonal variation of depressive symptoms in Antarctica. Acta Psychiatr Scand 1995, June;91(6):423–9.

[205] Reed HL, Burman KD, Shakir KM, O'Brian JT. Alterations in the hypothalamic-pituitary-thyroid axis after prolonged residence in Antarctica. Clin Endocrinol (Oxf) 1986, July;25(1):55–65.

[206] Xu C, Zhu G, Xue Q, Zhang S, Du G, Xi Y, et al. Effect of the Antarctic environment on hormone levels and mood of Chinese expeditioners. Int J Circumpolar Health 2003, September;62(3):255–67.

[207] Hodgdon JA, Hesslink RL, Hackney AC, Vickers RR, Hilbert RP. Norwegian military field exercises in the arctic: cognitive and physical performance. Arctic Med Res 1991;50(Suppl. 6):132–6.

[208] Maes M, Mommen K, Hendrickx D, Peeters D, D'Hondt P, Ranjan R, et al. Components of biological variation, including seasonality, in blood concentrations of TSH, TT3, FT4, PRL, cortisol and testosterone in healthy volunteers. Clin Endocrinol (Oxf) 1997, May;46(5):587–98.

[209] Palinkas LA, Reed HL, Reedy KR, Do NV, Case HS, Finney NS. Circannual pattern of hypothalamic-pituitary-thyroid (HPT) function and mood during extended antarctic residence. Psychoneuroendocrinology 2001, May;26(4):421–31.

[210] Wehr TA. Effect of seasonal changes in daylength on human neuroendocrine function. Horm Res 1998;49(3–4):118–24.

[211] Savourey G, Caravel JP, Barnavol B, Bittel JH. Thyroid hormone changes in a cold air environment after local cold acclimation. J Appl Physiol 1994, May;76(5):1963–7.

[212] Hackney AC, Hodgdon JA. Thyroid hormone changes during military field operations: effects of cold exposure in the Arctic. Aviat Space Environ Med 1992, July;63(7):606–11.

[213] Paakkonen T, Leppaluoto J. Cold exposure and hormonal secretion: a review. Int J Circumpolar Health 2002, August;61(3):265–76.

Chapter 7

Space

Hanns-Christian Gunga

Professor, Center for Space Medicine and Extreme Environments, Institute of Physiology, CharitéCrossOver (CCO), Charité University Medicine Berlin, Berlin, Germany

7.1 INTRODUCTION

From operational, physiological, and medical points of view, long-term missions in space can be characterized by (i) flight duration; (ii) high mission autonomy; (iii) varying gravitational loads (hypergravity during launch and landing, weightlessness during interplanetary transit, hypogravity during presence on the moon, artificial gravity loads); (iv) psychological, physiological, and social problems; (v) cosmic radiation; and (vi) hypomagnetic environment.

These exceptional natural and artificial environmental conditions during a long flight in space in a space vehicle or during a stay at a lunar or Mars base can cause physiological, anatomical, biomechanical, psychological, and social changes and have a lingering influence on the organism in complex ways. These include performance, health, and the general sense of well-being of the astronaut (e.g., sense of taste, tactile sense, reaction time, coarse and fine motor functions, muscle strength) as well as crew safety, health management, and the training program. The major physiological changes occurring in an astronaut during a stay in space are shown in Figure 7.1.

Here it must be emphasized that seemingly marginal medical, physiological, psychological, or social changes that occur during short flights can develop into pathological problems during long space flights, thus threatening entire missions.

7.2 MISSION SCENARIOS

Human exploration of space requires that the health and performance of astronauts be assured in different mission scenarios: in near-Earth orbits (International Space Station—ISS, Low Earth Orbit—LEO, mission scenario 1 (MS 1)); during an interplanetary flight (transit, M2); on the moon (MS 3);

Human Physiology in Extreme Environments. http://dx.doi.org/10.1016/B978-0-12-386947-0.00007-1

273

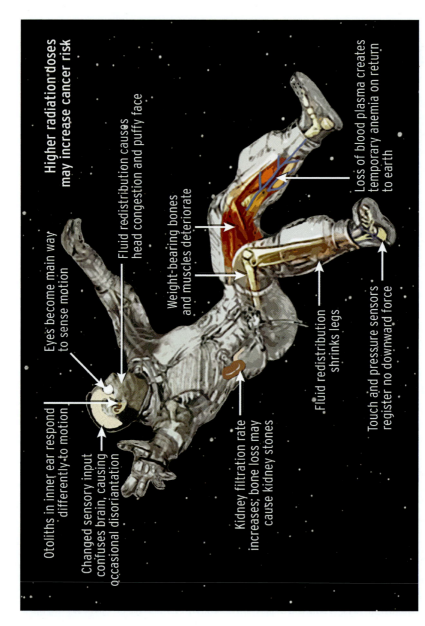

Higher radiation doses
may increase cancer risk

Fluid redistribution causes
head congestion and puffy face

Loss of blood plasma creates
temporary anemia on return
to earth

Eyes become main way
to sense motion

Weight-bearing bones
and muscles deteriorate

Otoliths in inner ear respond
differently to motion

Changed sensory input
confuses brain, causing
occasional disorientation

Fluid redistribution
shrinks legs

Touch and pressure sensors
register no downward force

Kidney filtration rate
increases; bone loss may
cause kidney stones

FIGURE 7.1   Major physiological changes in organ and organ systems under gravity-free conditions. *Adapted from Refs. [1,2].*

and on Mars (MS 4). On the basis of these scenarios, the following parameters can be identified that fundamentally influence the mission and that must be taken into account to varying extents depending on the nature of the particular mission: (i) distance from the Earth, (ii) changes in g-forces (10^6 g in LEO, 0 g during an interplanetary transit, 0.16 g on the moon, 0.38 g on Mars), as well as (iii) the probable duration of a mission. These mission scenarios lead to various load profiles related to health, safety, and the well-being of the crew. In addition, they significantly influence operational planning and technological issues down to the construction and design of the requisite space vehicles or planetary space stations (Habitat) [3,4]. In general, autonomy, as well as safety and health problems, substantially increase from scenario 1 to scenario 4 because (i) a real-time telecommunication system becomes more and more inefficient with increasing distance from the Earth. For scenario 4 (Mars), there is a 40-min time delay, so a natural dialog between the crew on Mars and the ground station on Earth can hardly take place; (ii) an evacuation of astronauts from LEO in an emergency would be possible within 24 h (Sojus capsule) and could be managed from the moon in the course of several days/weeks, but timely evacuation is as good as impossible in scenarios 3 (transit) and 4 (Mars), which means that the highest degree of autonomy is required for these scenarios, and (iii) the radiation risk in scenarios 2-4 substantially increases without appropriate shielding measures (see below).

7.3 MAJOR PHYSIOLOGICAL AND MEDICAL LIMITATIONS DURING LONG-TERM SPACEFLIGHT

In this chapter the emphasis will be on the problems connected with very long manned missions, extending beyond the ISS. As can be seen in Figure 7.6, a rather good database exists for approximately the first 30 days of habitation in space. In the meantime, we know from numerous investigations how the physiological progression of various processes takes place during this phase, for example, in the cardiovascular system and concerning the salt/water balance. Afterwards, a stable equilibrium is reached after about 6 weeks (Figure 7.2).

However, because of the scarce amount of relevant data, we know relatively little about the muscle and bone skeletal system during long periods in space, and this despite the fact that for very long human stays in space it will likely play a significant role. The same is true for long-term effects of increased radiation exposure in space. The progression and extent of recovery from this radiation load and the changes to other human organ systems after reentry into Earth's gravitation field are also open questions. It had been observed already that it takes approximately the same time to re-adapt to Earth's environment after space for astronauts and their different organ systems, as illustrated in Figure 7.3.

It can also be expected that during long-duration flights in space, psycho-physiological problems and interpersonal conflicts will be an increasing problem. A number of stress factors affect humans in space, including

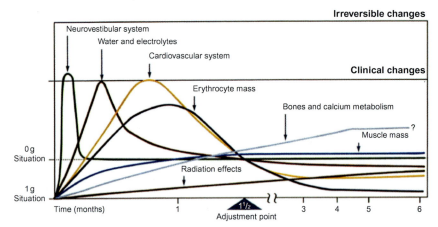

FIGURE 7.2 Temporal progression of physiological changes during adjustment to weightlessness. The 1-g condition represents the physiological situation on Earth. The 0-g condition is full physiological adjustment to space conditions, which probably could only be achieved by those born in space. An astronaut requires an average 6 weeks to reach a point of adjustment, at which there is at least partial adaptation to the new environment [7].

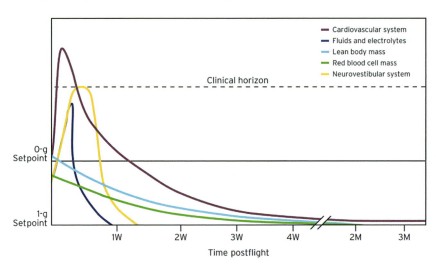

FIGURE 7.3 The time span for re-adaptation of different organs and organ systems of astronauts/cosmonauts after a long-term space flight to the Earth's environment; W, week, M, month [7].

isolation, confinement, lack of sources of stimuli, dependence on technical systems, and lack of comfort, as well as individual (lack of privacy, separation from family members, stress management, motivation) and interpersonal (territorial and dominance behavior, group structures, group thinking, group dynamics) conflicts, but also changes of circadian, lunar, circalunar, and annular endogenous rhythms. During long flights in space these changes are combined with an increasing feeling of monotony the longer the flight lasts.

All these factors contribute to making a long-duration flight in space, even after years of training, an unusual burden for astronauts that is far beyond their range of experience on Earth [1].

7.4   LIFE SUPPORT SYSTEMS ON ISS

Life support systems (LSS) developed in the past and already in operation on the ISS include CEBAS (Closed Equilibrated Biological Aquatic system), BLSS (Bioregenerative Life Support System), ARS (Atmosphere Revitalization Subsystem), ECLSS (Environmental Control and Life Support System (Shuttle)), ECLS (Environmental Control System), CELSS (Controlled Environmental Life Support System), and CEEF (Closed Ecology Experiment Facility). These LSS are designed to recycle as much as possible to reduce the uplift payload and to save energy consumption, bottlenecks for all current and future space missions [8]. As can be seen, they vary in nature according to their purpose in space. As a typical example, the LSS for the production of O_2 and the removal of CO_2 from the environment on ISS is shown in Figure 7.4.

The daily requirements for a single astronaut for oxygen, water, and dry food are substantial (Figure 7.5).

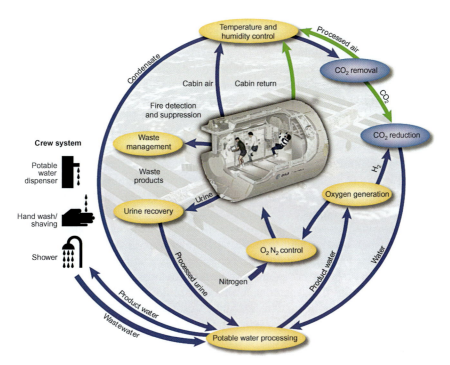

FIGURE 7.4   Typical LSS on the ISS for O_2 production and CO_2 removal. *Adapted from www. nasa.gov/sites/default/files/104840main_eclss.*

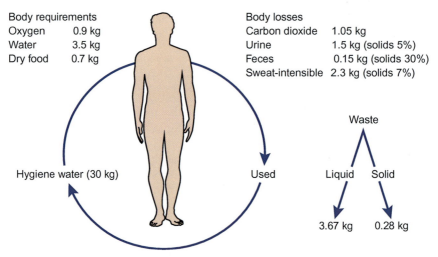

Body requirements
Oxygen   0.9 kg
Water   3.5 kg
Dry food   0.7 kg

Body losses
Carbon dioxide   1.05 kg
Urine   1.5 kg (solids 5%)
Feces   0.15 kg (solids 30%)
Sweat-intensible 2.3 kg (solids 7%)

Waste

Hygiene water (30 kg)

Used

Liquid   Solid

3.67 kg   0.28 kg

FIGURE 7.5   Daily body requirements and body losses (Piantadosi, 2012).

In the future, such a BLSS must be even more complex. A really sophisticated system will include photosynthetic activity of green plants, which in the presence of light convert CO_2 and water into carbohydrates and oxygen, providing chemical energy. To be fully effective, in addition to sufficient concentrations of CO_2 and O_2 in the ambient air and the right humidity, temperature, and barometric pressure, light of a suitable wavelength must also be available in a future space craft or habitat on the Moon or Mars. The circadian and seasonal rhythms of plants must also be taken into account. Cultivation in these artificial systems is usually in the form of monocultures, so the plants lack a natural social milieu (biotope), which, as is sufficiently well known from terrestrial experience, involves risks. The lack of gravity is the biggest disadvantage, because all systems developed to date use gravitropic land plants that do not prosper under weightlessness, if they grow at all, and do not produce edible fruits [8].The cause is likely the lack of thermal convection around plants growing under weightlessness. For this reason, the CEBAS system, among others, was developed, which produces biomass in an aquatic milieu using nongravitropic plants and provides animal protein as well as plant material. Despite all these problems, such systems have to be further developed until satisfactory solutions are found. Self-contained, completely closed systems with products that can be harvested will probably not exist for some time; in most cases, something either has to be added to the system somewhere or removed from somewhere else. The systems are semi-open, which means it costs energy to maintain them. One can, however, develop energy-saving concepts for them. From the perspective of physiologists and medical doctors, it should also be noted that for long-term presence in space, a good variety of energy producers and protein-rich nourishment to guard against certain monotony in diet must be assured. Otherwise, dietary deficiencies and malnutrition can be expected in the long term.

7.5   EXPERIENCE TO DATE

In October 2000 the first ISS crew (William Shepherd, Yury Gidzenko, Sergei Krikalev) was launched and stayed 141 days in space. In May 2014, the fortieth and forty-first expedition crew (Makin Suraev, Reid Wisemann, Alexander Gerst) was sent into space and will stay on ISS until November 2014. With the ISS in operation, the number of astronauts/cosmonauts and the frequency of long-term space flights has markedly increased. Yet, in 2005, Clément [5] published an interesting graph that illustrated nicely the fact that prior to the ISS, the overwhelming number of space flights were short-term missions (Figure 7.6).

According to his data in 2003, about 460 people had been in space at that time, most of them male astronauts; the female proportion was only about 10%, and the average flight duration was only 10 days. Furthermore, in 2003 only about 40 astronauts had flight experience longer than 6 months, and only four people were in space longer than 1 year [5]. The record for the longest continuous presence in space is still held by the Russian physician Dr Valery Polyakov, who flew for 437 days on the MIR station from 1994 to 1995. The cumulative records are also held by Russian cosmonauts, namely Valery Polyakov (679 days, 2 flights) and Sergei Avdejev (748 days, 3 flights). However, today, according to U.S. government information on astronauts, several U.S. astronauts can now be found on the hit list of long-term space flight, including Sunnita Williams (322 days), Donald Petit (370 days), Michael Foale (374 days), Peggy Withson (377 days), and Michael Fincke (282 days) [9]. In total, 330 American

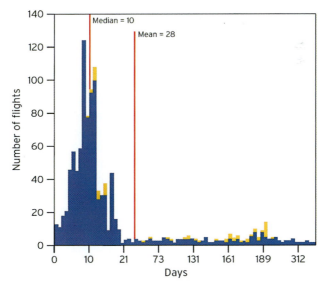

FIGURE 7.6   Frequency and duration of manned space flight until 2003 (blue) and from 2004 until 2012 (yellow) thereafter. The two red lines indicate the median (left) and the mean (right) duration of space flights until 2003. Note the rare number of long-term flights before 2003, that is, until the ISS was completed. *Adapted from Refs. [5,6].*

astronauts have flown into space, 282 males and 48 females, of which 201 had military and 129 civilian backgrounds. Furthermore, according to the NASA Johnson Space Flight data files, in 2012, a total of 124 astronauts and cosmonauts stayed on board and 35 expedition crews stayed longer than 120 days, including expedition crew 14 (Mike Lopez-Alegria, Mikhail Tyurin), who stayed 215 days [9]. As indicated in Figure 7.6 the database for long-term space flights has markedly increased since the ISS has been in operation (yellow bars), because in 2003 the median duration mission length had been only 10 days and the arithmetic mean duration length 28 days, respectively. This graph does not include the taikonauts from China. Since May 2014 a total of 8 taikonauts have been in space for up to 3 weeks.

It has often passed unnoted that research is always supplemented with accompanying terrestrial studies. These studies have always been a major component of space medicine, and it can be safely assumed that their significance will increase for the planning and execution of future long flights into space. It should be emphasized here that the European Space Agency (ESA) critically analyzed at an early date the various medical, physiological, psychological, social, and technical problems that could play a role during lengthy human presence in space. This subject, the human factor in manned missions, has been exhaustively treated in ESA studies, at special meetings, and in numerous technical reports [10–17]. By definition, studies of the human factor are an interdisciplinary research field that integrates insights from medicine, physiology, psychology, sociology, biomechanics, engineering, and other disciplines and complements them with habitability studies, which concern human abilities and maximum potentials. They undergird the design of equipment (ergonomics), machines, systems, workflows, tasks, and ambience so that people find environments as safe, comfortable, and effective as possible. Because the necessary information cannot be adequately appreciated and dealt with only on the basis of theoretical considerations, ESA, for example, established the interdisciplinary Simulation Mission Study Group (SIMIS) several years ago. It was the task of this working group to identify simulation facilities (e.g., pressurized diving chambers, submarine stations, bunkers) or other isolated places on Earth where people in small groups can live together to simulate the living quarters of long-duration flights in space. It turned out that polar stations, underwater research stations, hyperbaric and hypobaric chambers, bed-rest facilities, submarines, immersion facilities, parabolic flights, and mock-ups (habitability test beds) can all serve as reasonably analogous environments. In the meantime, ESA has conducted numerous isolation and confinement studies, and results are available from simulation studies making use of various facilities with different crew compositions and crew sizes (ESA reports: POLAREMSI, ANTEMSI, HYDREMSI, ISEMSI, EXEMSI, HUBES, SFINCSS). They make far-reaching recommendations on planning and conducting long-duration missions from medical, psychological, and technical perspectives [10–12,14,15,17–19]. These terrestrial simulation studies will remain a significant element of all space missions in the future—for example, for crew selection issues [20,21]. It is astonishing that there is hardly any data available from the field of clinical medicine specifically on problems related to isolation and confinement, although similar problems must arise in cases of long-term immobility (fractures), for example. A current example for the clinical relevance of results from such simulation studies are the findings of Heer et al. [22] and Titze et al. [23], who could prove that the human body can store osmotically inactive sodium. This offered a completely new perspective on the endemic disease high blood pressure (hypertension). It was significant that the discrete retention of salt was only noted after observation periods of longer than 3 weeks. As shown in Figure 7.7, earlier ESA isolation and confinement studies, such as ISEMSI conducted in 1990 and EXEMSI in the year 1992, gave first indications that this might be the case in humans [12,24–26].

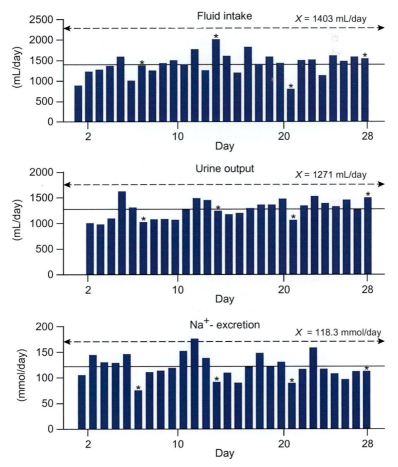

FIGURE 7.7 Salt and water turnover in six men during a 28-day isolation and confinement study. Please note the weekly rhythm of sodium-excretion (Na⁺) (ISEMSI) [24].

Short-term observation periods of 3-5 days were inadequate to reveal this effect. In the meantime it could be experimentally shown that the surface tissues (cutis and subcutis) play a central role in the osmotically inactive storage of sodium. Furthermore, most recently this research group showed during MARS500, a long-term isolation and confinement study with an international crew conducted in Moscow at the Institute of Biomedical Problems (IBMP) between June 2010 and November 2011 (Figures 7.8 and 7.9), that a continuous reduction in salt intake could decrease systolic blood pressure significantly [27,28].

Furthermore, these studies indicated that (i) weekly (circaseptan) patterns in sodium excretion exist that were inversely related to aldosterone and directly related to cortisol, and (ii) total body sodium is not dependent on sodium intake but instead exhibited far longer (≥ monthly) infradian rhythms independent of

FIGURE 7.8   The MARS500 facility at the IBMP in Moscow.

FIGURE 7.9   One main chamber of the MARS500 isolation and confinement facility at the IBMP in Moscow/Russia. Six male subjects from four different countries (three from Russia, one from China, two Europeans) were completely locked up for a period of 520 days to simulate a flight from Earth to Mars and back. *Photograph by courtesy of ESA.*

extracellular water, body weight, or blood pressure [28]. As mentioned above, problems of isolation and confinement have been the subject of intensive research in space medicine for years and are leading to results that could easily be relevant for clinical problems, considering that almost half of the population in industrialized countries lives in single-person households, either by necessity (sickness) or voluntarily. Here, too, medical physiological space research on humans is setting the pace in describing the physiological phenomena and pathophysiological changes that arise. Furthermore, a methodological repertoire from space research is made available that can later find practical application in clinical medicine, such as noninvasive technologies, telemetric transmission, and real-time data analysis programs.

7.6   SPECIAL MEDICAL PHYSIOLOGICAL PROBLEMS ARISING FROM RESIDENCE IN SPACE

Various authors [5,29] have summarized the most relevant medical physiological experiences, problems, and risks that have become known in the course of U.S. and Russian short- and long-term flights in space, as well as in work with terrestrial simulation models produced over the last 30 years [30]. The conclusion is that when planning, designing, and realizing missions in space, the following medical problems must be taken into account: loss of appetite (anorexia); nausea (space motion sickness); exhaustion; sleeplessness; dehydration; dry and inflamed skin;muscular-skeletal pain in the back, legs, and feet; respiratory tract infections; cardiovascular problems including arrhythmia; headaches; diarrhea; constipation; barotraumas (the formation of free gas bubbles in blood and tissues causing embolisms, limb, and joint pains, "bends"); chemically induced lung infections as a consequence of long, strenuous extravehicular activities (EVAs); deficient sensory perception; seborrhea; allergic reactions; fungal infections; and dental disease, including the inhalation of foreign bodies. In the compilations made by Billica et al. [29] and Clément [5], considerable attention is given to psychological problems that reduce the mental performance, the ability to concentrate, psychic stability, and resilience of astronauts (e.g., asthenia, stress, fear, nervousness, depression). The entire spectrum of possible problems is probably still not yet discovered. Based on the compilations [5,29] regarding medical problems during U.S. and Russian short- and long-term spaceflights, and additional data from earth-bound simulations (analogous) such as those in Antartica and submarines, the following common health hazards have to be taken into consideration when defining the main requirements for design, construction, and medical equipment: loss of appetite (anorexia); space motion sickness; facial fullness; sinus congestion; fatigue/tiredness; insomnia/sleeplessness; dehydration; inflammation/dryness of skin and subcutaneous tissue; musculoskeletal pain (back, leg, foot); upper respiratory infection (sneezing, coughing); conjunctival irritation, including foreign bodies in the eye; superficial injuries; genitourinary infections; cardiovascular problems (cardiac arrhythmias, low heart rate); headache; muscle strain; diarrhea;

constipation; barotitis; bends (decompression-caused limb pains); chemical pneumonitis (lung inflammation from EVA); sensory changes (tingling, numbness); fever; nosebleeds; psoriasis; folliculitis; seborrhoea; allergenic reactions; fungal infections; dental caries; inhalation of foreign bodies; glossitis; and psychological problems and mental problems/disorders (decreased concentrations capabilities, mood evelations, asthenia, stress/tension, anxiety). In addition, Comet [10] has identified the following research areas that at least will require more work before a human flight to Mars lasting 500-1000 days: the performance of the cardiovascular, immune, muscle and skeletal systems; nutrition under complete weightlessness $(0, \ldots, 0.1\,g)$; changing gravity loads (1/10, 3/10, 5/10, 8/10 g); toxicological and pharmacological kinetic studies under weightless conditions; studies on the occurrence of kidney stones caused by accelerated bone reduction in space; the dynamics of healing processes; as well as research on the influence of various types of artificial gravity on human, animal, and plant organisms. These special medical-physical problems and methodologies to detect or counteract them during a long-term space flight will be the focus in the following sections.

7.6.1   Changes in Body Composition

Liquids and electrolytes, fat, and bone mass all undergo considerable change during the first 6 weeks under simulated terrestrial (isolation studies) and in real weightlessness [7,25,31]. Biochemical equipment and technologies must be available on board to identify these bodily changes, such as body impedance devices, whole body plethysmographs, or whole body scanners. These data are of special importance for the planning, initiating, and follow-up control of special exercise as countermeasures such as exercise on a vibration platform or training exercises with a flywheel ergometer.

7.6.2   Cardiovascular System

It has been confirmed in several studies that the cardiovascular system undergoes rapid changes during presence in LEO on short-term (shuttle) and long-term (Skylab and MIR) manned missions [32]. Fluid displacement toward the head [33–36] occurs, which leads to a loss of liquid in the lower extremities of about 2 L and reduces the astronauts' plasma volume by up to 20% [37–39]. The production of red blood cells also decreases significantly during the first weeks in space, because erthropoietin production and release from the kidneys is reduced [37,39]. The reduced plasma volume and lower red blood cell mass—the latter is critical for oxygen transport capacity—reduce the maximal exercise astronauts can undertake, especially after returning to Earth. In addition, reduced plasma volume as well as an altered sympathetic nerve receptor density and threshold in the blood vessels and the heart can play a key role in orthostatic hypersensitivity. Also, the shift of volume along the body axis from the body's lower to upper half leads to noticeable changes in heat emission from the skin and probably affects the entire human

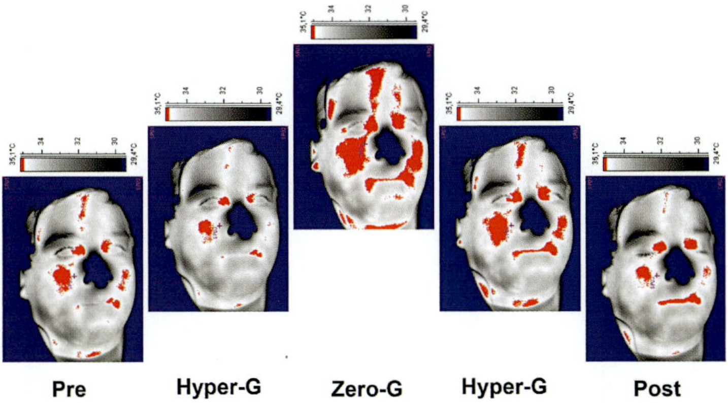

| Pre | Hyper-G | Zero-G | Hyper-G | Post |

FIGURE 7.10   Infrared images of facial skin temperature (male test subject, 20 years old) in the various phases of a parabolic flight during DLR's eighth parabolic flight campaign in 2006. The rapid dynamics of skin temperature changes that are caused by a fluid shift along the body axis and during weightlessness lead to a striking increase in heat emission [6].

heat balance under weightlessness. This is supported by recent data from relevant human physiology experiments in connection with parabolic flights (Figure 7.10).

Under terrestrial conditions, the head and neck are responsible for over 30% of the heat exchange between the body and the environment. Forced convection to ventilate the space vehicle or the ISS could cause heat emission from this part of the body to be much higher. This has been empirically confirmed by the experience of numerous astronauts who reported cool to cold extremities, to the extent that some of them wear socks although the room temperature is relatively high (ISS ca. 24-27 °C).

7.6.3 Muscles and Skeletal System

Both the muscles and skeleton require enough stimulus in the form of mechanical loading (strain) to be able to maintain their structure. Under weightless conditions this strain is reduced to a minimum [1,2,40,41]. Muscle atrophy is noticeable after only a brief time in space, accompanied by structural and functional changes in the muscles as well as a negative nitrogen balance. This atrophy is especially pronounced in the posture muscles that support the body to counteract gravity. Without corrective measures, astronauts lose up to 20% of their muscle mass during a short mission and up to 50% during long-term missions [5]. This loss is higher in the group of extensor muscles compared with the flexors. The leg muscles have the highest loss because astronauts move through their environment mainly by using their arms. Together with the changes in the neurovestibular and cardiovascular systems, such as dizziness and orthostasis, the loss of strength and stamina lead to a considerable reduction in performance for astronauts returning to an environment with gravity—which in the case of a landing on Mars could hinder the success of the mission. Changes in muscles

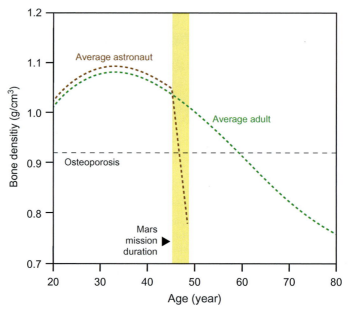

FIGURE 7.11   Estimated bone density loss in astronauts during a 3-year Mars mission. Please note that at the end of the mission the astronaut's bone density would be comparable to an 80-year-old adult. *Adapted from Ref. [2].*

are closely linked to changes in the skeletal system. Without adequate strain, there is an imbalance between bone development and breakdown with a global loss of bone mass of 1-3% per month on average. As illustrated in Figure 7.11 [2], this estimated bone density loss in astronauts during a 3-year mission is so considerable that an astronaut would have a bone density comparable to an 80-year-old man.

In areas experiencing especially low strain, bone density is reduced by up to 10% [5]. The same range of change was also observed in earth-bound long-term bed rest studies (LTBRs) [40,41], shown in Figure 7.12. However, the changes in astronauts observed thus far are generally faster and more pronounced. During this LTBR countermeasures that might counteract muscle and bone loss most successfully were tested as well during bed rest. Rittweger et al. found that (i) flywheel exercise had the potential to prevent muscle atrophy; (ii) both pamidronate, a bisphosphonate that initiates a pharmacological inhibition of bone resorption, plus flywheel exercise may be beneficial to preserve bone mineral content during bed rest; and finally, (iii) for some unexplained reason(s), efficacy appears to be more easily achieved in the tibia diaphysis than in the epiphysis [41].

Loss of bone mass is expressed in hypercalciuria. Although other factors in addition to increased calcium levels in the urine are relevant (pH value, uric acid, oxalate), there is an increased risk of post renal stone formation [5,7].

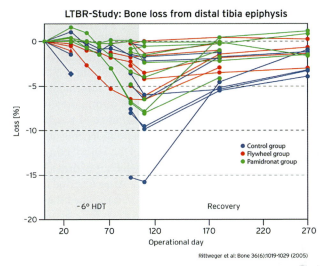

LTBR-Study: Bone loss from distal tibia epiphysis

Rittweger et al: Bone 36(6):1019-1029 (2005)

FIGURE 7.12 Results from a long-term head-down tilt (−6°) bed rest study (90 days). Please note that control subjects showed no changes, whereas the subjects on bed rest lost bone mass up 15%. The high interindividual differences can be mainly explained by the size of the endocortical surface. *Adapted from Ref. [41].*

Bone decomposition and the associated increased risk of fractures, particularly of the lower extremities, and possible renal stone formation seriously threaten the health of astronauts and the success of a mission. Appropriate countermeasures must be taken. How rapidly the hypercalciuria in astronauts progresses is shown in Figure 7.13: It starts during the first days in space [5].

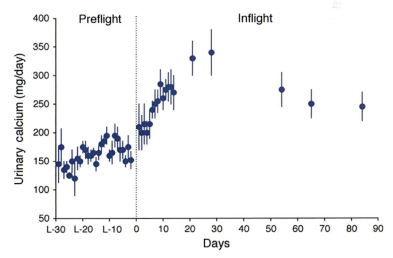

FIGURE 7.13 Compiled data from Gemini 7, Skylab 2-4, and Shuttle missions (each data point represents the mean ± SD of *n*=2-14 subjects). *Adapted from Ref. [5].*

7.6.4   Sensory Systems

The integration of visual, vestibular, and kinesthetic stimuli makes orientation through the environment possible. The individual systems are interconnected. Thus, proprioceptive and visual information flow to the vestibular system and from there back again to involuntary control of eye movement. The lack of gravitational stimulation leads to an incongruence of information in the semicircular canal of the inner ear, which only reacts to a change of position and the acceleration information from the saccule and utricle independently of gravity [42]. The discrepancy between this information and the kinesthetic and visual signals can lead to so-called space motion sickness, which is expressed in various types of dizziness and feeling ill or nauseous [7,31]. After a period of adjustment of a few days, these symptoms disappear but can return when the astronaut returns to a 1 g environment [42].

Some astronauts reported problems with nearsightedness, which may be caused by fluid displacement toward the head and changes in the shape of the eyeball and lens in a 0 g environment [5].

Because of the various permanent sources of noise on a space vehicle (motors, experiments, air conditioning equipment) noise pollution reaches levels of 60-100 dB [5], which in addition to causing stress also carries the risk of causing partial deafness and makes appropriate acoustic and auditory protection necessary.

7.6.5   Diet

Regarding an adequate diet and nutrition for astronauts/cosmonauts [43,44], since the first space flights, it has been recognized that special care has to be taken [45] and numerous factors must be considered. The food must be nonperishable (for months to years), the drinks powdered, and the meals freeze-dried so that only water is added on board and mixed into the food. Many meals are heated to high temperatures to make them long-lasting. All efforts must be made to avoid food poisoning, which could be life threatening for the entire crew on an interplanetary flight. In addition—as is known from crews overwintering in Antarctica [46]—it has been shown particularly during long-term residence on MIR and the ISS that astronauts love "special event meals" to counteract the monotony caused by stimulus deprivation [5]. Therefore, attention must be paid to taste and even to the type of packaging. Of course, nutritional value and trace element content must be precisely known and adequate. Food that is too salty is to be avoided because it fosters an increased removal of calcium from the bones [47], which is accelerated because of weightlessness [48]. These days, NASA has a collection of about 150 menus, to which another 150 menus from the Russians can be added on the ISS, so the astronauts have about 300 different menu variations available [43]. Alcohol is not permitted. The day's menu is divided into four segments, breakfast, lunch, dinner, and a snack. However, the consumption of cookies is problematic because crumbs are not easy to remove in conditions of weightlessness and can have unpleasant consequences, such as crumbs in the eyes, in other body openings, and interference with technical and scientific equipment. Carbonated drinks must also be avoided because they lead to unpleasant flatulence. The latter can also sometimes lead to spontaneous vomiting under weightlessness. Astronauts and cosmonauts have a marked preference for strongly spiced food. In part, the reason could be the liquid displacement along the body axis caused by weightlessness [33], which causes the oral mucosa in the mouth and the tongue to swell (edema). Also, the distribution of food in the mouth changes because of weightlessness, which leads to an altered sense of smell. All these factors taken together are probably sufficient to show that nutrition under conditions of weightlessness—particularly for long flights—is a problem that should not be underestimated [44]. It has already been well confirmed in isolation studies that a balanced, varied diet is a decisive factor for the overall psycho-physiological well-being of astronauts, whereby communal meals involving the entire crew also play an important role in the daily schedule (the social components of meals), independent of energy supply and taste [5,30,31]. As far as energy requirements, it was surprising to discover that the caloric demand under weightlessness hardly differs from that under 1 g conditions for the astronauts. This could be, for example, because there are increased energy requirements under weightlessness due to the daily muscle exercises instituted as a countermeasure to muscle and bone loss and/or the catabolic process in respect to the described adaptations of the muscle mass in space. Changes in digestion (resorption) and taste could also play a role [49,50].

7.6.6   Radiation

The radiation risk to humans in space is significant—it is one of the major threats for astronauts/cosmonauts, especially during long-term space stays [51]. Different kinds of radiation sources have been recognized, mainly solar X-rays, solar flare neutrons and gamma-rays, solar flare electrons, protons and heavy irons (solar particles), Jovian electrons, and galactic and extragalactic cosmic rays (Figure 7.14). The latter are of special concern during long-term space flights [52–54].

Depending on the kind of radiation, among others factors, the kind of shielding and the strength of the dosage short-term effects of a radiation exposure might be radiation sickness, involving symptoms such as nausea, vomiting, or even diarrhea; reduced blood count; hemorrhaging; or death. Long-term effects of radiation can damage the DNA and lead to uncontrolled cell growth and finally to cancer if compensatory mechanisms such as apoptosis (programmed cell death) or DNA repair (p-53 mechanism with premitotic pause and restoration) are ineffective [5,54]. Those radiation effects and their downstream consequences at cellular level are schematically illustrated in Figure 7.15.

Specifically, for missions near the Earth (LEO), crew members are well protected from ionizing radiation by the Earth's magnetic field. But further out,

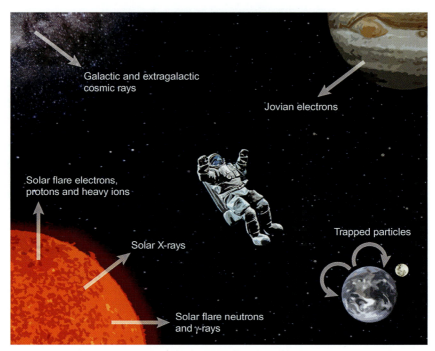

FIGURE 7.14   Different kinds of cosmic rays. *Adapted from Ref. [52].*

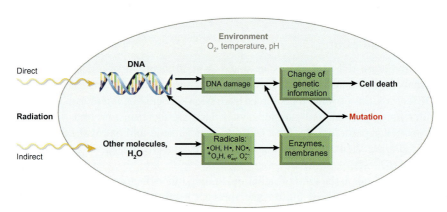

FIGURE 7.15   Radiobiological chain of events that starts in a microbial cell after exposure to ionizing radiation, with two alternative pathways of interaction, resulting in either direct or indirect radiation damage. *Adapted from [102].*

the crew is exposed to far higher radiation levels. In the Van Allen belt, which extends up to an altitude of 30,000 km, radiation composed of high-energy protons and electrons (ca. 1 keV to several MeV for electrons and hundreds of MeV for protons) prevails (Figure 7.16). Cosmic radiation (also known as cosmic background radiation) consists of high-energy ionized nuclei, from hydrogen to

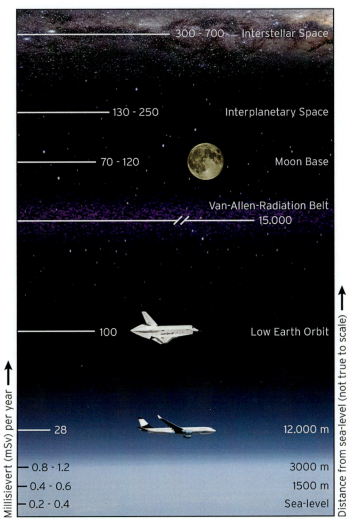

FIGURE 7.16 Radiation exposure on Earth and for different mission scenarios: LEO, interplanetary transmit, moon, and Mars [6]. *Adapted from Kent Snodgrass.*

uranium nuclei (up to 1000 MeV), whose origin is outside our solar system. The solar particles mentioned above primarily consist of high-energy protons that are ejected into space during a solar eruption (solar particle event, SPE). This radiation can significantly increase the energy of cosmic particles up to several hundred MeV. Neither the temporal occurrence and duration or strength of such an eruption has been so far reliably predicted; the radiation from SPEs represents a serious danger for interplanetary long-term missions [4,5]. Radiation dosage (*D*) is measured in Gray (Gy), where 1 Gy is equivalent to absorbed energy of 1 J/kg. Because the various radiation sources also differ in the strength of biological effects, the so-called equivalent dosage (De) in Sievert (Sv) is used. Depending on the radiation source, the radiation dosage D is multiplied by a factor for its significance in each case (1 for gamma rays, ca. 20 for alpha particles, ca. 10 for high-energy ions) [4]. Figure 7.16 shows the supposed radiation contamination in millisievert (mSv) for the various mission scenarios (MS) such as LEO (MS 1), interplanetary transit (MS 2), moon (MS 3), and Mars (MS 4), whereby the maximal dosage per lifetime according to present knowledge should not exceed 250 mSv. It can be seen that radiation exposure in mission scenarios 2-4 is substantial and that measures to protect the space vehicle, Moon, or Mars habitat must be taken [31,34].

7.6.7 Psycho-Physiological Problems Arising from Residence in Space

7.6.7.1 Biorhythms

In space, humans are almost completely isolated from the physical and social environmental factors that normally determine their biorhythms, thus a reduction in zeitgeber occur (for details, see Chapter 6) [18,55–59]. The terrestrial alternation between day and night is mimicked by controlling the artificial illumination in the space vehicle, and the daily fluctuations in ambient temperature are also artificially maintained, to mention only some of the relevant factors. Social factors, whose origins are family, workplace, and friendships, are lacking. Endogenous biorhythms must adjust to a large extent to the mission's predetermined work and relaxation rhythms [18]. The duration and quality of sleep have to be so optimized that performance readiness and a subjective feeling of well-being are maintained. However, the need to more or less stringently organize the daily routine means that individual needs have to be adapted to the needs of the entire group, and this for the duration of the entire mission. Negative consequences are inevitable because over long periods of time people's psychological states and work capacity fluctuate considerably. In this respect observations made during terrestrial isolation studies do not differ significantly from Russian experience on long missions [11]. The daily schedule during the Russian isolation experiences is as follows: work period 9-17 h, resting 17-23, sleeping 23-7, getting up phase 7-9, with meals four times a day. Presumably, there is accordingly little opportunity for any individual style that significantly varies from this routine. Measurements of activity patterns with the help of accelerometers during a 4-week ESA isolation study produced fairly similar results on workdays for all six test subjects [14]. Measurements on weekends showed, as expected, more individualized behavior patterns, especially that the distribution of sleep periods differed from the normal weekday pattern. However, strict regulation of the daily schedule should mean respecting natural sleep requirements, which during the day are especially pronounced after the main meal. It is known that forced, artificially extended periods of wakefulness negatively influence cognitive abilities, something that should definitely be avoided on long missions [55–57]. But there is no general strategy that can be applied to the situation in space, and appropriate ground-based experimental simulation is also missing. Perhaps the new established research facility envihab at the DLR in Cologne (Germany), which is highly equipped, could serve as a better analog for such long-duration term space flight simulation [60,61]. Isolation conditions and the length of the particular mission influence the sleep-wake rhythm as well as the cognition of the astronauts. Russian experience has shown that after an initial 14-day adjustment period, a mission phase follows in which there is full adaptation until about the 90th day. Afterwards, sleep problems can be expected, the activity of the crew members is reduced, and a reduction in the range of their interests can be observed. Increased nervousness and tiredness become evident. Because these phenomena were found during terrestrial isolation studies as well as in space, it is hard to identify the role played by weightlessness. It is currently unknown how weightlessness affects biorhythmic processes at the cellular and molecular levels and what the consequences are for the entire system, including important immunological issues [62]. Therefore, as a first step, currently long-term studies on ISS are ongoing to determine the time course and extent of changes of body core temperatures in astronauts, which are taken as an indicator of any circadian rhythm changes in the human body (Figure 7.17).

7.6.7.2   *Isolation and Confinement of Movement*

In the course of dealing with the physiological consequences of long periods in space, medical doctors had to consider the problems of isolation, solitude, loneliness, and confinement of movement [63]. Confinement is often connected with a lack of exercise, which plays a role in the discussion of countermeasures (see Section 7.6.8) as well as in this section. But isolation and confinement are not only problems limited to space flight; they can be observed every day in

FIGURE 7.17   An astronaut performing body core temperature with the Double Sensor system (yellow probe at the front of the head) and intraocular pressure measurements during long-term stay on ISS. *Photo courtesy by ESA/NASA.*

society. In the large cities of Western Europe almost 50% of the population lives in single-person households, either by choice (singles, primarily young people) or because of external circumstances such as the death of a partner. Increasing longevity reinforces this trend. In order to identify the physiological effects of this change in ambient conditions, new, noninvasive technologies must be developed for continuous data collection. They should have low energy requirements, not get in the way when they are worn, and be easy for a layperson to use. This continuous flow of signals produces large amounts of data, which can only be managed with the help of modern information technology. Time-series analyses are particularly valuable. When recording and processing data, the spectrum of normal stimuli on Earth should be kept in mind (Table 7.1).

For example, if gravity is lacking in space, then 40% of the body's stimuli disappear. In the case of the other parameters, noise, smell, taste, and colors of the living habitat (Antarctic station, spacecraft, Moon base), no large variations due to physical confinement are expected. Living together in close quarters always means consideration for others—not what everybody wants, but what is the least disruptive has to be determined. This leads to an overall leveling down and to monotony. Isolation and confinement therefore mean a reduction or deficit in stimuli [64–66]. In the best case, this can lead to relaxation followed by recovery, particularly if one has been living in an overstimulated environment. If the removal of stimuli is unavoidable, such as for long residence in space, the first step is light deconditioning leading to a vicious circle as described in Figure 7.18.

In a LEO this can be combated by the indicated countermeasures, which provide psychological and social support such as telephone calls with family members. Physical refreshment can, for example, be produced by thermal stimulation in an otherwise thermally monotonous environment. As another option, the environment in which the people are living can be specially designed, because the requirements for habitability, that is, the level of environmental acceptability, change dramatically with circumstances. For short periods, any environment may be taken as "acceptable," but not for a long mission [67–70]. Such stimulus substitution measures are absolutely necessary in a monotonous environment to break out of the vicious circle, which ends in total deconditioning

TABLE 7.1 Normal Stimulus Range on Earth [6]

Parameters	Whole Spectrum of Parameters (Hypothetical)
Gravitation	40% proprioceptors
Light	30% eyes
Acoustic noise	10% ears
Smell and taste	10% nose, mouth
Temperature	10% skin, CNS (central nervous system)

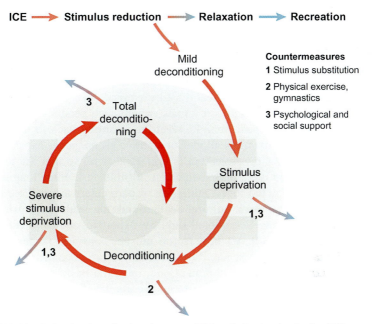

FIGURE 7.18   Isolated and confined environment (ICE) and stimulus deprivation [66].

[64–66]. Total deconditioning can lead to being unable to readjust to terrestrial conditions. Stimulus substitution not only helps to make the stress associated with residence in space bearable, it also improves the chances for smooth re-adaptation to terrestrial conditions. Considering the experience gathered under space conditions, rehabilitation measures on Earth, such as after long immobility and sickness, need re-evaluation. The effects of a reduced social spectrum should not be underestimated. A crew consists of four to eight members at most, and they have to get along. The communication within such a group can vary considerably during a mission, as illustrated in Figure 7.19. Here, the number

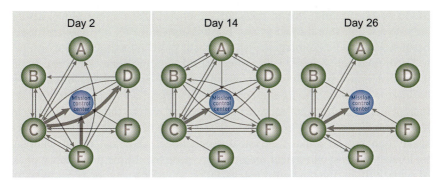

FIGURE 7.19   Communication diagram of an international crew of six men being isolated and confined for 28 days (ISEMSI, 1990); A–F different subjects inside the isolation chamber. *Adapted from Ref. [71].*

of verbal communications among group members during the joint meal during a 28-day period of isolation was recorded. On the second day it can be noted that the communication of person D with commander C dominates; the other members are almost excluded. But by the end of the mission (26th day), person D is totally isolated during the meal. This social isolation within the group can have somatic effects that must be avoided [71]. One reason for isolation might be modern computer games, which each person can play alone in some corner of the space vehicle, resulting in sitting back to back during leisure time without communicating. Competitive group games can counteract such a development. Crew selection is important here, so that the team has the right composition [18]. Only then can it fulfill its task in the long term. In summary, it can be noted that a multitude of parameters affect the cohesion and performance of a team: cultural background, the leadership qualities of the commander, sex, age, emotional stability, expertise, and cooperativeness, to name only a few [5,30]. All these factors must be considered when a crew is selected for a long stay in space. Physiological performance factors also play a role, but not necessarily a dominant one, once the basic parameters have been achieved (well trained at least to an average level). Psychologists have an important role in deciding the right data to consider.

7.6.8 Countermeasures

It should first be pointed out that the physiological changes noted so far under weightlessness are adaptations that are appropriate for the space environment. The Russian cosmonaut Polyakov lived, as mentioned above, in the Russian MIR station for periods up to 14 months. Medical physiological problems usually occur when the astronauts are again subjected to a higher level of gravity, whether on Earth or in the future on the Moon or Mars. But here, particularly in the case of EVAs in the context of a mission, physical fitness is the *conditio sine qua non*. Appropriate corrective measures must be taken, especially those assuring that the proper functioning of the cardiovascular, skeletal, and muscle systems is maintained. Providing emergency medical care for missions to the Moon or Mars could, just to consider the operational reasons alone, always only be possible to a limited extent. For missions in LEO, it is another matter, because in a medical emergency an astronaut or cosmonaut can be evacuated to Earth relatively quickly (via rescue capsule or emergency descent), something that is not possible for lunar and certainly not for Mars missions. In the narrower sense, among the corrective measures, all those treatments and therapeutic measures that serve to guarantee or restore the physical, mental, and spiritual well-being and health of astronauts before, during, and after a space flight can be counted. Physical, biochemical, chemical, pharmaceutical, biological, and psychological measures are possible. In the preparatory phase of a space flight, they include crew selection in order to assure a crew who are as efficient as possible and in which there is the lowest risk of interpersonal conflict [18]. Physical and mental training programs have to be made available, and care must be taken to achieve the required physical fitness and high, continuous mental performance levels and alertness, not to mention programs that prepare the organism for altered circadian rhythms. During the flight, most of the corrective measures concern providing and carrying out an exercise program of several hours with the help of customized training equipment such as treadmills, flywheels, rowing machines, a "penguin suit," or a vibration platform. After return from space, intensive medical and psycho-physiological training programs are necessary to accustom the astronauts to the terrestrial environment and to their social milieu (post-flight disorders) [72,73]. It has been shown that this post-flight phase to re-establish the original preflight situation requires almost as much time as the space flight itself (Figure 7.3). An exception in the physiological realm has to do with the changes that occur in bone metabolism, which, according to the present state of scientific knowledge, cannot be completely returned to initial values.

7.6.8.1   Cardiovascular System

In order to avoid orthostatic hypersensitivity and physical deconditioning on the part of the astronauts during their presence in space [5,31], a variety of different physical training devices should be available for the daily training periods such as a bicycle ergometer, treadmill, and/or rowing machine. This variety is not only necessary for physiological reasons, but also from a psychological point of view, particularly to avoid boredom. During the training units, basic cardiovascular and pulmonary data need to be noninvasively monitored, stored, and analyzed (e.g., $\dot{V}O_2$max, heart frequency, blood pressure, breathing rate, skin and body basal temperature, oxygen saturation). Before landing, astronauts need to be supplied with liquid in order to increase their plasma volume [74,75]. This kind of liquid replenishment can still be carried out after orthostatic intolerance has developed. For this treatment, practical devices for intravenous infusions and increasing plasma volume must be at hand for emergencies, and the astronauts have to be trained in the technique of intravenous infusion. During the critical phase of re-entry, antigravity suits are worn [75] that support blood flow toward the upper body by applying pressure to the legs [75]; in addition, horizontally oriented seats are used [7]. Despite these countermeasures, orthostatic intolerance remains for the most part as an unsolved clinical and mission-relevant problem [76,77].

7.6.8.2   Muscle and Skeletal System

Power training to maintain muscle and body strength as well as bone mass is the second pillar of physical exercise. The immediate motivation here, too, is to preserve the strength of the hands for work outside the space vehicle [5,30,31].

In addition to various elastic band and stretching systems for resistance training, the procedures mentioned above for cardiovascular training can also be used, such as rowing machines or a treadmill, in which the trainee is pulled toward the machine via a system of elastic resistance bands (high-impact training to maintain bone structure). It has been shown that a slight amount of vibration exerts sufficient stimuli to maintain bone mass [78–81]. The application of vibration plates during the flight could help shorten the current tedious exercise regime. Pharmacological approaches to maintain muscle tone and the skeletal system include doses of amino acids to stimulate protein synthesis and of biphosphonates to reduce global calcium loss and maintain bone density [79,82].

7.6.8.3   Neurosensory System

The lack of gravitational stimulation in space leads to incompatible information coming from the visual, vestibular, and kinesthetic systems, which can lead to what is known as space motion sickness [30,31]. Although symptoms like dizziness, nausea, and feeling ill recede after a few days, they often return upon entry into a gravitational field, which could hinder the continuation a mission on Mars. Unfortunately, there are large differences in the nature and particular expression of symptoms, both between individuals and for the same person on different occasions [83]. These often do not agree with observations made in experiments on Earth, making it difficult to predict susceptibility to space motion sickness. Selecting individuals with a high tolerance for movement, acceleration, and deceleration did not produce the desired results. The use of biofeedback and autogenous training was successful in reducing some of the symptoms. Medication (e.g., promethazine) can also reduce the severity of space motion sickness [84], but it also can lead to a drop in performance (affecting reaction time and pattern recognition, for example) and can negatively influence sleep and overall mood.

7.6.8.4   Radiation Protection

Selection of crew members based on the lowest possible number of changes in particular gene loci known to be oncogenes, as well as the extraction and storage of bone marrow as a regeneration possibility, can also be carried out before the flight as precautionary measures to avoid later malignant illness. Pharmacological radiation protection is available in the form of free radical scavengers as well as innovative radioprotective substances (such as DNA-binding WR-33278). Protease inhibitors and substances such as dimethylsulfoxide can hinder the evolution of an already radiation-damaged cell into a cancerous cell cluster [85–87]. This development gives rise to the hope that pharmacological approaches can be developed to counter radiation risks. On the technical side, the hull of the space vehicle itself supplies the most protection but as well induces special radiation problems due to the creation of new particles [88]. However, because most of the risk comes from high-energy solar particles moving slower than the speed of light, a satellite-supported early warning system could be put into operation [89]. As soon as a solar event (eruption, solar wind) has been identified, the information can be relayed to the crew on board the space vehicle, which would give them enough time to get to a specially shielded area [31,34,90]. An innovative technical approach would be to use a magnetic deflector to divert solar particles from the space vehicle [91]. New superconducting electromagnetic coils that no longer require supercoiling have shown promising results and are financially affordable. But data are not yet available on the long-term effects of comparable magnetic fields on human tissues.

7.6.8.5 Personal Databases for Astronauts

To summarize, there are great individual differences in the progression and extent of adjustments that astronauts make to their new environment (e.g., concerning the neurovestibular system, liquids and electrolytes, cardiovascular system, red blood cells, bone and calcium metabolism, lean body [5,30,92]). It is essential that these changes be recorded and monitored for each individual astronaut as completely and as long as possible before (baseline data), during, and after each one's presence in space. These data are, for example, essential for establishing guidelines for exercise programs designed to counteract cardiovascular deconditioning. Such data is also needed in case of emergencies. Preventive care starts with a most complete long-term individual database on the physiological, psychological, and medical history of the participating astronaut/cosmonaut in the mission. During and after the mission this database has to be both kept up to date and completed. In a second step, these individual data/changes in the course of the mission should be compared with the overall database from earlier astronauts/cosmonauts.

7.6.9   Emergency and Rescue Mission Scenarios

Possible emergency and rescue scenarios for different kinds of space missions have been discussed by several authors [5,93–95] and were a very special topic in the frame (i) of the HUMEX study on the survivability and adaptation of humans to long-duration exploratory missions, as well as (ii) in case of IHAB, an additional study on inflatable habitats [6,53,96–99]. Mainly, the descriptions, conclusions, and recommendations from these publications are compiled in the following subsections.

7.6.9.1 Scenario 1: In Low Earth Orbit

The guidelines for the handling of emergencies in space parallel those found on Earth, which can be divided into five categories according to (i) severity of illness and injury, (ii) capability of the onboard medical equipment, (iii) ability of a surgeon/physician to perform/assist during the medical event, (iv) level of skill and training of the crew medical officer, and (v) ease and feasibility of medical evacuation to Earth.

For life-threatening emergencies on the ISS, a health maintenance system is available that consists of (i) cardiac defibrillator, (ii) ambulatory medical pack, (iii) respiratory pack, (iv) advanced life support pack, (v) crew medical restraint system, and (vi) crew contamination protection kit. A cardiac defibrillator is necessary because especially during EVAs cardiac arrhythmias have been described [7].

The ambulatory medical pack consists mainly of those standard instruments and medical equipment necessary to perform first aid in case of injuries. Therefore, the medical kit is equipped with injectables (medications to be administered by injection), emergency items such as minor surgery equipment (e.g., sterile gloves, scalpel, needle holders), instruments for inspecting the body (e.g., stethoscope, tourniquet, tongue depressors, otoscope, binocular loupe), oral medications (e.g., pills, capsules), bandages, and noninjectables, which are medications to be used for topical application (e.g., bleeding, wound healing). Taken together, this package allows basic capabilities for surgery, dentistry, and/or anaesthesiology.

The respiratory pack consists of a device to allow continuous ventilatory support and oxygen administration in the case of a severe pulmonary dysfunction in flight either due to an infection of the lung—for example, pneumonia—or due to a thoracic injury of the astronaut/cosmonaut.

The advanced life support pack allows "bed-side" monitoring in space combined with the capabilities of ventilatory therapy and cardiovascular monitoring of the patient.

The crew medical restraint system is necessary in order to perform, in case of an emergency (e.g., a spinal injury due to contusion), proper transport (evacuation from LEO) or medical care of the astronaut/cosmonaut under micro-g conditions.

In any case where toxic substances are liberated, the crew contamination protection kit can be worn. In life-threatening situations this suit should ensure a hermetic sealing of the body and therefore is equipped with a respiration kit as well.

Finally, telemedicine equipment is on board the ISS for the purpose of providing for and monitoring the well-being of the astronauts/cosmonauts during their stay. Furthermore, an environmental health system functions on the ISS to monitor of the internal environment for early detection of biohazards and contamination risks in the fields of toxicology, water quality, microbiology, and radiation. Permissible exposure levels as well as maximum allowable concentration authorized for each material/substance in human spacecrafts are known as SMACs (Substance Maximum Allowable Concentration). They can be classified into three groups: (i) noncarcinogenic (e.g., acceptable, slight irritation, mild headache), (ii) noncarcinogenic (e.g., unacceptable, anaesthesia, blindness, disability), and (iii) carcinogenic (lifetime risks).

7.6.9.2   Scenario 2: On the Surface of the Moon

For a moon mission, the problems are still less critical as compared to scenarios 3 and 4 because (i) a telemedicine system will be efficient enough because of a nonsignificant signal delay in telecommunications, (ii) most diseases with a significant probability of occurrence can be prevented or treated using medical facilities similar to those available on ISS, (iii) in case of a any life-threatening situation a rescue to Earth can be performed in a reasonably possible time frame (days/weeks). Therefore, the astronauts/cosmonauts in mission scenario 2 should be (i) equipped and trained to perform basic medical routine procedures, including blood sampling for routine biochemical checkups of the crew members in the habitat on the moon's surface and (ii) able to perform basic medical emergency care with advanced emergency kits with support from ground-based experts on Earth via telemedicine/teleconferences. Data on the long-term effects on the human body exposed to a reduced gravity force (0.16 g) on the moon surface are very rare. Only during the APOLLO missions could some data be collected for up to 3 days on the moon's surface [7]. One can assume that the low gravity force on the Moon is not sufficient to prevent deconditioning in the long run, especially if a Moon mission is planned with a half-year stay on the surface of the moon. As in the LEO scenario, under these circumstances, to maintain crew health various countermeasures (dealing with, e.g., cardiovascular and skeletal-muscular systems) have to be implemented in the habitat on the moon's surface. Therefore, those medical and countermeasure devices listed for scenario 1 (LEO) also apply to the moon scenario. Different from scenario 1 might be that on the Moon, (i) even more advanced telemedical equipment will be needed because it would take at least several days to evacuate an ill or injured astronaut/cosmonaut, and (ii) surgical equipment should be more advanced than in scenario 1 because a higher rate of injuries can be assumed due extensive EVAs on the moon's surface. Because an evacuation from the Moon would take at least several days, it can be foreseen that in the case of an infectious/and or parasitic disease, the infected astronaut/cosmonaut would need to be treated in the medical facility at the habitat on the Moon's surface. Therefore, in addition to supplying the standard antibiotics, such a Moon medical facility should be designed and equipped to treat the patient in a separate, isolated medical room of the habitat. From a crew safety standpoint, this would minimize the risk that the other members of the crew would get infected. In light of the possible occurrence and treatment of psychological problems/disorders in the Moon scenario, the main difference between the LEO and Moon scenarios would be that the transport of letters and personal food items will be impeded. All the other channels to avoid crew member boredom and homesickness on the moon's surface (e.g., teleconferences, various possibilities for recreational activities during the leisure time) could and should be available in the Moon habitat facility. Even, in case of a severe psycho-medical issue, with pharmaceutical support (e.g., valium, haloperidol), an evacuation of the astronaut/cosmonaut to Earth could be realized. Definitely—and different from the scenario 1—on the Moon precautions must be taken to avoid dust contamination of the habitation (e.g., cave inhalation, toxic reactions). Furthermore, radiation shielding for the habitat and individual dosimetry have to be assured for the astronauts/cosmonauts due to the high radiation exposure on the Moon (see Section 7.6.6). A probable suitable solution for high radiation exposure might be habitats below the surface, at least to a certain extent.

7.6.9.3 Scenario 3: During Transit from Earth to Mars

During transit from Earth to Mars (scenario 3) the physiological, medical, psycho-physiological, social, and logistical problems are markedly increased. Whereas for mission scenarios 1 and 2 some data are available from flights to MIR, the ISS or the U.S. APOLLO program, there are no data on how humans will adapt to, perform, and socially interact during such a real long-term interplanetary transit from the Earth to Mars. Therefore, the flight route, the duration, and the distance from Earth have a much more critical impact, especially in view of psychological problems, as compared to the scenarios 1 and 2. This is mainly because (i) no immediate rescue scenario will possible due to the flight route and (ii) telecommunication/telemedicine systems will not be very efficient because of a significant signal delay. This will (i) increase the feeling of isolation and confinement and (ii) probably lead to social stress affecting group functioning. From a medical-physiological point view, the changes, adaptations, complications, and possible countermeasures during a transit from Earth to Mars seem to be comparable to those mentioned in scenarios 1 and 2, however, to higher degree because basically any acute emergency situation has to be performed without support from ground-based experts on Earth due to the time delay, which allows—at least because of the far distance from Earth—no effective communication. As in the LEO and Moon scenarios, to maintain crew health various countermeasure facilities have to be implemented in the transit spacecraft. Because an evacuation during transit from Earth to Mars is impossible, in case of an infectious/and or parasitic disease, the infected astronaut/cosmonaut has to be treated in the medical facility of the transfer spaceship. Therefore, different from scenario 1 but comparable to scenario 2, in addition to the standard antibiotics the medical facilities on the space ship should be designed and equipped to allow the treatment of the astronaut/cosmonaut in a separate, isolated medical room. In view of the possible occurrence and treatment of psychological problems/disorders, precautions must be taken especially to avoid boredom, homesickness, claustrophobia, asthenia, and anxiety on board during transit time. Therefore, an increasing variety of recreation facilities/opportunities than in scenarios 1 and 2 has to be implemented. This is especially important because in the case of a severe psycho-medical disease of an astronaut/cosmonaut, no evacuation to Earth can be realized. Then, so as not to endanger the whole mission, it might be necessary that the ill astronaut/cosmonaut be kept under pharmaceutical (e.g., valium, haloperidol) management for the whole mission. Furthermore, the radiation shielding for the transit spacecraft and individual dosimetry must be even more advanced and should include a "safe haven" rescue room because it can be assumed that, different from scenarios 1 and 2, during transit the astronauts/cosmonauts will eventually encounter markedly higher radiation doses (see Section 7.6.6).

7.6.9.4   Scenario 4: On the Surface of Mars

The Mars exploration mission, scenario 4, is the most challenging one because (i) an emergency return is impossible, (ii) real-time telemedicine communication to Earth will not be efficient due to the delay in transmission, (iii) only intelligible dialogs/telecommunication between the crew on ground and the crew in the orbit of Mars will be available, (iv) the 0.38 g load experience on the Mars surface after a long-term 0 g exposure might cause difficulties in gait and posture, leading to (vi) an increase of related injuries, (vii) crew members will experience maximum feelings of isolation and confinement, which (viii) will probably lead to social stress affecting group functioning and well-being of the astronauts/cosmonauts. From a medical-physiological point view, the changes, adaptations, complications, and possible countermeasures during the stay on the surface of Mars seems to be comparable to those that have been analyzed during the APOLLO missions, however at a much higher level, and to those that have been discussed in scenario 3. Different from scenarios 1, 2, and 3 might be that in scenario 4 (i) Mars has a 0.38 g force that has to be taken into account in the design and construction of countermeasure devices; (ii) still even more advanced self-automatic medical equipment is needed compared to scenarios 1 and 2 and similar to scenario 3, because no telemedical support from Earth will be available. Therefore, the crew should have at least one full physician on the surface of Mars. The surgical and diagnostic equipment should be more advanced than in scenarios 1 and 2 and similar to scenario 3 (highest autonomy). Due to the fact that the emotional and activity levels as well as the working schedules will be very high in contrast to mission scenario 3, it can be expected that feelings of boredom especially will be diminished, but feelings of loneliness and anxiety might increase, leading to an activation of the sympathetic-adrenal-medullary system and the hypothalamic-pituitary-adrenal axis. Due to the uniqueness of the mission itself, the psychological stress levels might be higher compared to the other scenarios; therefore, teleconference systems, although handicapped by time delay, and recreational facilities have to be implemented in the daily schedule. Precautions must be taken to avoid contamination from Mars dust and possible hostile Martian chemical/life forms; as on the Moon, radiation shielding on the surface of Mars and individual dosimetry must be improved. Psychological and social training programs should be performed routinely to avoid feelings of boredom, isolation, and confinement and to ensure group coherence. To counteract these changes during space flights a treadmill, lower-body negative pressure (LBNP) devices, and probably a short-arm centrifuge should be available. The treadmill, compared to an ergometer, allows a higher degree of training of the lower extremities and the circulatory system, especially if it is combined with a penguin suit. This penguin suit has been frequently used during Russian space flights and consists of a special elastic material that provides a passive stress on the antigravity muscle groups of the legs and the antigravitational muscles of the upper parts of the body (back). The LBNP device partially reverses the headward fluid shifts that occur under micro-g conditions due to LBNP. Therefore, the daily training sessions of the astronaut/cosmonaut with this device will counteract, at least partially, the cardiovascular deconditioning of humans in space. A short-arm centrifuge, similar to the LBNP device, will lead to a redistribution of body fluids to the lower parts of the body and will counteract the headward fluid shift under micro-g conditions. Therefore, currently, multinational research programs are ongoing to determine which duration and level of g-forces are required in humans to achieve the best countermeasure results to prevent cardiovascular deconditioning for astronauts/cosmonaut under micro-g conditions. Nevertheless, the most promising proposals to prevent cardiovascular deconditioning during the whole space flight seem to be not a single countermeasure device but instead a combination of the different training devices (treadmill, LBNP, and short-arm centrifuge). To counteract impairment of the musculo-skeletal system, most promising are the flywheel ergometer and a vibration plate. The flywheel ergometer is constructed for seated isometric, concentric, and eccentric uni- or bilateral leg press exercises. Currently, the flywheel ergometer's dimensions for height, width, and length are 0.95 meter (m), 0.55 m, and 1.3 m, respectively. It is composed of a fixed frame with a seat and a movable arm. A pivoting horizontal bar, perpendicular to the moment arm, functions as a footplate. Seat and footplate can be adjusted to allow different lower limb lengths. Polymer (PVC) flywheels (380 × 25 mm), 3.5 kg each, are used to provide resistance. The resistance level can be adjusted by using one, two, three, or four flywheels, where resistance increases with the number of flywheels used. The total weight of the flywheel ergometer (with four flywheels installed) is 55 kg. While seated, the subject pushes horizontally against the footplate. From a starting position of about 80° angle, the subject initiates flywheel rotation through the pull of a cord that is connected to the footplate. The cord will begin to unwind, and energy is imparted to the flywheel. Once the pushing phase (concentric muscle action) is completed to an almost full knee extension, the cord will start rewinding by virtue of the kinetic energy of the rotating flywheel and, thus, returning the footplate. While attempting to resist the force produced by the pull of the flywheel recoiling the cord, an eccentric muscle action is now performed. Force can be measured using a strain-gauge mounted on the footplate and attached via a ball-and-socket joint to the moment arm. The horizontal displacement of the footplate is measured by means of a linear potentiometer (source: http://lsda.jsc.nasa.gov/scripts/cf/hardw.cfm?hardware_id=855; P. Tesch device (http://lsda.jsc.nasa.gov/scripts/cf/exper.cfm?exp_index=322)). Ground-based studies such as the Berlin Bed Rest Study 2003/2004 have suggested that vibration *per se* has a beneficial effect on preserving bone mass. In the study mentioned above, vibration training sessions were performed twice daily five times per week. Each exercise session takes 30 min, plus 30 min during which the subjects recreate. Each morning session comprises the following units: (i) squatting exercise with maximum force, held by shoulders, hips and waist to maintain the thigh musculature; (ii) toe-standing to maintain the foot extensor musculature; (iii) heel-standing to maintain foot-dorsiflexors; and (iv) "kicks," rapid extensions of the entire leg, with a pause of 10 s in between. The rationale behind this unit is the observation that sub-maximum mechanic osteogenic stimuli become maximum when rest periods of 10 s are inserted. In the afternoon session, units 1-3 are performed over 60 s, two times each, without any break in between, but with the peak forces reduced to about 60% of those in the morning session. According to the capability of the subjects, exercise time should be 60 s or longer for each unit. The vibration frequency of the system obviously has to be between 20 and 25 Hz, and the peak forces should be 2.5 times the body weight.

Resuscitation, health stabilization, and transportation for a quick return to Earth can be realized in the near Earth orbit and probably also still during missions to the Moon. Actually, for Mars missions, the highest level of autonomy in respect to resuscitation, health stabilization, and therapeutic capability has to be guaranteed because evacuation to Earth will demand probably several months. The autonomy level can be compared to some extent with South Polar stations during the winter season. Based on the knowledge from the stations, basic surgery equipment for a Mars base has to cover the whole spectrum of basic surgery and treatment operational skills. Therefore, the participation of a physician is mandatory for long-term missions, at least in scenarios 3 and 4. Telemedicine could give limited support. Advanced telemedicine, or better e-Health device, should be able to transfer, in addition to basic physiological and medical data, pictures from ultrasound devices and computer and/or magnetic resonance tomographies. The ultrasound technique in the LEO and the noon mission scenario plays a very helpful role in avoiding unnecessary evacuations from there to Earth. In case of the scenarios 3 and 4, the telemedicine device is less efficient due to signal time delay, but it might be useful, as outlined above, for telemedical consultations between the Mars orbiter crew and the ground crew of the Mars surface. For the LEO and Moon scenarios, it seems advisable to have an echographic system that can be controlled from an expert center on Earth via a 2-D echograph guided by a teleoperated robotic arm [100]. In this system a robotic arm holding onto the patient with a real ultrasound probe is remotely controlled from the expert site with a fictive probe. It reproduces on the real probe all the movements of the expert hand via satellite. In detail, at the expert center, the medical expert moves a fictive probe, connected to a computer system that sends the coordinate changes of this probe via satellite line to a second computer, located in the LEO or Moon, that applies them to the robotic arm holding the real echographic probe. Meanwhile, several recent field tests have already shown that the expert was able to perform the main views (longitudinal, transverse) of the liver, gallbladder, kidneys, aorta, pancreas, bladder, prostate, and uterus the same as during direct examination of the patient. The heart and spleen seem to be more complicated for medical ultrasound examinations. Most interesting, the mean duration of the robotized echography procedures was approximately only 50% longer than direct echography of the patient.

A real concern during long-term-space flights far from Earth are rehabilitation countermeasures after injuries which have affected bone and connective tissue because we have to deal with a general depressed osteoblasty under micro-g/0-g conditions, combined with an increased bone resorption especially at the lower limb and lower parts of the spinal cord, and probably impaired connective tissue repair mechanisms. In view of the long-term mission to Mars (scenarios 3 and 4), it can be assumed that injuries affecting the skin, muscle, connective tissue, or even bone (fracture) have to be taken seriously into account and accordingly surgical diagnostic and therapeutic tools and capabilities have to be an integrative part of the mission planning. After any acute surgical and medical support of injuries, a rehabilitation program has to be started. Assuming a high risk of fractures of the lower legs, a device should be available in scenario 4 to allow an impulse-free ergometer exercise (in contrast to the flywheel device) with a direct and indirect biofeedback system for individual adjustment of force and speed. These biofeedback systems for rehabilitation programs seem to be especially necessary for mission scenarios 3 and 4, because in the case of a complicated fracture, an evacuation to Earth from LEO and the Moon seems to still be realistic. If such an injury should occur during transverse to Mars and on the surface of Mars, an evacuation to Earth would be impossible.

7.6.10 Outlook

Space medicine research is characterized by innovative science approaches and pioneering technological applications. Although an apparently very exotic speciality, it has a frequently underestimated high clinical relevance. Specifically, practical questions of high-altitude, climate, diving, sport, and occupational medicine, as well as rehabilitation and even isolation research are subjects of interest (e.g., osteoporosis, cardiovascular illness, countermeasures, and exercise regimes). Space medicine research thus extends far beyond the narrow realm of space physiology and space medicine: It is an advanced preventive medicine in its best sense. Terrestrial simulation studies with their customized and sometimes unique simulation models and equipment (e.g., bed rest, lowered head-down tilt positions, isolation and immersion, centrifuges) are superbly suited for addressing interdisciplinary human physiology concerns in great scientific depth and breadth. At the same time, such studies serve to introduce the next science generation to this field of investigation and to interest them in the subject, in order to assure that research continues. This problem should not be underestimated in the typically extremely long planning periods for space activities. For scientific, technological, and economic reasons, research in space medicine and space physiology must be regarded as an integral, essential component of any space policy in the future.

REFERENCES

[1] White RJ, Averner M. Humans in space. Nature 2001;409(6823):1115–8.

[2] Long M. Surviving in space. National Geographic Mag 2001;9–29.

[3] Harland DM, Harland DM. The MIR space station: a precursor to space colonization. Chichester: Wiley; 1997.

[4] Messerschmid E, Bertrand R, Freyer F. Space stations: systems and utilization. Berlin, New York: Springer-Verlag; 1999.

[5] Clément G. Fundamental of space medicine. Space technology library, Springer-Netherlands; 2005.

[6] Gunga HC, Steinach M, Kirsch K, Steinach M, Wittmann K. Space medicine and biology. In: Ley W, Wittmann K, Hallmann W, editors. Handbook of space technology. Wiley and Sons; 2009. p. 606–21.

[7] Nicogossian AE, Leach-Huntoon C, Pool SL. Space physiology and medicine. Philadelphia, PA: Lea & Febiger; 1989.

[8] Rahmann H, Kirsch K. Mensch-Leben-Schwerkraft-Kosmos. Perspektiven biowissenschaftlicher Weltraumforschung in Deutschland. Heimbach-Verlag; 2001.

[9] US Government. Astronaut fact book. Astronaut fact 2013.

[10] Comet B. Limiting factors for human health and performance: microgravity and reduced gravity. Study on the survivability and adaptation of humans to long-duration interplanetary and planetary, environments; 2001. http://ecls.esa.int/ecls/attachments/ECLS/Perspectives/humex/tn2.pdf

[11] Gushin VI, Kholin SF, Ivanovsky YR. Soviet psychophysiological investigations of simulated isolation: some results and prospects. Adv Space Biol Med 1993;3:5–14.

[12] Maillet A, Normand S, Gunga HC, Allevard AM, Cottet-Emard JM, Kihm E, et al. Hormonal, water balance, and electrolyte changes during sixty-day confinement. Adv Space Biol Med 1996;5:55–78.

[13] Schmitt DA, Schaffar L. European isolation and confinement study. Confinement and immune function. Adv Space Biol Med 1992;3:229–35.

[14] Tobler I, Borbély AA. European isolation and confinement study. Twenty-four hour rhythm of rest/activity and sleep/wakefulness: comparison of subjective and objective measures. Adv Space Biol Med 1993;3:163–83.

[15] Vaernes RJ, Bichi AF. 135 days in isolation and confinement: the HUBES simulation. Power 1995;200:706–12.

[16] Vaernes RJ, Bergan T, Ursin H, Warncke M. The psychological effects of isolation on a space station: a simulation study. Training 1996, 200107–01. SAE Technical Paper 921191, 1992, http://dx.doi.org/10.4271/921191.

[17] Værnes RJ. EXEMSI: description of facilities, organization, crew selection, and operational aspects. Experimental campaign for the European manned space infrastructure. Adv Space Biol Med 1996;5:7.

[18] Kanas N, Manzey D. Space psychology and psychiatry. Springer; 2003.

[19] Manzey D. Human missions to Mars: new psychological challenges and research issues. Acta Astronaut 2004;55(3):781–90.

[20] Nicholas JM, Penwell LW. A proposed profile of the effective leader in human spaceflight based on findings from analog environments. Aviat Space Environ Med 1995;.

[21] Sandal GM, Leon GR, Palinkas L. Human challenges in polar and space environments. Springer; 2007.

[22] Heer M, Baisch F, Kropp J, Gerzer R, Drummer C. High dietary sodium chloride consumption may not induce body fluid retention in humans. Am J Physiol Renal Physiol 2000;278(4):F585–95.

[23] Titze J, Maillet A, Lang R, Gunga HC, Johannes B, Gauquelin-Koch G, et al. Long-term sodium balance in humans in a terrestrial space station simulation study. Am J Kidney Dis 2002;40(3):508–16.

[24] Gunga HC, Maillet A, Kirsch K, Röcker L, Gharib C, Vaernes R. European isolation and confinement study. Water and salt turnover. Adv Space Biol Med 1992;3:185–200.

[25] Gunga HC, Kirsch KA, Röcker L, Maillet A, Gharib C. Body weight and body composition during sixty days of isolation. Adv Space Biol Med 1996;5:39–53.

[26] Maillet A, Gauquelin G, Gunga HC, Fortrat JO, Kirsch K, Guell A, et al. Blood volume regulating hormones response during two space related simulation protocols: four-week confinement and head-down bed-rest. Acta Astronaut 1995;35(8):547–52.

[27] Rakova N, Jüttner K, Dahlmann A, Schröder A, Linz P, Kopp C, et al. Long-term space flight simulation reveals infradian rhythmicity in human Na(+) balance. Cell Metab 2013;17(1):125–31.

[28] Titze J, Dahlmann A, Lerchl K, Kopp C, Rakova N, Schröder A, et al. Spooky sodium balance. Kidney Int 2014;85(4):759–67.

[29] Billica RD, Simmons SC, Mathes KL, McKinley BA, Chuang CC, Wear ML, et al. Perception of the medical risk of spaceflight. Aviat Space Environ Med 1996;67(5):467–73.

[30] Nicogossian AE, Huntoon CL, Pool SL. Space physiology and medicine. Philadelphia, PA: Lea & Fibiger; 1994.

[31] Pappenheimer JR, Fregly MJ, Blatties CM. Handbook of physiology: environmental physiology. Oxford University Press; 1996.

[32] Thornton WE, Hedge V, Coleman E, Uri JJ, Moore TP. Changes in leg volume during microgravity simulation. Aviat Space Environ Med 1992;63(9):789–94.

[33] Kirsch KA, Baartz FJ, Gunga HC, Röcker L. Fluid shifts into and out of superficial tissues under microgravity and terrestrial conditions. Clin Investig 1993;71(9):687–9.

[34] Moore D, Bie P, Oser H. Biological and medical research in space: an overview of life sciences research in microgravity. Berlin: Springer; 1996.

[35] Smith JJ, Porth CM, Erickson M. Hemodynamic response to the upright posture. J Clin Pharmacol 1994;34(5):375–86.

[36] Watenpaugh DE, Hargens AR. The cardiovascular system in microgravity. In: Comprehensive physiology. Hoboken, NJ, USA: John Wiley & Sons, Inc.; 2010.

[37] Alfrey CP, Udden MM, Leach-Huntoon C, Driscoll T, Pickett MH. Control of red blood cell mass in spaceflight. J Appl Physiol 1996;81(1):98–104.

[38] Aubert AE, Beckers F, Verheyden B. Cardiovascular function and basics of physiology in microgravity. Acta Cardiol 2005;60(2):129–51.

[39] Gunga HC, Kirsch K, Baartz F, Maillet A, Gharib C, Nalishiti W, et al. Erythropoietin under real and simulated microgravity conditions in humans. J Appl Physiol 1996;81(2):761–73.

[40] Rittweger J, Gunga HC, Felsenberg D, Kirsch KA. Muscle and bone-aging and space. J Gravit Physiol 1999;6(1):P133–6.

[41] Rittweger J, Frost HM, Schiessl H, Ohshima H, Alkner B, Tesch P, et al. Muscle atrophy and bone loss after 90 days' bed rest and the effects of flywheel resistive exercise and pamidronate: results from the LTBR study. Bone 2005;36(6):1019–29.

[42] Nooij SA, Vanspauwen R, Bos JE, Wuyts FL. A re-investigation of the role of utricular asymmetries in space motion sickness. J Vestib Res 2011;21(3):141–51.

[43] Perchonok M, Bourland C. NASA food systems: past, present, and future. Nutrition 2002;18(10):913–20.

[44] Smith SM, Zwart SR, Block G, Rice BL, Davis-Street JE. The nutritional status of astronauts is altered after long-term space flight aboard the International Space Station. J Nutr 2005;135(3):437–43.

[45] Lane HW, Feeback DL. History of nutrition in space flight: overview. Nutrition 2002;18(10):797–804.

[46] Rivolier J. Man in the Antarctic: the scientific work of the International Biomedical Expedition to the Antarctic (IBEA). London, New York: Taylor & Francis; 1988.

[47] Heer M. Nutritional interventions related to bone turnover in European space missions and simulation models. Nutrition 2002;18(10):853–6.

[48] Iwamoto J, Takeda T, Sato Y. Interventions to prevent bone loss in astronauts during space flight. Keio J Med 2005;54(2):55–9.

[49] Stein TP, Leskiw MJ, Schluter MD, Donaldson MR, Larina I. Protein kinetics during and after long-duration spaceflight on MIR. Am J Physiol Endocrinol Metab 1999;276(6):E1014–21.

[50] Stein TP, Leskiw MJ, Schluter MD, Hoyt RW, Lane HW, Gretebeck RE, et al. Energy expenditure and balance during spaceflight on the space shuttle. Am J Physiol Regul Integr Comp Physiol 1999;276(6):R1739–48.

[51] Cucinotta FA, Kim M-HY, Chappell LJ, Huff JL. How safe is safe enough? Radiation risk for a human mission to Mars. PLoS One 2013;8(10):e74988. http://dx.doi.org/10.1371/journal.pone.0074988.

[52] Hellweg CE, Baumstark-Khan C. Getting ready for the manned mission to Mars: the astronauts' risk from space radiation. Naturwissenschaften 2007;94(7):517–26.

[53] Horneck G, Facius R, Reichert M, Rettberg P, Seboldt W, Manzey D, et al. HUMEX, a study on the survivability and adaptation of humans to long-duration exploratory missions, part II: missions to Mars. Adv Space Res 2006;38(4):752–9.

[54] Reitz G, Facius R, Sandler H. Radiation protection in space. Acta Astronaut 1995;35(4–5):313–38.

[55] Basner M, Rao H, Goel N, Dinges DF. Sleep deprivation and neurobehavioral dynamics. Curr Opin Neurobiol 2013;23(5):854–63.

[56] Basner M, Dinges DF, Mollicone DJ, Savelev I, Ecker AJ, Di Antonio A, et al. Psychological and behavioral changes during confinement in a 520-day simulated interplanetary mission to mars. PLoS One 2014;9(3):e93298.

[57] Goel N, Basner M, Rao H, Dinges DF. Circadian rhythms, sleep deprivation, and human performance. Prog Mol Biol Transl Sci 2013;119:155–90.

[58] Gundel A, Drescher J, Spatenko YA, Polyakov VV. Changes in basal heart rate in spaceflights up to 438 days. Aviat Space Environ Med 2002;73(1):17–21.

[59] Gundel A, Polyakov V, Zulley J. The alteration of human sleep and circadian rhythms during spaceflight. J Sleep Res 1997;6(1):1–8.

[60] Koch B, Gerzer R. A research facility for habitation questions to be built at the German Aerospace Center in Cologne: future challenges of Space medicine. Hippokratia 2008;12(Suppl.):191–6.

[61] Rabbow E, Koch B, Gerzer R. Envihab: Neuartige Großforschungsanlage des DLR eröffnet-Ort des Fortschritts: von der Idee bis zur Realisierung. Flugmedizin extperiodcentered Tropenmedizin extperiodcentered Reisemedizin-FTR 2013;20(04):180–5.

[62] Crucian B, Simpson RJ, Mehta S, Stowe R, Chouker A, Hwang SA, et al. Terrestrial stress analogs for spaceflight associated immune system dysregulation. Brain Behav Immun 2014;39:23–32.

[63] Holland AW. NASA investigations of isolated and confined environments. Adv Space Biol Med 1993;3:15–21.

[64] Palinkas LA. The psychology of isolated and confined environments: understanding human behavior in Antarctica. Am Psychol 2003;58(5):353.

[65] Palinkas LA, Johnson JC, Boster JS. Social support and depressed mood in isolated and confined environments. Acta Astronaut 2004;54(9):639–47.

[66] Suedfeld P, Steel GD. The environmental psychology of capsule habitats. Annu Rev Psychol 2000;51(1):227–53.

[67] Clearwater YA. Space station habitability research. Acta Astronaut 1988;17(2):217–22.

[68] Clearwater YA, Coss RG. Functional esthetics to enhance well-being in isolated and confined settings. Springer; 1991.

[69] Harrison AA. Spacefaring the human dimension; 2002. Available from: http://site.ebrary.com/id/10064740.

[70] Harrison AA. Space habitability and the environment. In: Philip RH (Author). Space Enterprise, Springer Praxis Books; 2009, p. 103-52.

[71] Bergan T, Sandal G, Warncke M, Ursin H, Værnes RJ. European isolation and confinement study. Group functioning and communication. Adv Space Biol Med 1992;3:59–80.

[72] Flynn CF. An operational approach to long-duration mission behavioral health and performance factors. Aviat Space Environ Med 2005;76(6):B42–51. Available from: http://www.ingentaconnect.com/content/asma/asem/2005/00000076/A00106s1/art00007.

[73] Kanas N, Sandal G, Boyd JE, Gushin VI, Manzey D, North R, et al. Psychology and culture during long-duration space missions. Acta Astronaut 2009;64(7):659–77.

[74] Diedrich A, Paranjape SY, Robertson D. Plasma and blood volume in space. Am J Med Sci 2007;334(1):80–5.

[75] Stenger MB, Brown AK, Lee S, Locke JP, Platts SH. Gradient compression garments as a countermeasure to post-spaceflight orthostatic intolerance. Aviat Space Environ Med 2010;81(9):883–7.

[76] Buckey Jr JC, Lane LD, Levine BD, Watenpaugh DE, Wright SJ, Moore WE, et al. Orthostatic intolerance after spaceflight. J Appl Physiol 1996;81(1):7–18.

[77] Meck JV, Waters WW, Ziegler MG, deBlock HF, Mills PJ, Robertson D, et al. Mechanisms of postspaceflight orthostatic hypotension: low alpha1-adrenergic receptor responses before flight and central autonomic dysregulation postflight. Am J Physiol Heart Circ Physiol 2004;286(4):H1486–95.

[78] Cardinale M, Wakeling J. Whole body vibration exercise: are vibrations good for you? Br J Sports Med 2005;39(9):585–9.

[79] Cardinale M, Rittweger J. Vibration exercise makes your muscles and bones stronger: fact or fiction? J Br Menopause Soc 2006;12(1):12–8.

[80] Rauch F, Sievanen H, Boonen S, Cardinale M, Degens H, Felsenberg D, et al. Reporting whole-body vibration intervention studies: recommendations of the International Society of Musculoskeletal and Neuronal Interactions. J Musculoskelet Neuronal Interact 2010;10(3):193–8.

[81] Rittweger J, Beller G, Felsenberg D. Acute physiological effects of exhaustive whole-body vibration exercise in man. Clin Physiol 2000;20(2):134–42.

[82] Smith SM, McCoy T, Gazda D, Morgan JL, Heer M, Zwart SR. Space flight calcium: implications for astronaut health, spacecraft operations, and Earth. Nutrients 2012;4(12):2047–68.

[83] Lackner JR, DiZio P. Space motion sickness. Exp Brain Res 2006;175(3):377–99.

[84] Davis JR, Jennings RT, Beck BG, Bagian JP. Treatment efficacy of intramuscular promethazine for space motion sickness. Aviat Space Environ Med 1993;64(3 Pt 1):230–3.

[85] Fang Y-Z, Yang S, Wu G. Free radicals, antioxidants, and nutrition. Nutrition 2002; 18(10):872–9.

[86] Pospíšil M. Pharmacological radiation protection. Springer; 1999.

[87] Weiss JF, Landauer MR. Protection against ionizing radiation by antioxidant nutrients and phytochemicals. Toxicology 2003;189(1):1–20.

[88] Durante M, Cucinotta FA. Physical basis of radiation protection in space travel. Rev Mod Phys 2011;83(4):1245.

[89] Turner DRE, Levine MJM. Orbit selection and its impact on radiation warning architecture for a human mission to Mars. Acta Astronaut 1998;42(1):411–7.

[90] Byrnes DV, Longuski JM, Aldrin B. Cycler orbit between Earth and Mars. J Spacecraft Rockets 1993;30(3):334–6.

[91] Benton Sr MG, Bamford RA, Bingham R, Todd T, Silva L, Alves P. Concept for human exploration of NEO asteroids using MPCV, deep space vehicle, artificial gravity module, and mini-magnetosphere radiation shield. In: AIAA space conference and exposition 2011; 2011 (conference paper). Available from: http://www.scopus.com/inward/record.url?eid=2-s2.0-84880585359&partnerID=MN8TOARS.

[92] Fomina G, Kotovskaya A, Arbeille F, Pochuev V, Zhernavkov A, Ivanovskaya T. Changes in hemodynamic and post-flights orthostatic tolerance of cosmonauts under application of the preventive device-thigh cuffs bracelets in short-term flights. J Gravit Physiol 2004;11(2): 229–30.

[93] Anzai T, Frey MA, Nogami A. Cardiac arrhythmias during long-duration spaceflights. J Arrhythmia 2014;30:139–49.

[94] Smart J, Bacal K. Space medicine: the new frontier. In: Auerbach P, editor. Wilderness medicine. Elsevier; 2012. p. 2172–202.

[95] Taddeo TA, Armstrong CW. Spaceflight medical systems. Springer; 2008.

[96] Bernasconi, M, Zenger R, Versteeg M. Crew health support on long-duration missions in microgravity: some considerations. IAC-06 Symposium 2006; IAC-06-A1.P.2.04.

[97] Gunga H, Zenger R. IHAB phase O medical support activities. HTS AG document 513TN0001 2004.

[98] Horneck G, Comet B. General human health issues for Moon and Mars missions: results from the HUMEX study. Adv Space Res 2006;37(1):100–8.

[99] Horneck G, Facius R, Reichert M, Rettberg P, Seboldt W, Manzey D, et al. HUMEX, a study on the survivability and adaptation of humans to long-duration exploratory missions, part I: lunar missions. Adv Space Res 2003;31(11):2389–401.

[100] Arbeille P, Poisson G, Vieyres P, Ayoub J, Porcher M, Boulay JL. Echographic examination in isolated sites controlled from an expert center using a 2-D echograph guided by a teleoperated robotic arm. Ultrasound Med Biol 2003;29(7):993–1000.

[101] Piantadosi CA, Mankind beyond Earth: the history, science, and future of human space exploration. New York: Columbia University Press 2012;29(7):993–1000.

[102] Horneck G, Klaus DM, Mancinelli RL. Space microbiology. Microbiol Mol Biol Rev 2010;74(1):121–56.

Index

Note: Page numbers followed by *f* indicate figures and *t* indicate tables.